BLACK APOLLO OF SCIENCE

BLACK APOLLO OF SCIENCE

The Life of
Ernest Everett Just

Kenneth R. Manning

New York   Oxford
OXFORD UNIVERSITY PRESS

OXFORD UNIVERSITY PRESS
Oxford London Glasgow
New York Toronto Melbourne Auckland
Delhi Bombay Calcutta Madras Karachi
Kuala Lumpur Singapore Hong Kong Tokyo
Nairobi Dar es Salaam Cape Town
and associate companies in
Beirut Berlin Ibadan Mexico City Nicosia

Copyright © 1983 by Oxford University Press, Inc.
First published in 1983 by Oxford University Press, Inc.,
198 Madison Avenue, New York, New York 10016-4314
First issued as an Oxford University Press paperback, 1984

Library of Congress Cataloging in Publication Data
Manning, Kenneth R.
  Black Apollo of science.
  Bibliography: p.
  Includes index.
  1. Just, Ernest Everett, 1883-1941. 2. Biologists—
United States—Biography. I. Title.
QH31.J83M36 1983 574'.092'4 [B] 83-4009
ISBN 0-19-503299-3
ISBN 0-19-503498-8 (pbk.)

Printing (last digit): 9

Printed in the United States of America

For

Harold John Hanham

Contents

PART I

CHAPTER I

The Early Years, 1883-1907

The sun shone brightly on the colorful decorations. With displays of bunting on King, Meeting, and Broad streets, and on the ships *Delaware* and *Sappho* docked at Union Wharf, the city made a pretty picture at the start of a splendid summer morning. Stores were decked out with miniature flags and Chinese lanterns. The American colors flew from the Chamber of Commerce buildings; the state colors fluttered from the windows at Main Station; and the municipal colors hung at City Hall. A handsome spread of American and French and German flags waved in the breeze on King Street between Beaufain and Calhoun. A large blue banner with a white palmetto tree at the center and the dates 1783–1883 stretched across the street, just below Wentworth.

Households awakened at early dawn to the music of St. Michael's chimes. "The Star Spangled Banner" rang clear and bright through the city, followed by "Way Down Upon the Swanee River," "Oh Susannah," and "Taffy Was a Welshman." Within minutes, a hundred-gun salute burst from the Lafayette and German Artilleries. The dawn of a new century was being ushered in—Charleston was celebrating its centennial of incorporation and Charlestonians, black and white alike, were hoping to catch new inspiration from the past and take new courage for the future.[1]

It was 13 August, a Monday morning, the tail end of a long holiday weekend. A cool breeze was blowing in off the water, making it a perfect day for an outdoor festival. Dignitaries and visitors from throughout the state gathered to celebrate in song and dance along the city sidewalks and in the streets. They mingled and chatted and laughed together in the downtown area all day. Around five o'clock in the afternoon, a small crowd came together at City Hall for the official ceremony, exhibiting a "quiet dignity . . . so becoming to [their] historic city."[2] After prayers and an unveiling of artworks, Mayor

William A. Courtenay delivered an inspiring address. At about the same time a larger crowd was gathering at the Rutledge Street lake to take part in a less formal affair. Fireworks, the grandest pyrotechnic display ever seen in Charleston, were scheduled to begin at half-past eight in the evening. The streets were jammed with people trying to reach the lake and, by quarter-past seven, a crowd of no less than five thousand had assembled. The lakeshore, piazzas, and windows were packed with people of "every shade of color, sex, and condition of life."[3] Some families lolled on the grass, others sat in boats on the lake. Everyone, even the people from the sea islands, had come for the celebration. As night drew on, the band began to play. People clapped hands and swayed to the music as they waited anxiously for the fireworks, the high point of the day.

An impoverished aristocratic city, Charleston had put much time and money into these fireworks. Some of the displays sought to illustrate a theme; others were designed simply to create excitement and give pleasure. The first set of fireworks, balloon-shaped, changed from striped to red to white as it rose and soared toward the north. Exploding in midair, it showered colorful sparks onto the crowd below and continued floating higher and higher into the dark night. A black child, a little boy, called out, "Deh, now! he done tu'n into a star."[4] This display, entitled "The Sunburst," was just one of a series of magnificent spectacles. "The Peruvian Glory," "The Casket of Jewels," and "The Jewelled Cross of Malta" stunned and dazzled the crowd. Between the displays, bombs and rockets and Roman candles were sent up hundreds at a time. There were large colored ones in the shape of peacocks' plumes, silver streamers, golden clouds, and eagles' claws, and an especially brilliant one in the shape of a huge chandelier hanging from mid-heaven like an aerial fountain of fire.

The fun grew fast and furious, but the crowd did not tire. Everyone was in a state of jubilant frenzy, which increased when the aquatic fireworks got under way. Glowing rockets imitated the antics of porpoises; lights on the surface of the lake changed hue and color constantly; silver and gold fountains bubbled up from the watery depths. The lake seemed as if it were on fire. With each successive display the crowd murmured in awe or shrieked with delight. A wild roar went up when "Charles Town 1670" flashed in jeweled letters across the sky, and again a minute later when "Charleston City" appeared against the same background, flanked by the dates 1783 and 1883. The spectacle was over. Reluctantly, people dispersed through the streets and headed home.

Mary Mathews Just was in the crowd. Though her labor pains had already begun, she did not want to miss Charleston's hundredth birthday. After the celebration her mind turned from the excitement of the city to the future of her family. What would this new child, her fourth, be like? In the late evening she summoned Dr. T. B. McDow and went through an easy delivery a little after midnight on Tuesday morning, 14 August.[5] It was a boy with light reddish skin, coarse brown hair, and sharply carved features. For Mary he was precious and special, a symbol of hope like the day just past.

Ernest Everett Just came into the world at a time of jubilation, but it would not be long before he would be brought face to face with the reality that political and social conditions for his race in the state were difficult and growing worse. A better time had been right after the Civil War, when blacks took a share in the political power of the state, joined the ranks of the Republican Party, began to vote and to enjoy public education, became legislators and officials in the South Carolina government. It was the time of Black Reconstruction. In 1876, however, things changed. The election that year brought in the period of the so-called White Redemption. At noon on 10 April 1877, United States soldiers filed out of the State House in Columbia by order of President Rutherford B. Hayes, symbolically ending the period of Black Reconstruction. Whites began to usurp power and enlarge the Democratic Party through political chicanery. Throughout the state, and especially in Charleston, blacks were disenfranchised, legislators lost office, and the practices of jim crow emerged. This trend of white oppression continued throughout the eighties and reached a peak with the policies of arch-conservative Benjamin Ryan Tillman, nicknamed "Pitchfork Ben," who initiated a program of wholesale black disenfranchisement in the early nineties. The political and social adversities aroused in Ernest's mother, Mary, the strength and courage to press forward, but they did not evoke a similar response from his father, Charles Fraser.[6]

Charles Fraser Just and Mary Mathews Cooper had hardly met before they were attracted to each other. Charles Fraser was a man with an eye for the women. Handsome, light-skinned and curly-haired, he was seldom without a good deal of female companionship. During his courtship of Mary, a local girl from a good family, he was carrying on with another—apparently less respectable—woman named Alice, who had taken "Just" as her surname. Mary became pregnant during the summer of 1878. Religious and moral, she insisted upon marriage and hoped Charles Fraser would mend his ways. They married on 19 September and began to await the arrival of their child in their small house at 103 Calhoun Street.[7] On 15 March 1879, the child, a female, was stillborn.[8] Mary was disappointed, but she hoped for other children.

Charles Fraser was also disappointed, but circumstances were different for him. He was not long without progeny. In less than three months, he and his mistress Alice began their "outside" family with the birth of a baby girl.[9]

After his marriage, Charles Fraser openly lived in two households—with Alice on Sives Street, with Mary at 103 Calhoun Street and later at 28 Inspection Street.[10] Maintaining two residences was difficult for Charles Fraser. He rarely kept a steady job, going from one to another every few months. Also, he loved his home brew as much as he loved his women. A good portion of whatever wages he earned went to liquor before the remainder went to household provisions. Many a weekend he barely escaped arrest for drunkenness, a misdemeanor rigidly dealt with in the city at that time.[11] Mary seemed to withstand it all and, while she did not always have the affection of her husband, she had the support and sympathy of his family. She bore two more children, a boy and a girl, at 28 Inspection Street before she had Ernest.[12]

In many ways, Mary mistook her husband for what she saw in his family, especially in his father, Charles Just, Sr., with whom the couple lived in the Inspection Street house. The elder Charles was an industrious and prosperous man. Born in 1805, he was a former slave of George Just, an immigrant from Germany who had come to the United States in the early part of the century, settled in Charleston, built a successful wharf-building business, and been naturalized as a citizen on 19 September 1808.[13] The senior Charles's parentage remains unclear, but there is reason to believe he was George's natural son. A mulatto, he seems to have occupied a favored position in George's household. He learned the wharf-building trade from George and became very accomplished. From all indications, his life as a slave was somewhat better than that of most slaves in the city and a great deal better than that of most slaves on the plantations.

Charles was well-respected in the black community of Charleston. Though a slave, he moved in Charleston society as a "free man of color." Many of his friends were part of the free community, making him indistinguishable from the upper-class blacks in the city. On 28 December 1842, for example, he attended the baptism of Augustus Charles Ryan, the newborn child of two of his very close friends, Augustus and Hannah Ryan, both free blacks.[14] The Ryans had in part named their baby after Charles, whom they had chosen as godparent, along with Carolyn Johnson and Richard Houston. The ceremony took place in St. Michael's Church, the Episcopal parish where whites and blacks worshipped together but sat in separate pews. At the same service, Charles also sponsored Rosa Johnson Taylor, daughter of Isaac and Elizabeth Taylor, both free blacks. Hannah Ryan shared the sponsorship of Rosa with Charles. The two couples, the Ryans and the Taylors, were members of St. Michael's. Charles may have been a member too, since George worshipped there regularly. It is important to note that Charles never acknowledged his slave status in the registry of St. Michael's.

Charles was not unique in his position. Many blacks in Charleston lived in limbo, halfway between slavery and freedom, because slaveowners sometimes permitted favored slaves to go about the city as if they were free. What was unusual about Charles was his strong tie with the community of free blacks, a close-knit and exclusive circle. He was a prominent member of the group which founded the Unity and Friendship Society, a mutual insurance club for free blacks. Chartered on 17 September 1844, the society emerged from the careful planning of Charles Just, G. McKinley, Samuel Vanderhorst, and fifteen other persons; it was established to help "widows and orphans . . . in their hours of distress . . . and time of need."[15] The group made no fuss about the color of its members. Men of varying colors from the free population were invited to join. Older organizations like the Brown Fellowship Society, founded on 1 November 1790, admitted only light-skinned blacks, and others such as the Society of Free Dark Men selected mostly dark-skinned persons. Though Charles was a mulatto and might have mingled exclusively with others of his kind in antebellum Charleston, he chose not to do so. Color seems not to have been important to him.

In the late thirties, Charles married Mary Anne, a free black of unknown origin. They had a son, Simeon, in 1840, and another, James, in 1845. The family lived on Calhoun Street, close to the docks where the father worked as a wharf-builder in his master's business. Charles apparently accumulated substantial savings by hiring out his labor after hours. Over the course of time he managed to build up a small estate. As a slave he could not own property, so his wife held most of what they owned in her name. It is interesting to note, however, that on 21 May 1852 he passed himself off as "a free person of color" in order to purchase a piece of land.[16] When she died in 1853, Mary Anne was listed as owning the property on Calhoun Street and two slaves, Lucy and Woodson. A year earlier she had willed the estate to Charles, still a slave, and appointed George Just her executor.[17] An illiterate woman, she signed the will with an X. She did not fully understand the implications of her husband's legal status, and no doubt Charles and George had drawn up the will for her. It was a difficult arrangement, designed for the benefit of Charles but dependent on the good will of George.

Charles had always lived on the beneficence of George, but with time he came to prefer blanket freedom to paternalistic servitude. Legend has it that he bought his freedom around 1853.[18] By then he had saved a substantial sum of money from his work and acquired the property and slaves of his deceased wife. He was living like a free man.

Other circumstances support the story. On 13 June 1853, less than a month before his death, George drafted an affidavit concerning the "descent" of Charles. It was filed in the office of the secretary of state in Columbia, and reads as follows:

> I George Just in consideration of doing Justice and discharging the duty which one man owes to another and from no other motive or inducement— Do Hereby Certify—that Sometime about the 27th day of May in the year Eighteen hundred and seven—I purchased from William Ellis and Peter Whiteheart a colored male child about fourteen months old of the name of Charles. Shortly after purchasing—I was informed by the said Peter Whiteheart (now deceased) that he had procured this child from one Gerishia Everett a free white woman who being in extremely destitute circumstances was compelled to sell him to procure bread that the said Gerishia was extremely ignorant and incapable of writing and that the said Gerishia Everett was the Grandmother of said child. And I do further certify that shortly after my purchase of said child a white woman called and asked to see said child and after she saw the child fondled him and used towards him terms of endearment. Persuaded from these circumstances—that the said Boy Charles was the issue of a white woman and coloured or black man—I have brought him up protected—and instructed him—using towards him only so much restraint and control as was necessary and proper, and that since his manhood. I fully informed the said Charles of his Condition— which was also known by the members of my family and others. I further certify that on account of my firm belief that Charles is and was always free I have allowed him to use my name as his own and that he is now and has always been known as Charles Just.[19]

According to this account Charles had never legally been a slave. The law held that a black was free in cases of mixed parentage where the mother was white and the father black, but not the other way around. By filing this affidavit George avoided having to submit a formal deed of manumission, as he would have had to do if he had admitted fathering Charles. Manumission was a difficult process. Legislation passed in 1820 and 1841 had put all sorts of obstacles in the way of a master who wanted to free a slave, and as the Civil War approached manumission became virtually impossible.[20] A good shortcut was for George to construct a story attributing a white mother to Charles, draw up an affidavit to that effect, and place the affidavit on record. Of course, the story George reported may have been true. Charles may always have been legally a free man—and Ernest Everett Just's great-great-grandmother may have been a white woman named Gerishia Everett. Whether the story was true or false, the end result was the same: freedom for Charles. George's motivation was identical in either case: if true, the story was exposed to free Charles; if false, it was made up for the same purpose. Despite George's leniency, Charles no doubt had been viewed as a slave in some quarters. After 1853, however, he had a form of protection in the affidavit; he could pay off his master and begin to live more proudly and comfortably, as a free man.

George died in 1853. When he had written his will the year before, he had urged his wife, Margaret, to assume some kind of trusteeship over Charles.[21] He had "earnestly" recommended his "servant Charles" to her "kind care," insisting that she not "dispose or sell" him "except at his own request and then only to such person as he may choose." If she needed money, she would have to procure it through other means than by selling Charles to a slave trader. Similar clauses were often placed in the wills of Charleston slaveholders, primarily as a way of expressing the hope that a slave be allowed to enjoy limited freedom and be rewarded for good service. A free man, Charles no longer needed this kind of paternalism.

In 1855 Charles married again. He was now fifty years old, and his new wife, Sarah Fraser, was a young mulatto girl only eighteen years old. They continued to live at the property on Calhoun Street which had belonged to Mary Anne. There they began a family. Their first child was Charles Fraser. Other children followed over the next twenty years. In 1880 Charles purchased 28 Inspection Street, previously owned by George and left to a distant relation, Margaret Furches.[22] He moved in there with his second family; Mary Anne's sons, Simeon and James, took over the property on Calhoun Street.

Charles was well on the way to making a success of himself in post–Civil War Charleston. He took the wharf-building business over from the white Justs, acquired a collection of work tools and household furnishings (including a piano), bought expensive personal articles such as a gold watch, invested in real estate and bought stock in the Enterprise Railroad Company of Charleston, and sat in a select pew in the Emanuel African Methodist Church on Calhoun Street.[23] He was fast becoming one of the most prominent blacks in Charleston.

The wharf-building trade was skilled labor; it paid well and carried prestige. In the 1850s and '60s, the famous black legislator Robert Smalls had worked on the wharves of Charleston as a stevedore and pilot.[24] Such positions were not easy to acquire, and Charles wanted to keep the trade in the family. To that end, he taught his three sons—Simeon, James, and Charles Fraser—the ins and outs of wharf construction and engineering. Unfortunately, the sons showed neither the industry nor the inclination of their father. They went their own ways in defiance of his wishes. Before long, Charles realized his lack of control over them. Hurt and angry, he wrote them out of his will:

I do not give any portion of my Estate Real or Personal to my sons, Charles Just, James Just, and Simeon Just or their heirs executors or administrators, on account of their undutiful and disrespectful behavior, and I therefore will desire and intend that neither they nor any of them

nor the issue of them or of any of them shall under any circumstances
receive any portion of my said Estate.[25]

But this action did little to alter their behavior.

Charles Fraser continued his extramarital affairs throughout his
marriage, even after he and his family moved in with his father, and his
drinking only got worse. By the time Ernest was born, the Just family
was in stress. Charles Fraser could not maintain a proper balance
between alcohol and women on the one hand, and a job and a family
on the other. It is ironic that during this time he served on the petit jury
of the U.S. District Court under Judge George S. Bryan to decide,
among other things, cases involving the illegal sale of liquor.[26]

Soon after Ernest's birth, the family was struck by tragedy. Mary was
just getting over the loss of her first child when the health of her other
three was endangered. Cholera and diphtheria epidemics broke out in
the late summer of 1883, and black children were especially hard hit.[27]
The dreadful diseases struck 28 Inspection Street in early December.
Ernest was not quite four months old. The youngest of the three
children, he was also the most vulnerable. Somehow he survived. The
other two were not so fortunate. On 28 December 1883 Ernest's two-
year-old brother, Norman Rutledge, died of "inflammation of stomach
and bowels," one of the most common symptoms of cholera in children.
The tension in the Just household had not helped little Norman's
chances for survival, and his death only added to the problem. Two
weeks later, on 9 January 1884, Ernest's four-year-old sister, Vivian,
succumbed to "diphtheria asphyxia." Both children were laid to rest in
the African Cemetery.[28]

In 1883 medical care was sadly lacking in the black community of
Charleston. There were few black doctors. Some, Dr. William D. Crum
for instance, were just beginning to come to the city after completing
their medical education at a newly founded black school in Washington,
D.C.—Howard University. Even so, a good deal of the effort of black
doctors like Crum went to making money from nonmedical sources,
usually real estate, and advancing in politics.[29] Also, rarely were their
fees lower than those of white doctors, and almost never did they have
access to the medical facilities—hospitals and clinics—available to
their white counterparts. Many black families, like the Justs, used the
medical services of local white doctors. By and large, treatment by
whites was considered both safer and more socially acceptable. Black
physicians were still not fully trusted.

Charles Fraser and Mary always sought treatment from a white
doctor, even though he practiced under a cloud of suspicion, especially
in the Charleston white community. Dr. T. B. McDow, married to the
daughter of a well-to-do Charleston family, delivered and looked after
all three of the Just children. He was one of the few white doctors who

would treat poor blacks; in a way they were his best clients, since well-to-do blacks like the elder Charles preferred to seek treatment from more reputable old-line city doctors such as Edmund S. Ravenel. McDow's status was never high, and kept sliding. He was called in to treat the victims of street fights, shootings, and stabbings, and he served in the thankless position of surgeon to the enlisted men of the Lafayette Artillery.[30] Trouble pursued him. In 1889 he was charged with the murder of Captain Francis W. Dawson, who had suspected the doctor of making advances to his servant, a Swiss girl named Marie Bardayron.[31] Dawson's body was found under the floor of a small closet in McDow's office. McDow was brought to trial immediately. He pleaded self-defense. The jury, which included seven blacks, acquitted him in response to the impassioned plea of his lawyer, Asher D. Cohen. White Charlestonians angrily repudiated the verdict, and the Charleston Medical Society expelled McDow from its ranks. They could never forgive McDow for practicing widely among the city's black residents, and naturally they linked his acquittal to the black component of the jury. Disgraced and worn, he died some years later, possibly a suicide. Though his death was a loss for blacks like Charles Fraser and Mary, it may actually have been a blessing in disguise. We cannot know how little Norman and Vivian would have fared under the attention of a more capable doctor.

With the death of his brother and sister, Ernest became an only child. He was the hope of the household, though his chance of survival was not great in a city where the mortality rate for black children was significantly higher than that for white.[32] Sanitation was poor. Spoiled meats, infectious mosquitoes, and dirty chamber pots posed dangers in many households; diseases such as cholera, typhoid fever, and tuberculosis wreaked havoc in black Charleston. The area around Inspection Street was close to the wharves and prone to tidewater flooding and rat infestation. Highly congested, it was a hotbed of infection and disease. Ernest managed to escape these perils himself, but he was in constant contact with illness and death throughout his infancy and early childhood. His cousin Joseph, the son of A. J. Chairs and James Just, was the next victim. The James Justs lived at 93 Calhoun Street, around the corner from Inspection Street. Only two years old, Joseph died in March, two months after little Vivian. Phthisis (tuberculosis) was the diagnosis.[33] The tragedy did not stop there. Edith, the youngest daughter of Charles, Sr., and Sarah, was stricken with typhoid fever. To complicate matters, the family was suffering severe financial difficulties because Charles, Sr. had retired from work. In an effort to find money for Edith's medical treatment, Charles mortgaged his Inspection Street home to the Hibernia Savings Institution for $1,550 on 6 January 1885.[34] It was all in vain. Edith's fever became complicated by pneumonia. She died, at seventeen years of age, on 13 March 1885, a year to the day after the death of little Joseph.[35] The family was distraught. Sickness and death dominated their lives, leaving little chance for Ernest to find attention and affection.

In 1885 Mary was pregnant again, and before the end of the year she gave birth to her third son, Hunter. Charles Fraser continued drinking heavily as his family grew in size. Ernest was too young to sort out the details of the household situation, but he could tell that an atmosphere of stress and struggle prevailed. His grandfather Charles was deeper in debt. On the day of Edith's death, he had to borrow fifty dollars from an old family friend, Hannah Ryan. Doctors' bills and funeral expenses had accumulated rapidly. To make ends meet, he obtained a second mortgage on his house from James M. Redding, a private lender with an eagle eye for real estate speculation. He was paid $436 on 9 January 1886.[36] At about this time Mary began to contribute to the household where she, her husband, and her children lived. In addition, she made loans to her father-in-law for his personal expenses. A dressmaker by trade, she took in more and more work as the situation became worse. Shortly after the birth of Hunter she became pregnant again, and realized that another child would require even more money. She worked at home day in and day out, cared for Ernest and Hunter, and awaited her sixth child.

There was no end to their troubles. As Mary approached the seventh month of her pregnancy, the city suffered a tremendous earthquake, now famous in local history. Shortly after ten o'clock on Tuesday evening, 31 August 1886, "frightened and horror-stricken" residents fled from their houses onto the streets to escape falling walls and tumbling roofs.[37] The only part of the city left undamaged was the waterfront, clear testimony of the brilliant workmanship of wharf-builders like the Justs. Everyone was affected by the tremor, but black residents suffered more because their homes were generally less substantial than those of the whites. Houses, family possessions, and keepsakes were ruined by the quake, rendering many blacks destitute. From the disaster's onset, fear had run deep among the blacks. They came together and held prayer meetings in an effort to comfort each other. Determined and hopeful, they prayed well into the night and morning. Their prayer vigils continued four nights in succession. The end of the world had come, so they thought, and rewards missed in this world would have to be sought in the next. Superstition and religion thrived together. The earthquake evoked visions of Doomsday, the Apocalypse, Sodom and Gomorrah; evangelists used the occasion to draw converts with their cries, "Down on your face! Down on your knees, miserable sinners!"[38] A pious woman, Mary no doubt prayed, clutched her children, and hoped to bring her husband a little closer to the family and to God. First and foremost, she needed to calm herself and not lose her unborn child. It was a frightening time, with heavy property damage and a high death toll. Order resumed within a day or two, as the thousands of homeless began to receive food and temporary shelter.

The Justs survived. Mary had already seen too much tragedy to be paralyzed by an earthquake. She restored order in her household, continued her work at home, and looked forward to the birth of her sixth child. Ernest, barely in his fourth year, would harbor the event within himself for life.

In the fall of '86 Mary had her last baby. Inez joined the world of her two brothers, Ernest and Hunter. They all continued to live together at 28 Inspection Street, but it was not long before Mary began to think about building a new house for her part of the family. The earthquake left behind a great deal of ruined property, that now was selling cheaply. Mary took advantage of this by purchasing the land at 99 and 101 Calhoun Street and constructing a one-story wooden house, half of which was to be a source of rent income and the other half a home for, herself, her husband, and her children. She was among those Charlestonians who, "courageous" and "undaunted," showed "unbounded confidence in the future of the old city."[39] According to the *Charleston News and Courier*, the rapidity of the rebuilding process was "without parallel in the history of the world." The city abounded with enthusiasm; industrious residents began reviving the economy by purchasing construction materials, hiring labor, and paying taxes. Mary put a good deal of her own money into the Calhoun Street property, at least two hundred dollars by the spring of 1887. She managed her finances remarkably well, striving as she was to maintain a stable household.

The situation at 28 Inspection Street was proving impossible. The house was too small to hold everyone, and Charles Fraser and his father could not get on. An even bigger problem was that the health of the three children was suffering. They struggled through the early part of '87, barely surviving one disease after another. Finally, the Charles Fraser Justs moved to the new house on Calhoun Street. This change upset the children—especially Ernest, who was just old enough to understand a little of what was happening. Then, shortly after the family had settled into their new home, both father and grandfather fell ill. Charles Fraser began to weaken quickly. On 20 June he died, of alcoholism, with Ernest not yet four years old and Inez barely four months old. He was only thirty-two. A little over a month later Charles died, at eighty-two. Both father and son were laid to rest in the Friendly Union Cemetery.[40]

Mary was left without a husband and Ernest without a father. There was now no income from Charles Fraser's odd jobs. Debts mounted, with funeral bills and expenses for food and other day-to-day necessities. Charles Fraser left no will and no money, Charles an extensive will and some property. As executrix, Sarah moved to probate Charles, Sr.'s will, only to find that the debts exceeded the liquid assets in the estate. She petitioned the court for permission to sell 28 Inspection Street, in order to pay off the numerous creditors. Nearly penniless, Mary also had to sell her new house. She needed money badly and was beginning to appreciate the past support of her husband and his family. Her last hope was for some sort of legacy from Charles, Sr., if not for herself then for the children. But little Ernest, Hunter, and Inez had all been written out of their grandfather's will because of their father's "undutiful and disrespectful behavior." Their only legacy was the reputation of their grandfather. When they were growing up, Mary most likely represented the accomplishments of Charles as those of Charles Fraser. In later life Ernest was often to portray his father as a prosperous businessman working on the wharves of Charleston with over one hundred men under his direction, a description that fits Charles better than Charles Fraser.[41]

After the sale of 28 Inspection Street by public auction, Mary collected a debt of two hundred dollars she had against the estate and moved from Charleston to James Island, just three miles away. She was hoping to find a job in the new phosphate industry that had begun to flourish on the sea island because of a major investment by Dr. St. Julien Ravenel, of the old Charleston Ravenels. The heavy rocks lining the ditches and ravines along the Ashley River were full of phosphate, a substance that was used to produce a commercial fertilizer cheap enough to sustain farming interests in the impoverished South after the Civil War. One major phase of the work involved gathering the rocks and stacking them for collection, a job that required such enormous effort that only black men could be persuaded to take it. The work was well paid relative to other jobs held by blacks; at that time a man could earn a daily wage of ninety-seven cents for phosphate mining, as much as three times what he could get for most other forms of unskilled labor.[42] Few women were hired, but Mary knew she was persuasive enough to get the job and strong enough to handle it. In need of some lucrative means to support her family, she moved to the island to work in the phosphate mines—a decision that no doubt would have horrified her father-in-law. It was hard work, but she managed well.

Mary invested her savings in real estate. At the time, several hundred acres of land known as "the Hillsborough Plantation" were being parceled out to blacks by General W. N. Taft and the widow of C. C. Bowen, a sheriff of Charleston County during the Republican era.[43] A number of settlements were springing up all around, with between four and five hundred building lots in each. The competition for the lots

was fierce, but Mary, astute and determined, negotiated a solid investment.

Her property was quite substantial. She became a strong community
leader, canvassing the inhabitants, mostly the men, and persuading
them to transform the settlement into a town. They called the town
Maryville, after its prime mover. It was one of the first purely black
town governments in the state, a model community for blacks not only
in South Carolina but throughout the United States.[44] The inhabitants
worked long and hard together, organizing town meetings and peti-
tioning the state government for public services.[45]

Most of the blacks on the island were poor. By and large, they were
farmers who had assimilated little of the culture of white America.
They retained African customs and traditions; their language, known
as GeeChee or Gullah, was a dialect structured something like English
but infused with a variety of African words and pitch intonations. The
islanders kept mostly to themselves and mingled with neither the
whites nor the blacks of Charleston and its environs. Illiteracy was
high, since few teachers would dedicate themselves to work in this
isolated rural setting. Mary Mathews Just found there a calling, a
people who needed her help; she was "a woman of high character and
fairly good education" who devoted the next decade of her life to
teaching the islanders and raising her children.[46] One of the first things
she did was to organize religious instruction classes for blacks on
Sunday and any other days off she could spare from the mines. Though
two lots had been reserved for constructing churches, a Centenary
Congregation and a Reformed Episcopal church, no place of worship
had yet been built.[47] Mary held Sunday school at her home.

Shortly after the family moved to the island, typhoid fever struck.
The children had not yet had a chance to settle into their new home
before Ernest fell dreadfully ill. He was in bed for more than six weeks.
Then, when the fever passed, he had trouble recuperating because his
memory had been badly affected. Earlier he had learned to read and
write in "a very unusual way of excellence" for someone so young, but
now he had to go through the whole process all over again.[48] Mary had
always been sympathetic in teaching him, but she was now very "hard."
No matter how much she tried, she could not overcome the effects of the
fever, and her disappointment in him was "keen." After a lesson, little
Ernest would seclude himself with a book, only to shed "bitter tears"
because the "simplest words" appeared as "funny marks." His writing
fared no better; he could not steady his hand to make a clear letter or
number. He could neither read nor write, and the fact that his mother
showed no sympathy made it even worse. Then one day he read his first
page—by himself, without the help of his mother. The moment seemed
miraculous, as if some deep mystery had been revealed. Ernest was never
to forget the moment, remembering it in late adult life as if it had been

"yesterday." He kept his newly found ability from his mother, who had, after all, hurt him by her unreasonable expectations. Only later, perhaps a month or so, did he share his regained ability with her.

The Bible became the main source of reading for Ernest.[49] There were other books to choose from, but he was influenced by his mother, who had established herself as a religious leader. While the other boys played together outside, little Ernest cloistered himself in a back room of the family's small wooden dwelling and fantasized about "priests, angels, and Christ" as most children do about "fairies." He secluded himself in a world of biblical characters "more real than real people." Often he lay awake at night wondering about the story of Samuel: did God speak to Samuel in his sleep, and did Samuel hear the voice of God? Ernest never quite believed the story, but he still envied Samuel; he wished that he too could hear the voice of God. Many parts of the Bible affected him deeply. He was fascinated by the story of the Red Sea, and inspired by the verse "And in the cool of the evening God was walking in his garden." This image of God captured a serenity and control he longed to impose on the natural world.

The island was full of birds and flowers, especially in the spring, when the wrens awakened to the smell of wisteria and dogwood. Azaleas and camellias blossomed along the ditches where the tadpoles swam, and Spanish moss gleamed from the trees where the raccoons bedded. Ernest explored the flora and fauna of the black forests of the lowlands and found delight and harmony in their natural beauty. Before going to school he would stretch out under his "favorite oak," watching the ships pass by on the "sun-flecked" river and waiting for "the trill of the mockingbird" to break "the stillness of the dewy morning"; after supper he would sit gazing at "the dark blue field of night with its harvest of stars, hushed and wistful."[50] The wide world of nature came to mean almost as much to him as the strange religious sphere he had built for himself, his "world of hidden shadows and of ghosts of a dream world."[51]

The family moved back and forth between the island and the city. Having decided to give her "life and means . . . to the cause of Negro education," Mary taught half a year in one place and half in the other.[52] In the winter she worked the phosphate mines and held Sunday school classes on James Island; in the summer she taught in Charleston. When the family first moved back to the city, Ernest found the new atmosphere so strange he turned again to books for refuge. He lost his link with nature and immersed himself in the Bible.

One of Mary's main struggles was against the high rate of illiteracy both in the city and on the island. At home she taught the basics of reading, writing, spelling, and arithmetic to children, teenagers, and adults together in one room. Every Saturday she taught hatmaking and dressmaking to black women: a knowledge of some practical art, she thought, was as crucial as the ability to read and write. To all her students—young and old, academic and technical—she stressed cleanliness first and foremost. In school she taught the importance of personal hygiene, and on Sundays she led Bible services on the theme of cleanliness and purity of heart. As can well be imagined, the Just children were "always in school."[53]

Ernest was aware of his mother's dedication, and her teaching affected him deeply. As a child he cherished the idea of cleanliness. This was his own private pleasure, a sense of "secret well-being" that seemed to protect him from the outside world.[54] It gave him the will to forge forward against the effects of typhoid, the stress of a nomadic lifestyle, and the loneliness of being without a father. He loved wearing "spotless" underclothes but had no concern for the appearance of "outside dress"; as long as his underclothes were clean, he felt as if he could keep his soul secret and pure. Concern for physical and spiritual cleanliness was to remain a dominant part of his character.

When Ernest was nine years old, Mary decided to give up teaching in Charleston and devote her energies entirely to the community on the island. This decision placed heavy responsibility on Ernest's shoulders, and "a man's share" of work.[55] Cooking, washing, and cleaning became routine chores which he performed quietly and dutifully. He also had to take care of Hunter and Inez while his mother worked in the phosphate fields. This was especially hard since Hunter was almost deaf, having permanently damaged his eardrum while diving as a small child. But Hunter could and often did help his older brother prepare the evening supper for the family while Inez, the baby sister, played around the house. Though there were times when Ernest felt lonely and neglected,[56] he never complained about the responsibilities his mother placed on him. He could see her, widowed and alone, struggling to provide for the family and setting an example not only for them but for the entire community.

Mary proved herself a remarkable businesswoman with a gift for strategic organization. Always quick to spot a new money venture, she pushed the local farmers into large-scale curing of moss fiber for mattresses.[57] Fiber was abundant and the demand for it great. The farmers did not know how to make the most of the situation, so Mary held meetings to show them the advantage of working cooperatively to maximize profits. Instead of each farmer hauling his own moss to the factory and selling individually, it was decided that the farmers would pool their resources, buy a wagon to carry all the stock, and bargain collectively. The plan worked, and the islanders began to receive good prices for the fiber.

The town had no school, and Mary moved to start one. Because no public support was forthcoming for this worthwhile venture, she sold her property to found the Frederick Deming, Jr. Industrial School, one of the first industrial schools for blacks in the state.[58] Organizing and running this enterprise was a full-time occupation, so she left her job in the phosphate fields. Every day she attended to the various administrative details and taught the three Rs, sewing, millinery, and dressmaking. In the same building used for the school she set up Maryville's first church and conducted services each Sunday, not as an ordained minister but as a lay leader with a calling.

Ernest attended the school and church until he was twelve years old. His entire early education, scholastic and spiritual, formal and informal, had been under the direction of his mother. By the summer of '96, however, he had reached the limits of what she could offer and needed more to sustain his interest and foster his talents. Mary saw that further education would be good for him and his future. Though aware of the lack of opportunities for blacks and the growing effects of jim crow, she still wanted her son to continue his education, go away and prepare himself more extensively at a higher level, and perhaps return to Maryville to engage in the teaching of the people. A possible choice was the Avery Normal Institute. Founded in 1865 in nearby Charleston, it had a reputation for excellence, preparing its students for college and training them "in the old New England standards of efficiency."[59] But it catered mainly to light-skinned, upper-class blacks rather than to the black masses and would probably not have looked on Ernest with favor. Mary did not want him to go there anyway. She had heard of an industrial school for blacks at Orangeburg, about sixty miles away, a school where he could continue along the practical lines of study she had always stressed. The costs were high, but she would make whatever sacrifices were necessary.

The Colored Normal, Industrial, Agricultural and Mechanical College, or South Carolina State College, had just become an independent institution in the spring of 1896. It was founded in 1872 as the College of Agriculture and Mechanics' Institute for Colored Students—part of Claflin University, a black high school and college in Orangeburg.[60] When the conservative Wade Hampton took over the governor's office in Columbia in 1876, he dissolved the college's original board of trustees, mostly white Northern missionaries, and appointed a new board composed of whites from South Carolina. No aspects of the "radical Reconstruction" were allowed to survive. The college struggled on with little moral or financial support. Then, twenty years later, it came into its own as a political concession to blacks at the time when Winthrop and Clemson were being established as land-grant colleges for poor white women and men, respectively. Segregation in public educational facilities had meanwhile taken a firm hold. Under "Pitchfork Ben's" administration laws separating the races were drafted and passed, and everything was finally settled when the federal government handed down the landmark decision in *Plessy* v. *Ferguson* (1896) declaring legal the concept of "separate but equal."[61]

Ernest was thirteen years old when he arrived at State College in the fall of 1896. After taking required examinations in reading, spelling, arithmetic, physiology, English grammar, United States history, and geography,[62] he was admitted to the Classical Preparatory Department to follow the Normal Course, that is, to be trained as a teacher. Unlike the industrial and mechanical departments, which taught the trades of brick masonry, blacksmithing, and shoemaking, the Classical Department concentrated on "first principles." It was not "classical" in the traditional sense because the faculty and administration were opposed to blacks spending "years of toil and expense fumbling around . . . Greek and Latin roots."[63] Instead it offered preparation, at the secondary level, for basic teaching to students with a fair level of academic ability. Even this modest goal was difficult to achieve, however, as many of the students who entered State College were underprepared at the primary level, coming as they did from rural areas with little opportunity for booklearning. The hope, in any case, was to give them a firm grounding in at least two subjects: oratory and composition. A good teacher had to be able to demonstrate above all else the skills of speaking and writing. Two of Ernest's exercises, a speech on "National Progress" and an essay entitled "Time's Noblest Offering," won him special praise from the faculty.[64]

Ernest completed the Normal Course in three years. In the spring of 1899 he received his Licentiate of Instruction and was thereby licensed to teach in the black public schools of South Carolina without examination. But full-time teaching did not appeal to him; after all, he was only fifteen years old. What else could he do, though? The island seemed to beckon. He missed his family, the church and school, the townspeople, the plants and animals in the cypress swamps. He would return to Maryville and teach for a while, but Just was beginning to think about pursuing his education in more depth than he had done at State College.

On his return to Maryville that summer, Just saw that the community had changed. His mother was struggling to keep things going in the church and the school. The townspeople were no longer cooperating with the collective enterprises she had initiated; they had begun cheating her out of what property she had left, despite all she had done to improve things for them and the community. Though frustrated and angry, Mary was determined to stay on as long as possible. But she and Just agreed that perhaps he should not devote his life to teaching in Maryville; perhaps he should go on to college, away if possible.

One of Mary's major links with the outside world was the *Christian Endeavor World*, the official publication of the Christian Endeavor World Unity group centered in Chicago. A branch of the movement operated out of the Circular Church in Charleston, and no doubt it was there that Mary got her subscription.[65] The journal had guided her in strategies of community organization, in combining religion and politics. She read and studied it with her students and "parishioners." In its pages had been advertised a school—Kimball Union Academy in Meriden, New Hampshire. Perhaps Just could find an opportunity to further his academic and religious education there.

In the late fall of '99 the Just family experienced the last in a long series of tragedies. The schoolhouse in Maryville caught fire and burned to ashes. Mary's work, hope, and dream were gone. She could do no more; her mission had ended. She had "poured out her means and her life" for blacks in the country districts, but when it was all over she felt "broken-hearted because of the uselessness of it all, the ingratitude."[66] All she could do now was inspire her son to go forward, obtain more education, and try to better himself and others. The Clyde Line would be leaving for New York regularly in the spring. Ernest could take one of the ships from the Charleston docks, work on board to pay his passage, and arrive in New York within a week.

☙

Ernest reached New York in April, he was always fond of recalling, with two pairs of shoes and a five-dollar bill.[67] He found himself a room and a job as a cook in a restaurant. The city seemed strange, "cold, self-centered . . . and bustling."[68] But it was merely a stopover, for a few months at the most. Ernest's goal was to further his education. During the summer his mother had written to Kimball Union Academy asking if he could go, but he sent off yet another letter from New York. Sometime early in September, before receiving a reply to either letter, he boarded a northbound night train at Grand Central Depot.

Early the next morning he was at the station on the Central Vermont Railroad in Windsor, Vermont. He ran to get the stage for the campus in Meriden. Though tired, he took in the scene around him. Autumn was fast approaching, and the foliage had begun to turn rust and orange. Before long the sleepy town came into full view. The school's three main buildings—Dexter Richards Hall, Bryant Hall, and Rowe Hall—rose above everything else, exhibiting a New England simplicity of design so unlike the gracious mansions along the Battery in Charleston. Ernest must have at once appreciated both the beauty of the countryside and the isolation of the town, "free from the distracting influences of large places" and conducive to "hard study."[69]

Kimball Union Academy was founded in 1812 by a council of New England churches to educate young men for the ministry. Major financial support had been pledged by the Honorable Daniel Kimball, a prominent citizen of Meriden, and the school opened its doors in 1815. Some years later a female seminary was established with funds from the bequest of Mrs. Kimball. The male and female seminaries came together in 1840. By 1900 the academy no longer functioned as a theological school, but it still prided itself on the "high moral and Christian sentiments" of the faculty and student body.

At Kimball, Ernest received one of the special scholarships awarded to a few "deserving" students. In addition he applied for a "Special Plan" that would help further defray the cost of tuition, room, board, heat, and light. The plan required the student to be of the proper "mind and character," and to spend one hour daily in the "cheerful performance" of some assigned work. The principal, Mr. Woodbury, accepted Ernest's application and assigned him to work in the kitchen as a cook and general helper.

For the first week Ernest kept to himself. He was in an environment very different from what he had known in South Carolina. His mother's school on the sea island and the college at Orangeburg drew only black boys and girls from similar backgrounds. Never before had he interacted with whites in the classroom and never before had he shared living quarters with them. It was natural for him, a thousand miles from home and surrounded by a sea of white faces, to feel uncomfortable. There were no other blacks at Kimball when he arrived, and by the time he graduated there was only one besides himself, out of a total of a hundred and seventy students. Most were New Englanders, though an occasional white came from places as far away as Japan and Jamaica. Somewhat provincial in its student body, the school was nonetheless broad in attitude and fortunate in setting. Ernest was received without incident, not only by the students and faculty but also by the townspeople. There were too few blacks in Meriden for race to be a troublesome issue. Before long Ernest was feeling quite at home. The family of one of his fellow students, Converse Chellis, showed him many "kindnesses" and even invited him over for Sunday dinner regularly.[70]

Kimball was acquiring a reputation for scholastic excellence. The program, rigorous and highly structured, prepared women for "Smith, Wellesley, Vassar, or other colleges" and men for "any college." A student could pursue either the classical course or the academic one, depending on his or her preference and career goals. The first consisted of prescribed training in English, mathematics, and classical literature; there were no electives until the last year, when a student could choose to learn the basics of French grammar. The second provided mandatory training in English and mathematics, with a broad range of electives in the classics, modern languages and literatures, the social sciences, and the natural sciences. Both routes took four years to complete. If a student was not sufficiently prepared to embark on either course of study, a year-long course of remedial work, taken concurrently, was required.

Ernest decided on the classical course—perhaps a conscious effort on his part to broaden his outlook, to taste a kind of educational experience he had missed at his mother's school and South Carolina State College. He knew he would have to do remedial work to bring his academic skills up to the level required by the school. But this did not discourage him. He looked forward to hard work and new ideas. He registered for the standard courses in English, algebra, and Latin, and attended remedial classes in English grammar, arithmetic, geography, American history, penmanship, and spelling. By the end of the first year he had not only caught up with his classmates but was actually well on his way to fulfilling the requirements for the second year.

As a sophomore Ernest continued in the prescribed courses in English, algebra, geometry, and classics. He picked up the fundamentals of Greek, having begun Latin the year before. There was no opportunity for him to get much instruction in modern languages and the natural sciences (only in his final year did he have a chance to learn some French), but this was not really a defect as far as he was concerned. Shortly after arriving at Kimball he had decided to become a classics scholar. He went on to prepare himself well in classical languages and to read widely in classical history and literature. Among the authors and works he mastered were Caesar, Nepos, Cicero, Ovid, Virgil, the *Anabasis*, the *Iliad*, and the *Odyssey*.

Ernest finished the rigorous four-year classical course in three years. Such an acceleration would have been impossible without the support and encouragement of the Kimball faculty, a small and cohesive group, mostly spinsters, with the energy and commitment of New England missionaries. The teachers recognized Ernest as talented and ambitious. Older than his classmates by two or three years, he had an aura of determination and strength unmatched by the younger and less mature students. His teachers were impressed with him, and he was inspired by them. Miss Henshaw laid his lasting foundation in Latin; Miss Mathews opened his eyes to the world of the Greeks—their language, literature, and history; Miss Dike introduced him to the French language, although not much in the way of French literature; she and Miss Miller sparked his interest in modern poetry, English literature, and drama. Mathematics was a field for which he never expressed any particular fondness, but he had to take algebra from Mr. Dow and do geometry with Mr. Woodbury, the only courses he took from men. After hours he further developed his skills in oratory and began to take an interest in physical training, including tennis, horseback riding, and swimming. Miss Stearns, Miss Sawyer, and Miss Hobson directed these extracurricular activities.

Religion was a major aspect of life on the Kimball campus. Each day began with a compulsory chapel service, run by the staff and the students. There were two mandatory sabbath services, a Bible service at 9:45 and a preaching service at eleven o'clock. Also, the school had a strong chapter of the Young People's Society of Christian Endeavor, with weekly meetings and special functions arranged by the students themselves. Ernest participated in all these activities. Religion continued to be an integral part of his life, and it provided him with a way of getting to know fellow students and of meeting teachers outside the classroom.

By the end of his first year at Kimball, Ernest's religious interests had begun to take on grander dimensions than ever before. He was reading all kinds of theology and history. His favorite theologian was Adolf von Harnack, whose work inspired him almost as deeply as the Bible had done ten years earlier.

Born in 1851, Harnack had emerged as one of the world's leading .church scholars by the turn of the century. He held the coveted chair in church history at the University of Berlin. A series of lectures he delivered there in 1899–1900 was translated into English and became a popular book that went into two editions in 1901 alone.[71] *What Is Christianity?* was all the rage at Kimball. The book traced the path by which the gospel of Jesus had become embedded in the doctrines of the Church. According to Harnack, the two have nothing inherently in common and their fusion is primarily a historical concern; the early Church had identified them simply as a means of insuring their survival in the Hellenistic period. The world had come a long way since then, however, and in Harnack's view the gospel would have to be freed from narrow doctrines in order for it to continue as a source of inspiration to mankind.

Harnack's work extended beyond theology into science. In 1900 he completed a three-volume *History of the Prussian Academy of Sciences*, in celebration of the academy's two-hundredth anniversary. Shortly thereafter he became the first nonscientist to be elected to membership in the Berlin Academy of Sciences. His scientific interests continued to broaden. In 1911 he became president of the Kaiser-Wilhelm-Gesellschaft in Berlin-Dahlem, and obtained substantial funds from government and industry to start research institutes in the natural and medical sciences. Ernest was to meet and befriend Harnack in 1930 at the Kaiser-Wilhelm-Institut für Biologie, and at that time he fondly recalled the profound impact Harnack's religious writings had had on him at Kimball.[72]

Liberal, historical, and scholarly, Harnack's views added an intellectual dimension to religious experience. Ernest found that through them he could reconcile his faith with his interest in academic studies, the classics in particular. Still, it was hard for him to abandon the fundamentalism he had acquired as a child. Throughout his stay at

Kimball he continued to have visions of God, Christ, and other biblical figures.

During the winter of his second year at Kimball, Ernest became homesick. He had not seen Charleston in two years. As spring approached, he felt more and more lonely. His mother was on his mind. There was nothing specific to worry about, but no letters had come from her for some time. Then one morning—so the story goes—while he was washing dishes and helping prepare breakfast in the Kimball kitchen, he had a sudden premonition that something was wrong at home. He knew not what, only that he had to get there. He threw off his apron and, without explanation to anyone, headed for Maryville. His brother Hunter had come to New York the previous year and was working as a chef. Ernest picked him up on the way. They arrived home early the next evening. A neighbor greeted them with the news that Mary had been buried no more than an hour or two earlier.[73]

Ernest felt confused his entire trip back to Kimball. He had neither the desire nor the energy to sort out the details of his past, isolate the events of his childhood, and decide what they meant for his future. His mother had been many things to him, and he would leave it at that. He had never really known his father, only her portrayal of him. From photographs, he knew that in a way he looked like Charles Fraser. His forehead, like his father's, slanted back ever so slightly, highlighting his dark, recessed eyes, which were shaded by heavy but unobtrusive brown brows. Always stern, his countenance revealed the same aura of dignity, perhaps arrogance, that was displayed in Charles Fraser's profile. His mouth, irregularly shaped, was unique; it had not come to him from his mother or his father. And his physique was different too. His parents had been short and stocky, while he was tall, slender, and wiry. His hair resembled a bronze toupee; coarse and well-groomed, with a part down the middle, it glowed against his light tan skin. By conventional standards he was strikingly handsome, carrying the racial mix well.

Somehow he could not fit everything about himself—his birth, looks, intelligence, upbringing—into his own mind. Everything had happened too quickly. He hoped that he would be able to feel and understand it all in the years to come, perhaps when he returned to Maryville. For the moment he looked to Kimball with a childlike feeling of euphoria, anxious as he was to pursue an education and achieve the kind of success his mother had wanted for him. He was never to see the lowlands of Carolina again.

Ernest had little trouble getting back to work on campus. As summer approached, he thought about involving himself more deeply in afterschool activities, seeing where he could go. He needed room to grow, things to do. The debating society was the perfect choice. Kimball laid special emphasis on this group. The art of debating was considered very important in secondary education at the time; based on the much-admired skills of rhetoric and oratory, it gave students a chance to present and defend a point of view, a forum in which to project themselves. This was what Ernest needed. He had kept to himself as a child and continued to be shy at Kimball. The debating society would help him overcome his shyness, he thought, as long as he gave it a good try. Advanced work in oratory was offered to any student wishing private lessons, so Ernest took advantage of the opportunity and made special arrangements with Miss Sawyer for coaching during his third year. His reward came at the end of that year, when he had a chance to perform during commencement week.

As his penchant for oratory developed, so did his interest in journalism. The *Kimball Union*, the school paper, was run by a staff of dedicated students who worked hard together and shared many sociable hours. Ernest expressed interest in joining the staff during his second year, but never really became involved. That summer he decided to commit himself to the paper, and in the fall he was elected editor-in-chief for 1902-3. The position involved endless work and heavy responsibility, the kind of organizational and administrative skills which were as natural to Ernest as they had been to his mother. Everything went well. Ernest published the customary two issues per term, performing duties ranging from the tedious tasks of typing and pasting up layouts to the subtle problems of negotiating and signing contracts with important officials at the Dartmouth Press. Some days he had to troubleshoot, others he had to provide intellectual leadership to his staff. Then there were the public relations duties—canvassing wealthy and influential alumni for funds, sending them copies of the paper, and so forth. Ernest must have swelled with pride when he got a letter in mid-February from Henry E. Burnham of the U.S. Senate, commending the paper as "a very creditable publication and . . . another evidence of the spirit of progress which now pervades the institution and gives assurance of a prosperous future."[74] The Burnham letter inspired him to end his career as editor-in-chief with a varied, elegant, and almost professional commencement issue.

Commencement week began on Sunday morning with the baccalaureate sermon by the Reverend Frederic A. Noble (a Bostonian, class of 1854) in the local Congregational Church, followed in the evening by a service for the Young People's Society of Christian Endeavor.[75] On Monday Ernest emerged as an important contributor to the occasion. That evening the school gathered in the Academy Chapel for the Francis E. Clark Extemporaneous Speaking Prize Contest. Ernest's diligent preparation in oratory was rewarded; he easily took first prize, with Alice Mary Henderson of Brown's Town, Jamaica, struggling for second. On Tuesday evening the seniors presented Goldsmith's *She Stoops to Conquer* at the Town Hall. Miss Sawyer, teacher of oratory and physical training, had worked hard as director and managed to include everyone either on stage or behind the scenes. Ernest had been given a special part, that of Tom Twist, a character not in Goldsmith's original cast but created by Miss Sawyer—no doubt a servant or footman to one of the play's gentlefolk. The production was a success, so much so that Ernest later decided to put on *She Stoops to Conquer* for his debut as a director of college dramatics.

Then on Wednesday afternoon, at the class day exercises, Ernest delivered the prophecy. He had been chosen for this task as a gibe at certain of his pretensions to wisdom and his belief in fate, but he took it all in good spirit and managed to make the occasion quite lighthearted. On Thursday, commencement day, the task before him was more serious: he had won the honor of delivering one of the few speeches. Before a crowd of students, parents, and alumni, he spoke on "Government Ownership of Monopolies." His aim was to show that this and all such "sordid socialistic theories" involving redistribution of wealth would result not in a better society but in a violation of the individual's right to earn according to his ability. The speech, originally a class assignment, had won the William P. Fiske prize for excellence in senior essay writing. It was published in the commencement number of the *Kimball Union*.[76]

Alone and without parents, Ernest felt proud of his performance during the week, even if there was no one else there to be proud of him. As he listened to the benediction that concluded the commencement ceremony, he felt a sense of pride in his overall achievement at Kimball. Also, he was looking forward to a new experience. In the fall he would be at Dartmouth. The college had been strongly recommended by his teachers because of its close connections with Kimball. At first he preferred Bowdoin in Maine,[77] but he chose Dartmouth because it was in Hanover, just fifteen miles away from Meriden. Perhaps, Ernest thought, he could return to Kimball to visit old teachers and friends from time to time.

❧

At the turn of the century life at Dartmouth centered around football and, to a lesser extent, baseball and track.[78] A newly formed faculty committee, the Athletic Council, was helping to tie sports more closely to the rest of college life. Football got a special boost. Before 1903 Dartmouth had competed mainly with Amherst and Williams, but its powerful team, the Indians, almost always won the games. A stronger competitor had to be found, and Dartmouth looked toward Harvard. The "dawn of a new era" arrived in 1903, the year Dartmouth first took on Harvard as a regular rival. Harvard had just opened its football stadium at Soldiers Field, and the game against Dartmouth was the inaugural event, the high point of the season. Bonfires lit up Hanover the night before; the faculty, administration, and student body cheered the rugged Dartmouth team forward. Next day the Indians joined the Crimson for the stadium dedication ceremonies in Cambridge. Dartmouth went on to tromp Harvard mercilessly, ending with a decisive victory—11 to 0. The team was to have winning seasons for the next four decades. The gridiron would continue to reign supreme at this little country school, and the Dartmouth colors, the "big green," would fly forever.

Originally founded in 1769 to educate American Indians, Dartmouth had very soon modified its aim to include "a greater proportion of English youth."[79] By the early nineteenth century the college was beginning to attract the sons of wealthy New Englanders, as Harvard and Yale had been doing for some time. But more than either Harvard or Yale, Dartmouth had a very provincial student body well into the twentieth century. Mainly white and Protestant, the students came from New England, New Hampshire for the most part, after having been properly "fitted" at local private schools. They were young men who wanted four years of rugged country living in the farm town of Hanover, amid the sights and smells of chickens and cows and pigs along Wheelock and Main Streets. They loved sledging through the snowcapped White Mountains, romping through the muddy Vermont pastures nearby, and bathing in the Connecticut River just down the hill. They came to this remote and scenic setting to meet others with similar interests, to join fraternities; at the end of four years they returned to family businesses and settled down to reap the rewards of their Dartmouth heritage. Theirs was a life of privilege and pleasure, only a small part of which involved study and booklearning.

Yet there were always a few students at Dartmouth who sought and received a rigorous, advanced course of study. Ernest came to Dartmouth in the fall of '03 with just this purpose, to be introduced to the highest level of thought in American education. Before settling down to serious work, however, he first had to locate a place to live. For financial reasons he decided not to live on campus, but to take a small room on the second floor at 17 South Main Street. Next year, finances permitting, he would move onto campus and perhaps join more fully in collegiate life.[80]

Just did not really participate much his freshman year; in fact, he retreated from the scene altogether. There were more people to know than at Kimball and more things to do, but he could not even begin to like the students and the place. The Dartmouth spirit, with its casual approach to academics and its emphasis on football, struck him as strange. Also, his classmates were at a different stage of life than his Kimball friends had been. In general, students at the secondary school had had few social prejudices and competitive urges, but those at the college were beginning to form close friendships that would later blossom into business relationships, for college was a crucial link to the professional world. The students tended to be self-centered and aggressive, whereas in high school they had been more relaxed, honest, and helpful—qualities that had to have been important to the only black in this Dartmouth class of 287 whites. Another sharp difference was that social life and school activities at Dartmouth were run by men who were not benevolent in quite the same way as Miss Stearns and Miss Sawyer.

Ernest felt all this acutely. For the first six weeks he spoke to no one on campus.[81] He did not try out for the debating team, the editorial staff of the *Dartmouth*, the dramatics club, the choir, or the football team. The only organization in which he expressed the slightest interest was the Young Men's Christian Association; he was elected to the rank of associate member on 15 November.[82]

Scholastic achievement was Ernest's primary goal. He settled down to work as quickly as possible. The freshman program, he found, was straightforward enough, with requirements in Greek, Latin, English, French, mathematics, "hygiene," and oratory. Because classics was his first love, he started advanced work in Greek. The reading for the course included Herodotus and Plato in the fall term, Euripedes and Xenophon in the spring. Ernest also enrolled in Latin courses that covered works by Livy, Horace, and Terence. The English course he chose was on composition, rhetoric, and criticism, geared to strengthening basic skills in expository writing. He signed up for introductory rather than intermediate French, perhaps feeling insecure about his quick course at Kimball. For mathematics, never his strong point, he took algebra in the fall, solid geometry and trigonometry in the spring. Hygiene and oratory, a semester of each, rounded off his program.[83]

Just's first year was mixed. In some things he did excellently, in others he barely passed. He got a perfect grade for hygiene, perhaps a reflection of the intense concern for cleanliness he had developed as a child. Oratory was a different story. Even though he had excelled in the art at Orangeburg and Kimball, the quality of his performance dropped drastically at Dartmouth. He received a final mark of 65, indicating satisfactory but hardly excellent work. Perhaps he felt uncomfortable delivering weekly declamations in this new and strange environment. In English, French, and Latin he performed well above average, ending with marks in the mid- and high 80s. He did not fare so well in mathematics. A necessary evil for him, the subject occupied little of his attention. He barely passed, receiving a mark of 53 for both terms. His grades in Greek, on the other hand, were outstanding. He received 93 the first term and 90 the second, the highest marks any freshman had ever received in Greek at Dartmouth. This feat made his reputation on campus.

By June Ernest was committed to a scholarly career in Greek literature. It was a difficult commitment to follow through on, however, in a school placing so much stress on athletic success. No doubt it was especially difficult for Ernest, who was one of the few blacks at Dartmouth not starring on a college team. His marks in Greek were considered less of an achievement than the latest touchdown pass by the school's black quarterback, Matthew Bullock.[84] Not everyone could be a football star, of course, but even if a student was not athletically active, he was expected to compensate by showing intense team and school spirit. The college newspaper, the *Dartmouth*, chided students who placed their studies above their loyalty to the school.

Despite all these pressures, Just kept at his books. He was most likely seen as a "grind," reading rather than cheering and drinking, studying rather than dating and dancing.[85] But scholarship was not the only reason he gave little of his time to campus life and school sports. Because he needed money, he had to work at various odd jobs on and off campus. There were room, board, and tuition costs; miscellaneous school charges, such as a football tax of fifty cents every month; savings sent to Inez and Hunter for their education and living expenses.[86] Any time off from study, his first love and commitment, had to be spent working.

Weekend dating, a prominent social activity in colleges at the time, does not seem to have been part of Ernest's life at Dartmouth. Most Dartmouth men selected their dates from Smith, Mount Holyoke, Vassar and the like, but for Ernest or any other black it would have been a waste of time and money to make the trip to these women's colleges. Interracial dating was taboo, and there were few if any black women on those campuses. By and large, black men from colleges such as Dartmouth chose to travel elsewhere, most often to Boston, where there was a small black community with a relatively wide choice of young women. Ernest did not join in, however, as he could neither afford the heavy travel costs involved nor spend an entire weekend away from his books.

Still, Just wanted to take more part in college life. He preferred living on campus to living off. In the fall of his sophomore year he moved into 26 Hallgarten—closer to the center of things, but still on the outskirts of the college's main buildings. Though constructed of brick, well lit, and heated with steam, Hallgarten was an undesirable place to live, for it had no janitorial services and few plumbing facilities. The students nicknamed it "Hellgate."[87] Previously called Conant Hall, the building had been the residence of the superintendent of an agricultural school in Hanover, then later a dormitory for students of agriculture and a boarding establishment for students in general. When the agricultural school closed down, Dartmouth bought and remodeled the building. It was never liked by the undergraduates,[88] even though the rent was cheap enough to attract boarders. In 1904 rooms throughout the college cost anything from $30 to $120 annually, and Just's room at 26 Hallgarten came near the bottom, renting for only $40. Over the years a high proportion of black students lived in Hallgarten, presumably because the rent was low; whites who needed cheap housing joined them. Ernest was to remain there, in single quarters, for the remainder of his undergraduate years.

Once in Hallgarten Ernest resumed his studies, but not as eagerly as in his freshman year. Too much was happening around campus. Lord Dartmouth was coming from England to take part in the dedication ceremonies of the new Dartmouth Hall, and the campus was in an uproar over preparations for the earl and his entourage, including his wife and daughter. The gloom that had set in because of the loss by fire of old Dartmouth Hall in mid-February changed to excitement at the prospect of a visit from a member of the English nobility. Classes were moved and rescheduled, some canceled altogether, and Ernest must have become more and more despondent about not being able to get down to work. Even after Lord Dartmouth came and went during the last week of October, there was no relief from the constant flow of activity. The campus remained chaotic through November owing to the demands of fall football. An unusually high number of out-of-town games played havoc with the academic schedule. Bonfires had to be built and cheers had to be sung well into the night each time the team was about to leave Hanover. Amid all this frivolity, Ernest found intellectual stimulation by joining two studious friends, Ray Spencer and Dwight Hiestand, whom he had met in Hallgarten, for discussions of subjects ranging from the genealogy of Alexander the Great to Edith Wharton's *The Valley of Decision*.[89] Not before mid-December did regular classes resume and the college settle back into its usual routine.[90]

The first semester resulted in disappointment, if not disaster. Ernest was carrying a heavy load of seven courses—two in Greek, two in biology, and one each in history, English, and German. Upset by the confusion, he missed several classes. By December he had accumulated a total of fourteen "excused" and five "unexcused" absences from class. According to school policy, he was docked four points of the final grade of a course per unexcused absence in that course. He had also missed chapel thirteen times, and the rules stipulated that each absence over seven per semester would result in the reduction of a student's general average by one point. Ernest had six excess absences, so his average was reduced by six points. Not surprisingly, he received a D (deficiency) in most courses. After much worry and serious study, he replaced most of the Ds with pass marks by sitting for makeup examinations in the spring. The one course he did not manage to pass was oratory. All in all, however, he finished up not too badly, with an average of 78.2.

The first semester of Ernest's sophomore year had one positive effect: he was introduced to a new field of interest—science. He took botany under Professor Lyman and did fairly well, ending with a mark of 76. But it was a course entitled "The Principles of Biology," offered by Professor William Patten, chairman of the biology department, that really struck his fancy. A Harvard graduate with a Ph.D. from Leipzig, Patten was an established scientist who had made his reputation in research on evolution. He was a dynamic teacher and enjoyed introducing undergraduates to the mysteries of biology. He stimulated lively discussions of the phenomena of nutrition, growth, reproduction, heredity, variation, distribution, natural selection, and, perhaps most important of all, evolution. Ernest was captivated. He was also impressed by Patten's publications, which he described as "studies of rare beauty and marvellous technical excellence."[91] He worked hard for the makeup examination and ended that difficult first semester with his highest mark, 85, in Patten's course. On the whole, his final grades in science were better than those in Greek. Perhaps he had found an interest to rival his devotion to the classics.

In the second semester, Ernest's program did not change. Signing up for courses in Greek, biology, history, and German, he was determined to study much harder in all subjects, improve his grades, and keep the small scholarship Dartmouth had been giving him. As soon as the term began, however, he lost perspective and focused all his attention on a course on Demosthenes. In addition to the course requirements, he did some independent research in philology and began a project to construct a concordance to Demosthenes' work.[92] Spirited and intent, he worked through April and into May before be realized that it had all been done before, by a German scholar in the eighteenth century. The experience made him want to give up Greek completely. Moreover, he had fallen behind in two of his other courses, German language and Greek translation, while going off on the Demosthenes tangent. In fact, he failed so badly in German that he never took another course in that language again—ironic in light of his later love for the Germans and their culture.

Eager to press forward, Just decided to major in science, biology in particular. The world of plants and animals held mysteries he wished to explore. He knew that biology at Dartmouth would be a difficult major—the professors were demanding, the subject matter was complicated at the advanced level, and the extensive laboratory work required would cut into the time he could devote to an outside job. But biology was a pleasure, worth every bit of hard work and financial sacrifice. Just longed for a chance to look more closely at nature, the world of living things which he had first come to know and love on James

Island. He would bring the same fervor to science that he had put into religion and the classics, perhaps even more. Science would be his; the joys and rewards would be private. No longer would it matter that as a black he could not join fraternity life; he would not be depressed by the constant talk of football in the dormitory and classrooms; he would not be disturbed that his classmates were, for the most part, ambitious social climbers rather than thoroughbred scholars.

Just also declared two minors, Greek and history, perhaps because he did not want to repeat the mistake of devoting himself too single-mindedly to one field. He was never to take another course in Greek, but as a junior he took five courses in history, spanning the period from the Renaissance through the nineteenth century and covering both Europe and America. The biology courses he chose, six in all, ranged from phanerogamic botany to invertebrate zoology. He received excellent grades in everything. In spite of his precautions against overspecialization, he put more effort into biology than history, and in biology he seems to have been captivated by zoology rather than botany.

One of the zoology teachers was John H. Gerould. A graduate of Dartmouth, he had recently returned to his alma mater to teach biology, having spent several years doing research at marine laboratories in Italy and France where Just was later to work. Gerould taught three courses in which Just was registered his junior year: invertebrate zoology, cytology, and the comparative anatomy and physiology of the nervous system and sense organs. He immediately recognized Ernest as a brilliant student with a talent for science, and went on to offer him "inspiration, kind sympathy and intellectual aid" during the remainder of Just's undergraduate years.[93] A "deep fraternal affection" grew between them and lasted long after Ernest left college. Gerould was always puzzled by Just's color, for he had never before taught a black who performed so superbly. Was Just black or white? Though he knew nothing of Just's parentage, Gerould always "credited" him with "some very exceptional German father or grandfather."[94] He preferred to see Just as white. It was "curious" to him that American society classified a man "as negro though he may be largely white." He felt Just had been wrongly classified, and wanted him not to be deprived of opportunity because of the "handicap" of his "partly negro blood." Gerould's generosity was genuine, even if his analysis was shallow.

But even more than Gerould, it was Patten who inspired Ernest to achieve academic excellence and to start thinking about a possible career in science.[95] Ernest got to know Patten well during the spring of 1906, when he was taking his course on the comparative anatomy of vertebrates. Feeling motivated to do some independent research, he sought permission to work on a special project. Patten granted the request. He teamed Ernest up with a fellow student, A. O. Kelly, and set them to work on a section of the book he was writing. Ernest devoted at least half of the course time to studying the development of oral arches in frogs, "a pretty difficult problem for a beginner."[96] He did "excellent" work, unraveling the history of the embryonic and larval jaws of the frog, putting together about a dozen plates to illustrate the text, and advancing a new view concerning the number of oral arches in vertebrates. The work was fitted into the chapter on the oral arches of primitive vertebrates in Patten's classic book, *The Evolution of the Vertebrates and Their Kin.* Patten gratefully acknowledged Ernest's assistance in a footnote.[97]

In his junior year Ernest performed exceptionally well, and ended with an overall average above 92. He was awarded the title of Rufus Choate Scholar, the highest academic award for an undergraduate at Dartmouth: any student who achieved a rank of 92 was so designated. The title carried no "pecuniary allowance," but it was a prestigious "term of honor" that only three or four students per class won each year. As a matter of fact, Ernest was the only Choate Scholar in the junior class of 1906.

There was just one year left to round out his studies. Though he had already met the requirement for the minor in history, he decided to take another course in the fall and yet another in the spring. Also, he chose to further his knowledge of botany by taking another course under Professor Lyman. Zoology remained his first love, however. There were only three zoology courses he had not taken, and he registered for them all: Patten's course on vertebrate embryology the fall term, his course on research in animal morphology, and his seminar in the spring designed especially for students who intended to go on to graduate work in zoology. Ernest planned to use the seminar to write his undergraduate thesis. His schedule would exhaust the offerings in biology and almost all those in history, but he wanted even more. He was looking for a way to bridge his work in biology and history. Sociology, a relatively new field, was used at Dartmouth for just this purpose, and, more generally, to tie together the natural sciences and the social sciences. A student could even pursue a major in sociology by combining two biology courses with a number of sociology courses. Ernest had no such plans, but the field seemed to offer the kind of broad perspective he was hoping to acquire. He took all four courses given by the sociology department during his senior year: "Ethnology" and "Anthropological Sociology" in the fall, "Biological" and "Psychological Sociology" in the spring. These courses sparked an interest in anthropology—one that Ernest would maintain later in his career as a zoologist.

Senior year was climactic for Ernest. He performed brilliantly in his courses, wrote a thesis, and even published two short stories and a poem in the *Dartmouth Magazine*, which issued high-quality literary pieces by undergraduates.[98] As the end of the year approached, however, he became more and more eager to leave Hanover. He had been there for four years without going home at all; he had been in New England for seven years, going home only once. The granite of the surrounding hills was beginning to depress him, to make him feel "hemmed in."[99] He wanted to return to the "broad, level stretches" of the Carolina lowlands, to be in Charleston again, to see the wharves and hear "the melting chimes of St. Michael's with their reserved quality of pensive sadness." New England was a strange world, one that he had not adjusted to fully. If he stayed much longer perhaps he would lose his ability to do what the South Carolinians did so well—feel and fear, treat life as a "joy" rather than a "business." His ultimate goal was to write, perhaps even to teach literature and drama, but at the moment he just wanted to revive fading memories. With a somewhat Miltonic fervor, he tried to articulate his sense of emotional urgency in a poem entitled "The Crucifixion":

> An Angel brought the tidings—wildly rushed
> Through streets of gold, his diamond lance dipt in
> The spilt blood of the Prince. Wild-eyed and dazed
> His fellows saw the symbol, straightway turned
> To marble caryatids. Wailed the saints.
> Job cursed the day, and marshalled all the Blest
> In arms. A tumult raged, the din of hell.
> The king upon his snowy throne enraged
> That His one Son such mortal agony
> Shouldst bear, in anger stood upright; one foot
> Upon his footstool weighed—it burst in seams
> And yawned in gulfs whence dead men came, and stalked
> About, instilling terror in the hearts
> Of mortal men; from seas new islands rose,
> New continents were made, the oceans found
> New paths; huge icebergs sweeping south were caught
> And held, to stone were changed. The hundred gates
> Of heaven smoked; the sun in sack-cloth hid;
> The moon did swoon, and could not mount her car,
> And ever and anon flashed from the hands
> Of God's most chosen knights, ten thousand swords:
> The voice of God appalled the universe.[100]

The religious joys and fears of his childhood swept over him. Strangely, however, they seemed detached and remote, a thing of the past. Had he lost touch? Perhaps he had grown up a little; perhaps he was becoming used to the world away from home.

Ernest was conscious of the intellectual benefits of his four years of study and hard work, but it was not until commencement day—the one time when the scholarly community paraded itself in public, when town and gown came together for praise and glory, pomp and circumstance—that he experienced the tangible rewards. On Wednesday, 26 June 1907, people from all over New England converged on Hanover to celebrate the 138th Dartmouth Commencement. There were official dignitaries, alumni, relatives and friends of seniors. Ernest had no family present to help him celebrate the occasion, but there he was nonetheless—prominent in the ranks as the top honors student in a class of 182.[101]

He had won everything. His superior grades ranked him as Rufus Choate Scholar for the second consecutive year, an honor he would often recall in later life. He also received the Grimes Prize for General Improvement, awarded to "that member of the Senior or Graduating Class who, in the judgment of the College Faculty, has made the most satisfactory progress during his College course, taking into consideration his preparation for the course when he entered." The prize carried the sum of seventy dollars, but it did not hold the same prestige as the Rufus Choate award. It is the one prize Ernest never mentioned in later life, perhaps because he detected condescension in its wording.

Every award and citation that Ernest received was won fair and square; there was no question of charity. Because of his stellar performance as an undergraduate he was elected to the ranks of Phi Beta Kappa. He was given a "commencement mark" for averaging above 85 percent in his course work for four years. He and only three other students received "honorable mention" in a subject; his was in sociology. He was the only student to be awarded "honors" in a subject; in fact, he received honors in three—botany, history, and sociology. He and only one other student captured "special honors" in a subject; his was in zoology, his favorite field. And it was he, and he alone, who graduated *magna cum laude*. Three classmates received the lower rank of *cum laude*, none the higher honor of *summa cum laude*.

Despite his achievement, Just attained neither valedictory nor salutatory rank; nor did he deliver any of the commencement addresses. The posts of valedictorian and salutatorian went to the students with the two highest averages. Just ranked tenth, so he was not in the running for those honors. On the other hand, he was a candidate for selection as a commencement speaker. There were to be six speakers in all, the two top students and four of the other fifteen having commencement marks. Just was not selected. For some reason, the faculty appointed six speakers privately but listed only five names publicly.[102] Perhaps they had chosen Just, then decided it would be a faux pas to allow the only black in the graduating class to address the crowd of parents, alumni, and benefactors. It would have made too glaring the fact that Just had won just about every prize imaginable.

The ceremony went without incident. After the benediction, the students went their separate ways. At twenty-three Just had reached a goal he had long been aiming for: he had earned a bachelor's degree from a first-rate college. The professional world lay ahead. He set his sights beyond Hanover.

CHAPTER 2

The Beginnings of a Professional Career, 1907-16

In 1907 there was no chance for a black college graduate to enter the white professional world, even if that black hailed from so prestigious an institution as Dartmouth. An educated black had two options, both limited: he could either teach or preach—and only among blacks. Having neither training for nor interest in the ministry as such, Just saw education as the profession to pursue. A Dartmouth graduate, he naturally wanted to teach at a good black university, preferably Morehouse or Howard. Morehouse had a dynamic president in John Hope, Howard an experienced dean in Kelly Miller. Both men wanted Just and perhaps competed for his services. Just thought hard before reaching a final decision.

Morehouse College provided a superior undergraduate education for black men primarily from rural Southern towns; Howard University offered excellent undergraduate and graduate education in many academic disciplines—as well as professional schooling in law, dentistry, and medicine—to black men *and* women from throughout the nation. Morehouse was in the deep South, in Atlanta, which was rife with jim crow practices, even lynchings; Howard was in Washington, where jim crow practices were less oppressive and lynchings almost nonexistent. Howard seemed the right choice.[1]

Wilbur P. Thirkield, president of Howard since 1906, was as eager as Miller to employ Just, "a brilliant young man about to graduate from Dartmouth who had gained high rank in English and science."[2] Aware of "the need of exalting English in the curriculum," he offered Just a position as instructor in English and rhetoric at an annual salary of $400.[3] Contrary to Thirkield's impression, Just had not received honors in English at Dartmouth. He did enjoy reading and writing, however, and the offer from Thirkield was too good to refuse. He packed his bags and boarded the train for Union Station.

It was his first real job. At Howard, he threw himself headlong into teaching an introductory course in literature and a basic course on rhetoric, given in the Teachers College and Commercial College respectively.[4] In the literature course he focused on simple selections of prose and poetry, stressing interpretation and appreciation for its own sake and ignoring complicated historical questions; in rhetoric he taught basic English grammar and assigned short themes daily. The students in the Teachers College and the Commercial College were less academically oriented than those in Howard's prestigious College of Arts and Sciences. Just had not only to instruct his students, but also to motivate them. He proved to be a "superb" teacher who "awakened the enthusiasm of his pupils."[5]

The following year Just was assigned to teach at a more advanced level, in the College of Arts and Sciences. He taught "Narration and Description," a required course for freshmen. The reading list included works of history, biography, fiction, and drama. Using classic writers as models, Just guided his students in constructing narratives on various themes. He stressed sympathetic interpretation of character and motive and at the same time demanded careful form and fluid style in their compositions. Notebooks, sketchbooks, pictures, and newspaper clippings were some of the materials he used to provide students with ideas for plot development. The students were fascinated by his approach. He had a sensitive and adept mind and a talent for description and narration, and could turn the most banal narrative into one of marvelous intrigue.

Just also taught "Exposition," a course required of sophomores. Here he substituted scientific treatises and essays for history and fiction and drama. His concern was with helping students develop skills of logic and argumentation. It was crucial, he thought, for everyone to achieve a level of expression capable of making a complex issue comprehensible to a "simple mind." Again he used classic writers, this time as models of stylistic clarity. William Harvey's *On the Motion of the Heart and Blood* and Henry David Thoreau's *Civil Disobedience* were both assigned. Just had read these works as an undergraduate at Dartmouth, and had been impressed by the literary as well as the scientific brilliance of Harvey's work in particular. Since this was a course about writing, not science, his lectures concentrated on the quality of Harvey's style.

The main impetus for Just to switch from English to biology came from Howard's president. In the spring of 1909 Thirkield was completing plans for a science building, scheduled to open in 1910. It was an attempt on his part to develop scientific education at Howard. In addition to the new building, he wanted to put together a new faculty, to increase the number of black scholars in the school's science departments and reduce the number of white liberal missionaries generally.

He recalled Just's "exceptional . . . record in biology" at Dartmouth.[6] While there were many black English teachers available, Just was the only one "well-equipped" for biology. Thirkield encouraged him to leave the English department and help Richard E. Schuh, "a high-class biological professor" from Pennsylvania State College, revitalize the biology department. Just was opposed to the change, but said nothing. He accepted the dual title of "Instructor in English and Biology," marking his first step in making science his career. Thirkield later congratulated himself on having "discovered Just," and called it "one of the greatest" discoveries he ever made.

When the new science building opened in 1910, Just joined Schuh in the Department of Biology and Geology.[7] He handled the zoology, Schuh the botany and geology. He taught an introductory course in general zoology, taken primarily by freshmen, with Hertwig's *Manual of Zoology* as the basic text. In the lectures, given four times a week, he tried to instill in his students the general principles of zoology as part of a liberal education and also in preparation for the study of medicine. In the laboratory sessions, conducted twice a week, he helped students analyze the structure, life history, and physiology of various animals. In his lectures he delivered his ideas with force, enthusiasm, and authority; in the laboratory he demonstrated the use of apparatus and methods of observation, always suggesting new and innovative approaches to laboratory technique and experiment design. Exposure to such an excellent teacher was rewarding for the freshmen, many of whom were taking a science course for the first time. They went away with a deeper appreciation for the world of living things and, in some cases, a desire to continue their study of science.

Just also taught two advanced courses in zoology, invertebrate and vertebrate, using Parker and Haswell's *Textbook of Zoology*. In the invertebrate course he dealt mainly with comparative anatomy, physiology, and embryology. But natural history came in for extensive treatment too. Just was, then and later, intent on having his students learn the habits of the organisms under consideration. In the vertebrate course he likewise stressed the comparative study of anatomy, with considerable attention to histology and physiology as well. The course was designed for students who wanted to go on to study human anatomy, physiology, and psychology. In fact, it became a requirement for premedical students and others continuing in allied fields.

In 1910 Just completed his switch from English; he acquired the title "Assistant Professor of Biology." Having given up all responsibilities in the English department, he added to his teaching load a course on the principles of animal histology. This was an advanced course, an elective for juniors and seniors who had had two previous courses in zoology. The focus was on the structure of the cell, with Dahlgren and Kepner's *Principles of Animal Histology* as a basic guide. Some attention was given to the comparative histology of animal tissues and to the structure of systems in various animals, including invertebrates. The course rounded out the offerings in zoology at Howard. Just was well on his way to instituting a rigorous and innovative academic program, and to establishing himself as a first-rate teacher of science.

His work did not go unnoticed by Thirkield's administration. He advanced through the ranks rapidly. In 1911 he was appointed associate professor at an annual salary of $1,500. Tireless, he examined ways of extending his work to the professional training of doctors. In 1912 he was appointed professor of biology in the college, at an annual salary of $1,250, and professor of physiology in the medical school with a supplementary stipend of $900, making his total salary $2,150. He was earning maximum salary and exerting maximum influence on science education at Howard.

Though he had climbed the professional ladder by virtue of his versatility as a teacher, Just no doubt hastened his ascent by contributing significantly to extracurricular activities on campus. At Kimball Union Academy he had been involved deeply in forensics and drama. On arrival at Howard in 1907 he was struck by the lack of art and culture in the college community, especially the absence of any organized performing arts. He sensed, in particular, the need for a drama society. In 1909 he set about the task of organizing a club to produce several plays a year at Howard.[8] He found a number of interested and talented students and convinced two other faculty members, Benjamin Brawley and Marie Moore-Forrest, to act as cosponsors. Less than a year later Goldsmith's *She Stoops to Conquer* came to the Howard stage. Just was familiar with the play from appearing in it at Kimball Union and had little trouble persuading his coworkers to produce it as the club's debut. His past experience reaped dividends: the Howard production of *She Stoops to Conquer* won tremendous acclaim. Howard went on to achieve a reputation for excellence in drama.[9] Just gave active support to the drama club until the 1920s.

During his early years at Howard Just became known on campus as an energetic young man willing to offer advice and guidance to students. A bachelor with few domestic obligations, he often spent his social hours with students, going on field trips, playing tennis, and swimming. He was the perfect person to serve as advisor to a group of young undergraduates interested in fraternity living. At Kimball he had been a member of the all-male Philadelphian Club, and he knew the social importance of such groups. So when Edgar A. Love, Oscar J. Cooper, and Frank Coleman descended on him in his office on the evening of 15 November 1911, he agreed to serve as faculty advisor to a group trying to establish a nationwide fraternity of black students.[10] But that was only the beginning. Much work would have to be done to convince the Howard administration to recognize such a fraternity. Fearful of the political threat a secret organization of young blacks might pose to Howard's white administration, the university's faculty and administration opposed the whole idea. Just was to play a major role in bringing them round. Viewed on all sides as a sensible young professor, he was chosen by the fraternity organizers to serve as mediator between the administration and the student body.

On 15 December 1911 the Alpha chapter of Omega Psi Phi was organized at Howard. Early the following year the members elected Just the first honorary member. It was not until 29 October 1914, however, that the faculty finally allowed the fraternity to be officially established on the Howard campus and incorporated in the District of Columbia. Though the students had done most of the organizational work, they always acknowledged Just as a founding member. His later scientific achievements were identified with his loyalty and dedication to the fraternity, and he was viewed as an embodiment of the fraternity's "ideals of fellowship, scholarship, and manhood."[11]

These were ideals that Just had set for himself early in life. By the 1910s he knew that he would excel. He wanted to be a success, perhaps even a notable in the field of biology. A mark of distinction, he thought, would be good for the race; he hoped it would be good for Howard, and he knew it would be good for him.

Now that Just was discovered, he had to be developed. Anxious to pursue a career in biology, he contacted his old teacher at Dartmouth, William Patten, about possible routes for graduate training. This was sometime in 1908 or early 1909. Patten advised him that medicine was perhaps a "wiser course" for a black interested in science, but he promised to arrange a contact if Just had his mind set on biology.[12] Just was adamant, so Patten put him in touch with Frank Rattray Lillie, head of the Department of Zoology at the University of Chicago and director of the Marine Biological Laboratory (MBL) at Woods Hole, Massachusetts. As a member of the board of trustees of the MBL, Patten had known Lillie for years. He was certain that Lillie, coming from a major institution granting graduate degrees, would be a good person to advise Just about career possibilities in science.

When Just contacted Lillie, he was warned again about the problem of race. He showed such determination, however, that Lillie advised him to come to Woods Hole for the summer as a research assistant, gain firsthand experience in laboratory technique, and acquire extensive theoretical knowledge in the biological sciences by enrolling for course work.

Just agreed. There were two courses he needed to take: invertebrate zoology under W. C. Curtis and embryology under Gilman A. Drew. His duties as Lillie's research assistant, however, would be so rigorous that he could not think of registering for them both. Wisely, he decided to take invertebrate zoology in 1909 and leave embryology for 1910.

Course work began in late June and ran through part of August.[13] Twice a week Just attended a morning lecture. Guest talks by eminent scientists, "the biological mighty," took place on a regular basis. Just's classes were led by Curtis and Drew, but many others—W. E. Kellicott, L. L. Woodruff, E. G. Conklin, R. A. Budington, T. H. Morgan, and E. B. Wilson—made regular contributions.[14] All experts in their fields, these men inspired Just to work long and hard. After the morning lecture he often hurried off to board a marine collecting vessel, either the *Cayadetta* or the *Nereis*, for a field trip to Cuttyhunk, Gay Head, Nonamessett Island, Tarpaulin Cove, or Vineyard Haven. Some afternoons were spent collecting and tentatively identifying well over a hundred specimens, everything from squid to horseshoe crabs, within two or three hours. Then it was back to the laboratory for an evening of hard work—identifying, labeling, and drawing specimens; dissecting and embedding eggs of the sea worm *Nereis* and the sea urchin *Arbacia*; locating microscopic ganglia in the marine snail *Busycon*; and watching the often curious process of cell division. On the whole, course work at the MBL was a demanding and enjoyable experience for him.

From all indications, Just's first two summers at Woods Hole were very successful. His interest in science flowered, capturing his multifaceted imagination. He completed the invertebrate zoology course "in a very creditable manner."[15] In embryology, he proved to be "a faithful, hardworking student who grasped the work presented thoroughly"; he was rated by the instructors as "one of the best" students in the laboratory. His best work, however, was done as research assistant to Lillie. He excelled in preparing slides for microscopic observation—sectioning, mounting, staining, and so on—and in collecting marine specimens, as well as in the routine chores of washing glassware, maintaining supplies, and cleaning floors and tables. Never before had Lillie observed such intelligence and devotion in an assistant.[16] It was the beginning of a student-teacher relationship that would blossom into full and equal collaboration.

From the very beginning Lillie took an interest in Just's development as a scientist. He arranged for Just to enroll at the University of Chicago *in absentia* in 1911 and begin work for the Ph.D. by taking courses at the MBL.[17] Many of Lillie's former students had done likewise, and he wanted Just to have the same opportunity. A year of residency in Chicago was required before Just could formally apply for the Ph.D., however. It was a difficult requirement for Just because of his numerous commitments and teaching responsibilities at Howard, but he would have to find a way to fulfill it somehow. Lillie also encouraged Just to undertake scientific research. In 1909 he set Just to work at Woods Hole on the problem of cell cleavage in the eggs of *Nereis* and *Arbacia*. This work laid the foundation for Just's lifelong interest in the study of marine eggs.

In the summer of 1911 Just hit upon an important discovery in the area of cell cleavage—one which he set forth a year later in his very first scientific publication, "The Relation of the First Cleavage Plane to the Entrance Point of the Sperm." He argued convincingly that biologists had been wrong to accept the notion that the cleavage plane invariably corresponds with the median plane of the embryo; his results demonstrated that the sperm entrance point is critical in determining the line of cleavage. At the time, the thrust and import of this discovery were generally recognized. The article came up for discussion in domestic and foreign journals, including *Biological Bulletin* and *L'Année Biologique*.[18] Years later it was to be cited by the renowned geneticist and Nobel laureate T. H. Morgan as the fundamental and authoritative study on the subject.[19] Publication of the paper increased Just's self-esteem and called attention to his work. Thereafter he was viewed around the MBL as a hardworking and intelligent young man with a possible future in science.

During his first years at Woods Hole, Just met many scientists, and his opportunities for graduate study and professional advancement increased. Though he went to Woods Hole primarily to learn science, he wanted also to make contacts that would extend his career beyond the gates of Howard. As early as the summer of 1911 there were indications that people were becoming interested in him and his work. At that time Thomas H. Montgomery of the University of Pennsylvania approached him with a scientific problem to solve.[20] A mere beginner, Just was flattered to receive overtures from such an eminent scientist. But although he wanted to work on the problem, he was too busy; he informed Montgomery that he could not undertake the research for at least a year because of heavy teaching commitments at Howard. Montgomery then advised him to apply for the Harrison Fellowship at the University of Pennsylvania. Once the application was processed, Montgomery said, Just could go to Pennsylvania and enjoy a year of research free from teaching responsibilities. Over the course of the 1911–12 academic year Howard administrators pressured him to accept the offer; they could not see why he should insist on the University of Chicago, which had few fellowship funds available. The reason, though Just never mentioned it to the people at Howard, was that he did not want to associate himself with anyone but Lillie. He had developed a good relationship with his mentor and would do nothing to jeopardize it. He put aside Montgomery's suggestion and kept up his registration at Chicago.

The matter did not end there. Just was concerned that Lillie would misinterpret what had happened. He was afraid that Lillie would feel slighted, and was anxious to explain that the initiative for the proposal had come not from him but from Montgomery. The fear was unfounded. Lillie understood, and he wanted Just to feel free to act independently when it came to offers of that kind. On the one hand, he appreciated Just's loyalty; on the other, he feared the consequences should Just carry that loyalty too far. He was comfortable with some sort of balance —something that Just, as the only black trying to enter a white community, could not easily maintain.

As spring approached Just began to consider a number of plans for the summer of 1912, not all of which included Lillie and Woods Hole. By April he knew that he would not be able to take leave for residency at the University of Chicago in the 1912–13 academic year,[21] so he resigned himself to the disappointment of not being able to obtain his Ph.D. for some years to come. This did not dampen his enthusiasm, however. He was ready for some hard work with Lillie at Woods Hole. But he thought that instead of devoting his entire summer to biological research, he might spend July and August traveling through France and Germany, in order to improve his facility with languages. With Lillie's approval, he would collect and study the sea worm *Nereis limbata* at Woods Hole through June and sail for Europe 1 July.

Lillie decided against the proposed vacation, despite evidence that Just was in need of a change of scene after an unusually hard year of work at Howard.[22] His idea was that Just should use his entire summer to consolidate the *Nereis* work. Though he stressed that the decision had to lie with Just alone, he requested that it be made as soon as possible, to facilitate arrangements for another research assistant if need be. The request was too much for Just; his feelings of loyalty overcame whatever impulse he may have had to act independently. Any response from Lillie that was less than enthusiastic, or that had even a hint of a reservation about it, would have made him alter his plans. Perhaps after three years Lillie did not know that Just's character was so composed, but he was fast finding out.

Just packed his bags and went to Woods Hole as Lillie's research assistant. He arrived in the new capacity of "beginning investigator."[23] His status had remained that of "student" since his first summer in 1909, but in 1912 he sought and received a promotion from Lillie. He had been loyal and hardworking, and now he was even giving up his plans for European travel to work with and for Lillie. His request seemed reasonable to him and, as it turned out, to Lillie as well.

The work that summer went well. Lillie and Just wrote an article together on the breeding habits of *Nereis limbata*. The article appeared in 1913. In 1914 Just published an independent article on the breeding habits of a related marine worm, *Platynereis megalops*, using observations he had made while working on the *Nereis* article in 1912. Both articles were important in setting the stage for Just's work as a beginning graduate student.[24]

🖋

Early in the summer of 1912, Just returned to Washington from Woods Hole to consummate a long-standing courtship. On 26 June he married Ethel Highwarden. The ceremony took place at the Highwarden home, and was attended only by close relatives.

On her wedding day, Ethel displayed a youthful charm and vigor that did not in any way betray her twenty-seven years. She was pretty, delicate, and petite, with a serious and subdued countenance. Her long straight hair, jet black, revealed traces of white ancestry. Unmistakably dark-skinned, she was blended of the lightest hues of brown with the darkest hues of charcoal gray. She carried herself in a gracious manner, which could often seem arrogant and haughty. Alongside her sophistication and refinement stood her determination and sternness of character—qualities befitting the daughter of a riverboat captain.

Ethel had grown up in Ripley, Ohio, a small town on the banks of the Ohio River. Her father, William H. Highwarden, was a man of means. He ran freight up and down the Ohio. Through him Ethel traced her ancestry back to William Whipple, a signer of the Declaration of Independence. Her mother, Belle Johnson, a woman of dignity, was one of the first black female graduates of Oberlin College. She worked as a teacher and businesswoman throughout Ohio. She was admired, in particular, for her management skills: the way she operated her store convinced people that women could "carry on business as successfully as men."[25] Her family had been among the early black settlers in Ohio, where they had enjoyed freedom and education even during the days of slavery. Though somewhat prosperous, the Johnsons had not been as well off as the older, more established black families in the northern part of the state. Black settlements along the river had generally been poor, comprised of recent runaway slaves and older free blacks without the means to remove themselves from the perils of slave hunters along the banks of the Ohio.[26] The sharp and dangerous boundary of the Ohio River was vividly portrayed by Harriet Beecher Stowe in *Uncle Tom's Cabin*. One of the novel's most memorable scenes is Eliza's flight with an infant across the water from Kentucky to Ohio, from slavery to freedom, on sheets of melting ice. Legend has it that a similar crossing actually took place at Ripley.[27]

By the 1880s and '90s Ripley was emerging as a vibrant community of blacks and whites, the latter mostly German immigrants. Ethel grew up in a bilingual community, speaking German and English interchangeably in the town and in the local grammar school. As a teenager she left Ripley to matriculate at Ohio State University in Columbus. There she pursued a broad course of study and developed a special interest in German literature. After graduation in June 1906 she took up an appointment at Howard, a year before Just arrived.

She began as an instructor in the College of Arts and Sciences and the Academy, a preparatory school run by the university. She and Miss Cook made up the Department of Modern Languages, offering German and French respectively at the college and preparatory levels. Ethel put special stress on literature in her classes. A demanding teacher, she introduced her students early to the works of German writers such as Goethe, Lessing, Vilmar, and Sturm.[28] German culture interested her, but it did not captivate her imagination as it would her future husband's over two decades later.

The two met in the fall of 1908 during a faculty tea at the home of President Thirkield. Just was attracted by Ethel's stately, even noble appearance, her knowledge of German language and culture, her relative aloofness from what he regarded as the philistine mentality of the black middle class. Ethel admired the cultivated demeanor of this Dartmouth man, his keen intellect, and, above all, his professional ambition. He courted her persistently thereafter, making calls at 1903 Third Street, where she lived under the watchful eye of her mother. Only after receiving his promotion to full professor, in the spring of 1912, did he propose and she accept.

Soon after their engagement, Ethel began to plan a honeymoon abroad. They both were anxious to visit Europe, especially Germany. She longed to see the art and architecture of the Continent's cities and towns, and he hoped to take a look at the universities and research institutes. Their plans were rare but not unique, for the black bourgeoisie had begun to take European tours in earnest by the early 1910s. Though Just and Ethel had more reason than most to visit Europe, it must have appeared a little pretentious to other young faculty to propose such an elaborate trip.

The honeymoon never took place. As summer approached, Just was forced to face facts: he needed to consolidate his work on Nereis and Lillie needed him at Woods Hole. He was in too formative a stage of his career to take the prime months of summer off from his research. A week's leave was about all he could afford. The marriage took place on a Wednesday. Just arrived from Woods Hole a day or two early and then spent the rest of the week in Washington with his bride before returning to his laboratory. Ethel no doubt was disappointed by the decision, but she gave up a honeymoon so that her husband could get on with his career.

Just did not focus on his work to the exclusion of starting a family, however. A child, Margaret, was born to Ethel on 28 April 1913, within a year of the marriage. At the time the family did not have a permanent residence; they were living with Just's sister Inez at 1853 Third Street. In the summer and autumn of 1914 they tried to find their own place. They moved twice before finally settling, around Christmas, at 412 T Street in the fashionable Le Droit Park section. The community was plush, made up mainly of black professionals and academics who had secure positions and incomes. The Justs began making payments on their new home, a stately three-story dwelling with a carriage house in back. A typically Victorian structure, rambling and spacious, it was large enough to accommodate two other children, Highwarden and Maribel, born in 1917 and 1922. The family lived in comfort, even elegance, while Just pursued his career in science.

ℜ

The summer of 1912 marked the beginning of many things for Just. It was then that he met Jacques Loeb, the eminent biologist of the Rockefeller Institute for Medical Research. This early encounter enhanced Just's reputation as a prominent black scientist. Just knew that Loeb was deeply concerned with social issues and the plight of the oppressed, but he could hardly have hoped that Loeb's interest would focus as intensely as in fact it did on him and the situation at Howard.

Born in Germany in 1859, Loeb moved to the United States in 1891.[29] He held a position at Bryn Mawr before moving to the University of Chicago in 1899 and the Rockefeller Institute in 1910. Even early in life he had been concerned with politics, especially the condition of oppressed peoples. As a Jew, he had endured the racial antagonism and anti-Semitism of the Germans under Bismarck,[30] and had left Germany for that reason. He understood the problems of the Filipinos and Japanese in the United States, and he fought the evils of the "yellow peril" mentality in American society.[31] Socialism, he believed, was the one political philosophy that would remove oppression from the world. Nothing pleased him more than to meet Just in 1912. Here was a chance for him to direct his political energies; here was a living specimen, downtrodden and oppressed, whom he could get to know and help firsthand.

If Loeb's concern was to help Just and black people generally, Just's was to make that help possible. Throughout 1913 and 1914 Loeb received many invitations from Just—as well as from the dean and the secretary of the Howard Medical School—to visit the campus, inspire the student body, and lend prestige to the effort to strengthen the financial status of the school.[32] Because of his failing health and chronic "throat trouble," he never made the visit, although he accepted several invitations to go.[33] He continued, however, to express "great

interest" in the Howard cause. In October 1914 he contacted Simon
Flexner of the Rockefeller Institute and Simon's brother Abraham, of
the Rockefeller Foundation, about what could be done to improve
things at Howard.[34] He also told Jerome D. Greene, a high-ranking
official at the foundation, that there was no question but that Rockefeller
funds should be channeled into support for Howard's medical program.
"A rather large amount of good," he assured Greene, could be done
with "comparatively little" financial outlay.[35] Moreover, he linked
Howard's success as an institution with Just's brilliance as a scientist.
Just, he felt, was "certainly a superior man."

Loeb's concern for the improvement of medical education went hand
in hand with his interest in the practice of medicine in the black
community. Well-trained physicians, according to his assessment of
the social problem, were "the best missionaries for uplift that could be
sent to the Negro."[36] Black doctors were effective because in the South
white physicians did not "care to practice among Negroes," and
Negroes, in many cases, did not "care to be treated by their white
oppressor." It was Just, or men like Just, who would train black
doctors; it was Howard, or schools like Howard, where they would be
trained. Loeb wanted to visit the Howard Medical School to see Just
and gather information for the Rockefeller Foundation, which he
hoped was "above the racial prejudice which would deny help to the
colored people."

Loeb's interest in black medical education was part of his larger
concern for the improvement of the general education of the Negro.
Through Just, his awareness of the predicament of Negroes increased,
and he came to the firm conclusion that help for blacks lay "purely in
educational directions."[37] One efficient and humanitarian method of
eliminating prejudice in America, he insisted, was "to uplift the Negro"
and thus remove the myth that "his low status" was due to "racial
inferiority."[38] The black man's status was the result not of any inherent
limitations but of "poor educational facilities" and the poverty that
existed in his home.

Though Loeb championed the black cause, Just was one of the few
blacks he knew personally. The two enjoyed long, intimate conversations at Woods Hole during the summers of 1913 and 1914. They
shared thoughts on what could be done to improve the lot of black
people. And more important, perhaps, Just gained through Loeb the
inspiration to try and become a first-rate scientist.

One evening as he was leaving the laboratory Just met Loeb on the
steps of the MBL. They once again began talking about the problems
of blacks in America, one of Loeb's favorite topics. This night, however, Just was tired and frustrated, and concerned about his work, not
his race. He tried to explain his anxiety over the results of some
experiments. Loeb said nothing about the experiments, but held forth

instead on the importance of Just's work for the uplift of black people. He had a vision of Just serving as a model in the black community and as an excellent teacher for future black doctors. In conclusion he assured Just of the general importance of his work for the advancement of science, and encouraged him to further heights in the world of biology. Impressed by Loeb's patience and sympathy, Just made the rather odd remark that "colored men" were like "dumb animals" in their gratitude for "kindness and interest" from others.[39] Through Loeb, he was beginning to view his work from a global perspective and to relate it to the larger cause of black people.

Loeb had successfully touched Just, one individual black, but he wanted to work on many levels at the same time. He would continue to lend his support to Just and medical education at Howard; there was much work to be done in that direction. He also wanted, however, to reach out directly to the black population. He was aware that the National Association for the Advancement of Colored People (NAACP) was the one organization concerned with social and political reform for blacks throughout the country and the world. In 1913 he began to support the causes of that organization.

Loeb was introduced to the NAACP at a reception for Miss Lillian D. Wald, a member of the association's board of directors, in the spring of 1913.[40] There he met the wealthy donors and the various board members, mostly Jewish, who had guided the organization from its birth in 1909. His attention was less on them, however, than on the famous black radical W.E.B. Du Bois, also a board member and the editor of the organization's official journal, *The Crisis*.[41] Du Bois's radicalism impressed Loeb deeply. The two men talked about politics and the evils of racism in America. They parted that night close comrades, united in a cause. Loeb was to be involved with two major NAACP campaigns over the next two years.

The NAACP had always held its national conferences in Northern cities such as New York, Boston, Chicago, and Philadelphia, but in 1913 it planned to convene in Baltimore, a stone's throw from the Southern states of Virginia and North Carolina. Important figures like Jane Addams, Franz Boas, William Hayes Ward, and John Dewey had been guest speakers in the past, and this year Oswald Garrison Villard and Belle Case (Mrs. Robert M.) La Follette were scheduled. While previous conferences had focused on questions of education, voting, and employment for the Negro, this one was to stress the "scientific side" of the race question. Not surprisingly, Loeb was invited to play a major role.

Du Bois asked Loeb to deliver a lecture on "The Theory of Race Inferiority as Modified by Recent Biological Knowledge." Loeb agreed to speak at the main session of the conference on Tuesday, 5 May, in McCoy Hall at Johns Hopkins University.[42] He was to be part of a

program to "impress the public by the weight of its scientific authority."
But when the time came he could not attend the forum owing to "a
very unpleasant attack of lumbago."[43] Instead, he wrote out a short
paper which Du Bois read to the association on the last night of the
conference and subsequently published in *The Crisis*.[44]

That year Loeb wrote several articles on the question of racial
inferiority, but never once did he appear on the NAACP podium.[45]
Perhaps he felt uncomfortable. He knew so few black people personally.
He had developed a friendship with Just, met Du Bois, and heard of
C. H. Turner, another black scientist engaged in biological research,
but that was the extent of his acquaintance with blacks. Still, he was
devoted to their cause. His commitment to the uplift of blacks was part
of his larger goal, dearly held, of eradicating racism internationally. He
would continue to take an interest in the NAACP as long as men like
Just were confronted with racism in American society.

In 1914 one of the NAACP's big concerns was D. W. Griffith's film
The Birth of a Nation. The film glorified the brutality of the Ku Klux
Klan in particular and white supremacy in general. Loeb was involved
from the beginning. Stunned and bewildered by the film, he emerged
from a theater in New York one cold evening in mid-March 1915
feeling as if "the most ghastly nightmare" had "taken possession" of
him.[46] His mind was spinning, so much so that he did not remember
being introduced in the lobby to Joel Spingarn, a pioneering leader of
the NAACP. Loeb felt tired and nauseated, anxious only to get home.
The film and the audience had made him sick. The film exhibited "a
display of homicidal paranoia with the special grievance against the
Negro," and, even more horrifying, the "homicidal paranoia" seemed
to take possession of the audience. To call "this diabolical appeal to
race hatred" *The Birth of a Nation* was, Loeb thought, "the worst
insult" that had "ever been heaped upon this country." But he would
not support the NAACP in its efforts to censor or suppress the film. He
felt that this would only be an advertisement for it. A better course, he
suggested, was to appeal to "the common sense and the decency" of
actors, directors, and producers "to unite against performances of this
kind in the future and to quietly drop 'The Birth of a Nation.'" No
doubt Loeb thought that these people were different from the authors of
Birth of a Nation's script.

Spingarn, chairman of the board of directors of the NAACP, shared
Loeb's feelings about censorship, but in this case he found himself
confronted by "a condition and not a theory."[47] The tangible results of
the film, in terms of social bitterness, unrest, even violence, might be
devastating. He intended to push forward against "this hideous libel"
disguised as historical truth. He only wanted permission to quote
Loeb's characterization of the film "in its pathological aspects." After

consulting legal authorities, Loeb agreed to allow the NAACP to make use of his remarks.[48]

Loeb's contact with the NAACP, by anchoring his theoretical politics to real situations, allowed him to see more clearly the pernicious effects of racism on the everyday lives of black Americans. He also became more aware of the feelings that a black like Just experienced when confronted by racist theories in science. He was inspired to fight racism within the scientific professions as well as in society at large. One of his goals was to rid the scientific literature of racist tendencies. Science was rational; racism was emotional. For Loeb, the two had nothing to do with each other.

In 1916 a relatively unknown scientist, J. E. Wodsedalek, published an article entitled "Causes of Sterility in the Mule." The article appeared in the *Biological Bulletin*, the official journal of the MBL. One paragraph compared the Negro with the ass:

> The cause of the difference of opinion is probably due to the fact that Guyer used Negro material in his investigation, while Von Winiwarter studied tissue obtained from a white man. In view of the difference between these two human races it is not at all improbable that a difference in the number of chromosomes also exists. The Negro is fully as far removed from the white man as is the ass from the horse, where a great difference in the number of chromosomes apparently exists. This vast difference in number appears to be, at least in part, responsible for the sterility of the mule. The difference between the white man and the Negro in regard to the number of chromosomes according to Guyer and Von Winiwarter is equally as great, but unfortunately the mulatto is fertile.[49]

Loeb was outraged. He felt that the editorial board of the *Bulletin* should disavow any association with the statement, especially since Just and C. H. Turner were regular contributors.[50]

Just's sponsor Frank Lillie, who was the managing editor of the *Bulletin*, felt otherwise. Though he admitted that the choice of words in the article was "extremely unfortunate," he did not believe that there was any "derogatory" intention on Wodsedalek's part.[51] Still, he apologized for not having read and edited the article more carefully. T. H. Morgan, to whom Loeb also appealed for help, agreed with Lillie. He doubted that Wodsedalek intended "to compare the Negro with the ass."[52] In his opinion, the phrasing was "merely a piece of stupidity" on Wodsedalek's part. It was best, in any case, to give Wodsedalek "the benefit of the doubt," since the statement could be interpreted in several ways.

Both Lillie and Morgan felt that Loeb's interpretation might have been erroneous and, further, that any notice given to Wodsedalek's remarks would only serve to publicize them too much. Lillie agreed, however, to contact Wodsedalek personally. His position was clear: he would submit the entire matter to the editorial board should Wodsedalek admit racial animosity. Wodsedalek admitted none, stating that he had not intended to insult the Negro in his article. Loeb found Wodsedalek's disavowal "a sufficient safeguard" against further "persecution of the Negro."[53] He dropped the issue; a rebuttal he had written was never published, and Wodsedalek's comparison received no further attention.[54]

Just, who would certainly have felt "the sting" of Wodsedalek's remarks,[55] never mentioned the article. He read it, just as he read every article published in the *Biological Bulletin*. But the matter was best left alone. Just was still feeling his way at Woods Hole, trying to learn the routes necessary for smooth traveling along his career path. Loeb could afford to criticize: he was in a different position, accomplished, recognized, entrenched.

As Just struggled to advance professionally, Loeb and the NAACP played a major role in helping him along. In 1913 the association developed plans to present an award the following year. Sponsored by Spingarn, the award would go to the man or woman of "African descent" who had performed "the foremost service to his race."[56] A selection committee was formed, comprising Bishop John Hurst (chairman), William Howard Taft, John Hope, James H. Dillard, and Oswald Garrison Villard. At first it seemed it might be difficult to find anyone worthy of the honor; but when the nominations began to pour in, members of the committee were astonished at the number of people who had made significant contributions.[57] The nominees—thirty in all—included Meta Warwick Fuller, the sculptress; William Monroe Trotter, publisher of the *Guardian* and militant agitator; Heman Perry, founder and president of the first major black life insurance company, in Atlanta; Major John R. Lynch, the Reconstruction statesman and author; Major R. R. Moton, principal of Hampton Institute; Isaac Fisher, the essayist of Tuskegee Institute; Howard Drew, world record-holder for the hundred-yard dash; and Cornelia Bowen, principal of Mt. Meigs School in Alabama. The nominations reflected a predominant interest in art, politics, social work, business, literature, education, and athletics—the fields in which black achievement had traditionally been concentrated.

Apparently the association decided that science should also be represented: a black scientist had to be found. Publicity for his or her achievements would demonstrate the intellectual depth of people of "African descent"; furthermore, it would help place black people and the NAACP in particular in the mainstream of Western tradition.

Science, unlike music, literature, and art, was held to be objective in its approach and method, culture-free and universal. Scientific truth could be denied by no one, white or black. If this was the first time the whole world was to know of the achievements of blacks, then those achievements should be as universal as possible. What better way to command the respect of all than to consider a scientist for the Spingarn Medal? No longer would the white scientific world find it easy to perpetuate the idea that blacks had never done creditable scientific work. Ironically, a close friend of Loeb's, J. McKeen Cattell, made a remark to that effect in a speech entitled "Science, Education and Democracy" at the annual conference of the American Association for the Advancement of Science in Atlanta, Georgia, in December 1913. Du Bois was quick to take issue with him in *The Crisis*, calling attention to the work of Just and C. H. Turner.[58]

Loeb was asked to recommend possible candidates for the medal. He explained that his acquaintance with "colored people" was so limited that a recommendation might do an "injustice."[59] He did suggest, however, that the medal be awarded to "one of the few colored people who, in the face of great difficulties," had done important scientific research. The person he had in mind was E. E. Just, professor of physiology at the Howard Medical School. He knew Just "personally very well," and considered him to be "a very able researcher." Just had "sacrificed a good deal for the advancement of the medical schools for colored people," and would do "a good deal more if . . . given a chance." Loeb knew of another black "by the name of Turner" engaged in scientific research, but he was unfamiliar with Turner's work. It was Just whom Loeb had come to know intimately, though personally more than professionally.[60] When asked by Villard for a letter recounting Just's scientific achievements, he said that he did not feel competent to provide detailed information. Instead of submitting an inadequate statement, he referred Villard to Lillie as the person most knowledgeable in the matter. His recommendation simply asserted a general impression that Just's "attitude of scientific subjects, his knowledge of biology, and his critical ability" were of an "unusually high and lofty order." Most importantly, Just seemed to be "guided by very high motives" in his work within the black community. He would not give up his position at Howard in order to increase his income. This Loeb particularly admired.

Just had championed the cause of the Howard Medical School and sought funds for its survival because he believed in its integrity and high quality. His motives were not, however, entirely altruistic. The medical school provided a substantial part of his income and even paid his room rent at Woods Hole during the summers. The exact amount of his salary depended, among other things, on the level of the medical school's enrollment. While most of its faculty members derived their major income from medical practice, he was dependent upon teaching alone and so had to "bear the hazards of a fluctuating salary."[61] Yet in any case, he was willing "to sink the personal for the larger cause."[62] This attitude impressed Loeb. Who was more deserving of the Spingarn Medal than Just?

The American public did not know Just as well as they knew some of the other nominees, especially Trotter. Newspapers at the time were filled with stories of Trotter's dramatic exploits: his involvement with the Niagara Movement; his opposition to Booker T. Washington; his confrontation with Woodrow Wilson on the issue of jim crow in Washington's federal office buildings.[63] When Just won the medal, everyone was surprised. The exchange of telegrams between the editor of the Chisholm News Service in Denison, Texas, and the secretary of the NAACP captures this:

EDITOR: Is Trotter winner Spingarn Medal?
SECRETARY: Ernest Everett Just winner of medal.
EDITOR: Who is Ernest Everett what?
SECRETARY: Ernest Everett Just is head Howard University Medical School.[64]

Perhaps Just was selected precisely because of his relative obscurity, because the NAACP wanted the public to recognize a black man of science, his great promise and lasting achievement.[65] Notification of the award reached Just at the beginning of February 1915. It was not the first time he had been honored by the association; three years earlier he had been chosen their "man of the month," and his photograph had appeared in *The Crisis* along with an article detailing his accomplishments.[66] But the Spingarn Medal was a far more prestigious honor, involving a great deal of publicity and an elaborate ceremony. Just informed the committee that he would be embarrassed to face a large audience, and he attacked publicity as incompatible with scientific endeavors.[67] Still, in the cold of winter he journeyed to New York to accept the award.

On the evening of 12 February 1915, Governor Charles Whitman presented Just the first Spingarn Medal. The ceremony took place in Ethical Culture Hall. There were crowds and crowds of spectators and several important speakers, including Charlotte Perkins Gilman and Oswald Garrison Villard. The hall was buzzing with talk of the new "flood of anti-negro legislation" in Washington and of how timely the Spingarn award was as a reminder to the country of "the spirit of brotherhood, fellowship, and Americanism."[68] Just watched and listened attentively. His speech was brief and to the point: "I thank the Association for the award not so much for myself as for the students whom I represent."

The NAACP managed to secure good press coverage for this ceremony in honor of black achievement. It did not matter that Just was relatively unknown. That a black, any black, had done brilliant work in marine biology was startling news in itself. Presses throughout the country, even Southern newspapers, commented extensively on the fact that a medal had been awarded a Negro for scientific achievement.[69] The *Courier-Journal* of Louisville, Kentucky, a white Southern paper, astounded the black community by referring favorably to Just's "scholarly research, original, not imitative work, in the field of biology."[70]

The award and the accompanying publicity made Just known to the American public. But the question in his mind was: How would the award appear to the scientific world? At first he feared that it represented a popular and perhaps vulgar watering down of his scholarly pursuits: it was an award given specifically to a black, unlike any he had ever received before, so different from the notion of general merit on which he believed science was based. After the ceremony, Lillie explained to him the value of the award and informed him that Loeb had been instrumental in calling the committee's attention to his work.[71] Only then did Just feel comfortable, and start to look at the award as an object of pride, as a possible factor in building his career and scientific reputation. That Loeb was interested in the medal eased his "deep-seated fear . . . that real biologists would rather deride the whole business and especially the paltry recipient."[72] Lillie's words of congratulation "removed a deep burden" from his mind.[73] He had thought that Lillie would consider the award "a travesty." Quite the contrary—Lillie was pleased that the award had gone to someone in pure science rather than a more practical field. He felt that the award was good for Just, the country, and the Negro race. Loeb, more narrow in his perspectives, considered it a harbinger of ambassadorial possibilities for Just in the black medical world.

Just perhaps never thought of the award in its larger context. To him it was mainly a stepping-stone to a successful career. He rarely mentioned it later, but he did feel its effects at the time. The award, he confessed, signified "a new day" in his life; it allowed him suddenly to become "a real human being alive and anxious to work."[74] He would continue to make sacrifices, at Howard and elsewhere, now that there was appreciation for his "striving."

The Spingarn Medal may have called attention to Just's scientific accomplishments and helped build his scientific stature, but a Ph.D. from a major university was nevertheless essential to expand opportunities, to maintain and further a reputation. This was especially true for Just, one of the first blacks to join the professional ranks of science. If he was to have a serious career in science, Just had to have a Ph.D. Lillie understood this and began to encourage him more strongly than before to obtain the "official stamp of approval."[75]

In June 1915 Lillie decided that Just had completed the research requirements for the Ph.D., and he agreed to accept previously published articles as a doctoral thesis. Only minor subject and residence requirements remained, all of which could be completed in three quarters at the University of Chicago. Eager to finish up, Just requested a leave of absence, with pay, from the Howard Medical School. Lillie lent support to this request, urging the leave as necessary for Just to establish a "scientific reputation" that would in turn be of "greatest advantage" to Howard.[76] He persuaded the administration that the reputation of faculty members was an important consideration for any medical school and that a leave of absence for Just would be an "investment" for the further development of medical education at Howard. The Howard administration recognized its indebtedness to Just for past loyal service and understood the wisdom of Lillie's argument of cost-benefit. The dean, W. C. McNeill, vowed that his office would "do its part" to find sufficient funds for the leave of absence.[77] Sacrifices would be made to help Just obtain the Ph.D.

Just began to make plans for his year of study in Chicago. His wife Ethel and their two-year-old daughter, Margaret, would remain in Washington so that he could focus full attention on his studies and avoid the distractions of domestic life. He would board with his kinsfolk, the Bentleys, who lived on East Forty-first Street within walking distance of the university. He would return to Washington from time to time, certainly at Christmas, but he could not promise how often. His priority was to take full advantage of resources at the university, of opportunities to meet people and to get to know the city. Sabbaticals were almost unheard of at Howard, and he could hardly hope to have another chance like this any time in the near future.

Just went to Chicago for the 1915–16 academic year and was awakened to the life of a big city. Washington was still an oversized town of sorts, whereas Chicago had already taken its place as America's second-largest and most flourishing metropolis. There he found "much-needed rest and mental refreshment."[78] The high-powered intellectual atmosphere of the University of Chicago was a welcome change from the dull routine of Howard Medical School. Opportunities abounded. Just gained perspective on his life and work by establishing contact with many high-ranking professionals in and out of his field in the black and white communities.

What inspired him most was the prosperity of blacks in Chicago. Cloistered in his Washington and Woods Hole laboratories, he had not realized that "anywhere in the United States colored could get even half a chance."[79] The large number of "highly respected, competent . . . Negro surgeons, dentists, and lawyers" in Chicago was a revelation to him. Also, he was surprised to observe that a young black Ph.D. had been working for some time under H. G. Wells at the Sprague Research Institute—probably a reference to Julian H. Lewis, the pathologist who later became the first black member of the faculty at the University of Chicago.[80] Of course, black professionals at this educational level were the exception rather than the rule. Law, medicine, and research may have attracted a privileged few, but business was the occupation of most middle-class blacks.[81] It did not demand long, expensive years of study at a university, just a little capital and a lot of hard work. No doubt Just was astonished by the number of black Chicagoans in proprietary and managerial positions. His daily activities must have brought him into contact with many who ran their own factories, restaurants, hotels, funeral parlors, printing plants, and real estate offices.

Just's kinsfolk the Bentleys were in the vanguard of black social and professional life in Chicago, part of a circle that included E. M. A. Chandler, chief chemist with the Dicks, David, and Heller Company; William E. Quinn, a well-known physician who maintained a plush residence on Indiana Avenue; Jesse Binga, president of the Binga State Bank; Edward H. Wright, the lawyer who reputedly grossed over thirty thousand dollars in fees each year; Willard Landry, the surgeon who standardized the method of childbirth by caesarean section; Robert S. Abbott, founder and editor of the weekly *Defender*; Charles S. James, the first black to be appointed dentist to the city's public schools; Sheadrick B. Turner, member of the Illinois legislature; Anthony Overton, head of the Overton Hygienic Manufacturing Company; Charles S. Duke, bridge-designing engineer for the city; and Daniel H. Williams, the first cardiologist to perform a successful operation on the human heart.[82] The list could go on. Chicago was one city where blacks had established a reputation for "their work, their education, their prosperity, and their political power."[83] Anyone who claimed that there was "not a single Negro" in Chicago, other than those who "black shoes, transport . . . luggage, or serve . . . in . . . restaurants," was sorely out of touch with reality.[84]

Just's hosts exemplified the best in cultured black life. C. E. Bentley had grown up in Cincinnati, where he studied voice and began a career in concert singing. In his early twenties he moved to Chicago and enrolled in the College of Dental Surgery. A diligent and independent worker, he opened private offices in the city shortly after graduation. Before long he numbered among his patients "some of the most distinguished and discriminating citizens of Chicago."[85] His social and professional activities came to cover a broad spectrum: he could be

found delivering an address at the Appomattox Club in remembrance
of the birth of Samuel Coleridge-Taylor, the black composer; lecturing
at Howard University on systemic infections due to pyorrhea and
inadequate root therapy; attending a dinner given in his honor by
administrators of Provident Hospital; organizing a dental congress in
St. Louis for three thousand members of the profession; and developing
a program of dental hygiene in the Chicago public schools.[86] Nor was
he the only distinguished member of the family. His wife, Florence
Lewis Bentley, was an experienced literary editor and "a strong social
force in the city," while his daughter, Vilma E. Bentley, was a highly
respected music teacher in the public school system.[87]

Yet if some blacks were able to achieve social status in Chicago,
others were plagued by insuperable social difficulties. Just did not
comment on the seamy side of Chicago life, but he could not have been
unaware of the problems faced by destitute blacks there. Decent housing
and worthwhile employment were difficult to come by.[88] Forced into
ghettos on the southern and western sides of the city, blacks had to pay
extortionate rents for tiny apartments in run-down buildings. Job
opportunities were by and large restricted to domestic work or manual
labor; the average black put in an eight- to nine-hour day for a one- to
two-dollar wage. Weekly earnings often fell short of covering the cost
of rent and food, and at times a choice had to be made between the
comforts of a warm room and a full stomach. The situation, dismal
indeed, was aggravated by racial hostility. Lurid stories of white violence against blacks appeared daily in the newspapers. Over the course
of a month, the *Chicago Tribune* reported six beatings (one fatal) of
blacks by police, one illegal execution, two lynchings, three attempted
lynchings, and assorted minor street confrontations.[89] Meanwhile the
trial of "Chicken Joe" Campbell, a black accused of murdering his
white employer, made headlines throughout October and November.[90]
After his conviction, unrest increased. The problem was getting out of
hand. Civic organizations decided to make a move. The Progressive
Negroes' League and the Chicago chapter of the NAACP arranged
lectures and conferences. Du Bois spoke at a meeting in February 1916,
sponsored in part by Julius Rosenwald, the Chicago philanthropist.[91]
But nothing could be done to prevent trouble. The situation simmered
for the next three years and finally bubbled over into near race war in
1919.[92]

How much Just was aware of all this is unclear, but he could not
easily have ignored such a potentially explosive setup. His evenings at
the Bentleys' usually were spent playing cards,[93] but there can be little
doubt that the conversation often turned from whist bids and poker
bets to the urgent social and political issues facing blacks. Bentley was
one of the original members of the Niagara Movement, "that first bitter
protest against the program of surrender and segregation," and one of

the founder-directors of the NAACP.[94] His passion for reform ran deep: "no cause that involved emancipation of the Negro failed of his enthusiastic support, so long as it did not involve any surrender of human rights." As chairman of the Chicago chapter of the NAACP, he spearheaded the investigation of certain racial incidents in the city: two of the most publicized cases involved the fatal beating of a black drayman in 1911 and the burning of the home of a black chauffeur in 1914.[95] More important, perhaps, was Bentley's firm stand against discrimination in general. He spoke his mind on the brazen efforts of state legislators to introduce anti-intermarriage and jim crow car bills, as well as on the more subtle attempts of city administrators to institute segregation in schools and hospitals.[96] Concerned by the unemployment and poor housing plaguing blacks, he was a leader in the effort to solve the problem of indigent and displaced families.[97] Another concern of black Chicagoans in 1915 and 1916 was whether or not the city would grant a license for the showing of *The Birth of a Nation*. Bentley led the NAACP's Committee on Grievances straight to the mayor's office each time a nickelodeon manager put in an application.[98]

Just did not talk about Bentley's activism, just as he did not mention any of the depressing aspects of black life in Chicago. He preferred calling the attention of people like Loeb to the status and accomplishments of black Chicagoans, perhaps as a way of suggesting possibilities for his own professional advancement. The bright side appealed to him; unlike Du Bois, he did not find the "murk and shadows"[99] intolerable.

Trouble in Chicago in 1915–16 was not exclusively racial; there were also well-publicized political struggles along class lines. On 8 December Just wrote Loeb about the "strike situation," noting in particular that Mrs. Frank Lillie had been taking "a great part in the movement to give the workers fair play."[100] He knew this would fascinate Loeb, who could never have imagined that a woman of Frances Crane Lillie's piety, wealth, and social standing would throw herself headlong into the most radical kind of political activism. She was a daughter of the multimillionaire Richard T. Crane and the wife of a conservative academician; it was indeed out of the ordinary that she should commit herself as she did—not only to the special cause of the strikers but also to a socialist philosophy, including public control of essential services. Just's interest in the strike perhaps was prompted more by Mrs. Lillie's participation than by any special concern about what was happening. The daily reports in the *Chicago Tribune* seem, however, to have aroused his political consciousness to some extent. He was beginning to wonder whether "freedom and justice" would ever be prevalent anywhere and whether "brute force" might not be the only viable political tool for the oppressed. This was a far cry from the anti-union sentiments expressed in his first article, "Government Ownership of

Monopolies," published thirteen years earlier in the Kimball Union Academy school newspaper.

The garment workers' strike began in late September. It was caused by the clothing manufacturers' refusal to negotiate a wage agreement and accept collective bargaining. From the beginning the workers received suppport from several groups, including the Socialist Party, the Hull House organization under Jane Addams, and "a number of wealthy Chicago women."[101] Mrs. Lillie first showed interest in mid-October, when she signed a resolution drawn up by Addams condemning the "suffering and waste" caused by the strike and demanding immediate settlement by arbitration.[102] Upset by the manufacturers' continued rejection of pleas for peaceful mediation, she joined the front ranks of the picket lines, alongside Ellen Gates Starr of Hull House. Their signs read, *We Are Picketing for the Pickets Driven from the Line*," a reference to the brutality with which the police had been implementing a court order prohibiting public protest. No matter what the time or weather, Mrs. Lillie could be seen marching to and fro with her placard outside Kuppenheimer's factory on Western Avenue and Twenty-second Street. In mid-November she loudly ridiculed a policeman who had arrested Starr, accusing him of being afraid to arrest someone rich. Her activism became more and more vocal. In early December she took out a full-page advertisement in the *Daily Maroon*, inviting all interested parties to a strike meeting in Kent Hall at the University of Chicago. She also sent out several hundred postcards to members of women's clubs, civic organizations, and church societies assailing the police as "obscene."[103]

Soon Mrs. Lillie found herself in direct conflict with the police. On 6 December (two days before Just wrote Loeb) she created a scene by accusing a policeman of allowing "sluggers," hired by manufacturers, to attack strikers.[104] The officer arrested her for interfering in the performance of his duty. Far from daunted, she protested all the way to the police station and the courtroom. Her language was radical and impassioned: "I protest against the mayor of Chicago using the police to crush the desire of liberty in the working people's breast through terrorism; . . . I protest against the power that seeks to perpetuate industrial slavery in America." The incident only served to reaffirm her commitment to the struggle for workers' rights. She returned to picket duty the next day.

As expected, her husband's attitude towards all this was diplomatically neutral. Lillie neither condoned nor condemned; he simply went about his business at the university.[105] Everyone knew that Lillie did not involve himself in politics. The contrast between him and his wife may have been amusing to some and upsetting to others, but it was a fact that had to be accepted. Just accepted it, and was probably amused. Unlike Loeb, he was not enough of a political activist to be either upset by Lillie's apathy or ecstatic over Mrs. Lillie's commitment. He was, after all, in Chicago to get on with completing the requirements for the doctorate. He had neither the time nor the inclination to think as deeply about political issues as he had done in his youth.

The academic atmosphere at the university was as much of an awakening to him as the social and political life of the city. The two were not, however, unrelated; the university's growth was undoubtedly connected with the flourishing aspects of the city. Having established itself as one of America's top universities, on a par with Columbia and Harvard, the University of Chicago had come a long way since it opened in the fall of 1892.[106] John D. Rockefeller, who gave the original gift of $600,000 to found the university, had given well over $35 million to its various parts by 1916. William Rainey Harper, the university's first president and its organizer, at once brought together a distinguished faculty—one that engaged itself in both first-class teaching and research—and managed to attract superb students from the start. Included on the faculty were nine former presidents of other universities who brought with them not only scholarly credentials but also administrative acumen. The benefactors and trustees were, for the most part, Chicago's leading businessmen, who gave the university the entrepreneurial force needed to take it to the forefront of education, men who took an active part in the affairs of the institution. In this respect the University of Chicago was different from older Eastern universities, whose donors and trustees were usually old-line families who observed school affairs from afar.

In general science research and teaching were riding high when Just entered for the academic year 1915-16, the closing year in the first quarter-century in the history of the university.[107] The university laboratory system, comprising laboratories from the various disciplines, had mushroomed under the direction of Julius Stieglitz, a renowned chemist. Most of the science departments had come to be among the best in any American university. The physics department in particular boasted A. A. Michelson, who had received the Nobel Prize in 1907, and R. A. Millikan, who was to receive the prize in 1923. The zoology department had distinguished itself, first under the direction of C. O. Whitman and from 1910 under Lillie, and had been associated in the 1890s with Jacques Loeb and T. H. Morgan. It was still a growing department, and in the winter quarter of 1915 it had just gained a laboratory of its own; previously the laboratories for the departments of pathology and bacteriology were housed in the zoology laboratory. When Just arrived on campus, he not only was inspired by the vibrant scientific community within the university, but he also felt a strong sense of identity with the upwardly moving zoology department.

The department required its graduate students to complete a rigorous program of course work for the doctorate. A major and a minor field had to be selected.[108] Fortunately, Just was permitted to use his previous course work at the MBL to satisfy a substantial part of the zoology requirement. He had passed the invertebrate zoology and embryology courses with highest praise. In Lillie's view, each of these courses, involving five hours of lecture, four hours of field work, and thirty hours of laboratory work per week, was the equivalent of two major graduate credits.[109] Just also received credit equivalent to five courses for the independent research he had done under Lillie's supervision at Woods Hole between 1911 and 1915. There was still much to be done, however, to complete the course requirements in his major field, not to mention the problem of selecting a minor. Physiology, a subject combining important aspects of biological research and medical teaching, seemed to be the right choice as a minor. Just was becoming as fascinated by the living organism's internal processes as he had been by its external habits; further, he was growing more and more convinced of the significance of his work at the Howard Medical School. With these factors in mind, he registered for one zoology course and three physiology courses in the fall of 1915.

Zoology 45 was perhaps the most stimulating of the courses he took at Chicago. Entitled "Physiology of Development," it considered a broad spectrum of data and views as material for a theory of development and heredity. Lillie was in charge, exuding his usual air of calm authority. Each week he delivered a two-hour lecture and presided over a one-hour conference session. A good thing about the course was its small enrollment. Only seven students registered, several of whom, including Elinor Behre and Tadachika Minoura, had been at Woods Hole.[110]

The course began with a consideration of the basic structure and function of the egg. In October the topic was fused eggs in the marine hermaphroditic animals Ctenophora and the roundworm *Ascaris*, and in November, germinal localization and nuclear determination in the eggs of the worm-like chordate *Amphioxus*, the slipper limpet *Crepidula*, and the frog *Rana*. Questions that would later figure prominently in Just's own research came up for detailed analysis: the rapidity of the cleavage process, the possibility of nuclear differentiation, the interdependence of nucleus and cytoplasm. Lillie was careful not to allow his students to become enmeshed in theoretical questions at the expense of practical problems. He took up next the topic of tissue culture, a subject requiring a sophisticated treatment of both the idea of cell specificity and the fact of physical development. These were the most inspiring classes. Just and his fellow students argued over the various interpretations by Hammar, Rhumbler, Harrison, Burrows, and the Lewises. Just argued for the position taken by Hammar, a German biologist who in 1847 had insisted on the significance of "ectoplasm." Later, in the 1930s, Just's own work would center on this theme.

From tissue culture Lillie went on to consider the law of genetic restriction, as prelude to an analysis of the development of the nervous system. He linked Hertwig's ideas on biogenesis to experiments with the optic vesicle of various animals, using Hans Spemann's *The Development of the Vertebrate Eye* (1912) as the starting point for further discussion and research. In early December the students completed individual research projects applying genetic theory to the problem of physiological development. Just followed up some of T. H. Morgan's work in 1902 on muscular regeneration in the earthworm.

The course concluded with a summary of the various approaches to *Entwicklungsmechanik*, developmental mechanics—Wilhelm Roux's theory of nuclear analysis, Hertwig's principle of correlation, vitalism, and so on. Lillie insisted on the importance of keeping an open mind, especially in so complex a field of research. He referred several times to Roux's opinion that the attempt to understand *Entwicklungsmechanik* is "very difficult, the most difficult task that the human intellect can undertake." Finally, he stressed that genetics and embryology had not yet been closely enough applied to each other. Scientists still seemed more interested in tissue processes than in the activity of the egg. Zoology 45 was supposed to correct that imbalance, to impress on rising young biologists the need to approach embryology with greater sophistication and respect.

Zoology 45 stood out in Just's mind as the most rewarding of all seventeen courses he took or audited at Chicago.[111] He vowed to do at least one important "piece of work" which would show how he had "caught part" of Lillie's teaching. It was a demanding course and assumed that the students were knowledgeable, independent, and motivated. It increased Just's interest in the physiological approach to embryology and sharpened his awareness of the international dimensions of science. Much of the source material came from German journals, especially Wilhelm Roux's *Archiv für Entwicklungsmechanik der Organismen*. Lillie brought up the theories and interpretations of no less than forty-seven scientists—twenty-three Germans, thirteen Americans, five Englishmen, three Frenchmen, one Chinese, one Italian, and one Pole.

Just felt he had a duty to prepare himself for medical teaching as well as for biological research. He looked for relevant courses, but was not impressed by the offerings. Though pathology and pharmacology seemed interesting in the catalog description, they turned out to be a complete waste of time. Just attended pharmacology once, "more than enough," and pathology twice, "three times too much."[112] Besides, he had decided on physiology as the perfect compromise between zoology and medicine. In the end he restricted his course selections to the physiology department. He registered for three physiology courses in the fall quarter, two in the winter, and three in the spring.

Most of the physiology offerings were undergraduate survey courses designed primarily for premedical students, but registration by graduate students with physiology as a minor field was also encouraged. There was no chance for theoretical discussion and independent research. Students simply read the assigned texts, took lecture notes, and performed required experiments. Though he was beyond this kind of regimen, Just took it in stride, keeping his duties at the Howard Medical School always at the back of his mind. He was one of a few graduate students surrounded by a crowd of eager undergraduates. For the most part, the graduate students received poor grades. Unlike the undergraduates, they had neither the time nor the desire to compete for top honors, especially in a field outside their major area of interest. Just always received the highest grade of any graduate student, but, like the others, he was less involved than in the conference courses in his major field.

In addition to the basic physiology courses Just registered for A. P. Mathews's advanced course, which involved a detailed analysis of the chemistry of cell constituents such as carbohydrates, lipids, proteins, and nucleic acids. The prerequisites for admission were general chemistry, qualitative and quantitative analysis, and organic chemistry. Though the chemistry prerequisite caused him some anxiety at first, he was assured by Mathews, a Woods Hole regular, that this would be no problem considering his experience and maturity. Mathews did, however, advise him to audit a few chemistry courses so as to fill in any gaps of knowledge. So Just attended courses in qualitative analysis and organic chemistry regularly and took detailed notes, a practice he was to continue in the winter and spring quarters. Of twelve students registered, ten were graduates; many were regular investigators at Woods Hole. Classes met in conference style for three hours each week, and each student was expected to put in a minimum of six hours per week of independent research in the laboratory and close reading of articles in foreign journals. Just enjoyed this type of course and its emphasis on the creative exchange of ideas.

Mathews's course, stressing the physiological properties of the cell, laid the foundation for Just's future work on the protoplasmic systems of marine invertebrates. Mathews later recalled that Just seemed to understand cell chemistry "wonderfully well," certainly "better than any other student in the class." He was impressed by Just's insight and ability; he knew that Just would go on to advance "original ideas" and accomplish "great things," to become "one of the most.original and creative men in zoology in the United States."[113]

Just's choice of courses for the winter quarter followed the pattern of the fall, with minor variations. In January he enrolled for two zoology and two physiology courses: William L. Tower's graduate course on organic evolution and C. M. Child's graduate seminar on reproduction;

an elementary course on the nervous system and the senses and Mathews's seminar on digestion. He also decided to audit A. J. Carlson's course on the physiology of the digestive tract.

The spring quarter, from April to June, was the last opportunity he would have for full-time study, satisfying requirements for the Ph.D. and acquiring knowledge for future research. He planned to go to Woods Hole in the summer, and he was obligated to return to Washington for the start of the fall term at Howard in September. The spring quarter in Chicago therefore had to be a time of relentless effort and concentration on his part. Besides taking three basic physiology courses, Just audited Carlson's advanced course on the physiology of blood circulation and respiration, Lillie's course on vertebrate embryology for medical students, and Kyes's anatomy course on immunity. Kyes and his course had special significance for Just.

Preston Kyes was professor of preventive medicine at Chicago, and there was a strictly medical side to his course involving consideration of the physiological mechanisms of disease production and prevention. Just noted Kyes's ideas on the relative susceptibility of blacks and Semites to diseases such as tuberculosis and diabetes. But he did not sit in on the course primarily for medical information; rather, he wanted to gain familiarity with the language of immunity, with the terminology Lillie had used to put forward the fertilizin theory which was to be so important to Just's work later. J. G. Adami's textbook on pathology provided the basic definitions for terms such as "agglutination" and "specificity." Just paid close attention to Kyes's analyses of Ehrlich's work, in order to grasp the full meaning of the terminology and how Lillie had used it in formulating his theory of fertilization.

But it was in Lillie's own seminar on the historical development and current status of problems of fertilization that Just gained a new perspective on his past work and his plans for the future. The seminar dealt with the questions on which he and Lillie had worked together for years at Woods Hole. Lillie led lively discussions on the mechanism of fertilization. Just listened carefully and read and reread Lillie's articles, seeking new answers to old problems. To do research in this area would place him close to Lillie; it would make his reputation as a scientist; and ultimately it would produce work he could call his own.

The spring quarter ended in early June. The department had already agreed to accept two of Just's articles as the main text for a doctoral thesis entitled "Studies of Fertilization in *Platynereis megalops*."[114] All he had left to do was receive his diploma.

One of a handful of blacks in the country seeking a doctorate in science, let alone receiving one,[115] Just could hardly be calm about his accomplishment; he was jubilant and afraid at the same time. He would certainly take part in the ceremony and mark the occasion. The convocation services were, however, to be a multifaceted celebration.

The university had set aside from 2 to 6 June to commemorate its quarter-centennial.[116] Just's self-pride was heightened and magnified by the numerous celebrations hailing the university for being the pioneering educational institution it was. Ceremonies took place in every department, and scholars from all over the country converged on the campus—renowned scholars such as William Henry Welch, the medical professor from Johns Hopkins, whom Just listened to with all attention as he dreamed of the future for the Howard Medical School.

On Tuesday, 6 June, Just attended the school's ninety-ninth convocation (the university graduated students after each quarter). The commencement addresses were made by representatives of various groups at the university, from the trustees to the alumni to the students, even one by the founder himself. John D. Rockefeller spoke briefly to the crowd, and Just, in cap and gown, was dazzled by the power and legend of the man. He hardly knew what to think, moving among the wealthy. He knew, however, that it was people like Rockefeller who had helped to make the university move, who embodied the American dream. What these people would mean for him he could not predict; he hoped no doubt that he himself would keep moving.

Badly worn out and in desperate need of a vacation, proud but tired, he passed up the opportunity to go to Woods Hole. He boarded the train on Sixty-third Street and headed home for a well-deserved two-month rest from the pressures and pleasures of academia. He left Chicago overjoyed that he had finally completed a rigorous program under eminent biologists at a leading university, had finally equipped himself with the technical training necessary for success in science. As he approached Washington, his doctoral diploma was uppermost in his mind. How would he use his "union card," as Loeb would have labeled it, his "official stamp of approval," as Lillie had called it earlier? He knew now he could be a more valuable asset to Howard, establish himself within the university system, and begin his career in the white world of science.

CHAPTER 3

The Expansion of Just's Scientific World: Woods Hole, 1909-29

Oh M.B.L., dear M.B.L.
We never can forget
Those rubber boots, those bathing suits
And that collecting net
Those songs and things will soon take wings
But through the coming years
What e'er the scene, dear formalin
Will fill our eyes with tears,
What e'er the scene, dear formalin
Will fill our eyes with tears.[1]

For nearly a hundred years, the Marine Biological Laboratory (MBL) has been central to Woods Hole, a small town on the southwestern tip of Cape Cod, Massachusetts. Founded on 20 March 1888, it was slated to be a "sea-side laboratory for instruction and investigation in Biology," catering to the needs and interests of students and scientists throughout the United States.[2] The obvious choice for the inaugural directorship was C. O. Whitman. He had demonstrated administrative skill as chairman of the zoology department at the University of Chicago and had learned much about the organization and operation of marine laboratories from two great zoologists, Louis Agassiz and Anton Dohrn.

Agassiz, professor of natural history at Harvard, had founded the first American marine laboratory on the small island of Penikese in Buzzard's Bay off the shoulder of Cape Cod.[3] His plan—one that turned out to be a classic innovation in the history of American science and education—was to bring teaching and research together under one roof. The doors of the Penikese laboratory opened for the first time in the summer of 1873, with Agassiz himself as director and main lecturer. The first group of students, forty-three in all, included Whitman, who was fascinated not only by Agassiz's lectures but also by his larger conception of a marine laboratory. Unfortunately, Agassiz's dream laboratory was short-lived. He died in December 1873, after the first session, and the school was to survive for only one more session, in 1874, before its doors closed permanently. Whitman attended both sessions.[4]

Seven years later, as a guest at the Stazione Zoologica in Naples, Whitman came under the influence of Anton Dohrn. Dohrn had in 1872 founded the Stazione Zoologica, which rapidly became the first great marine laboratory in Europe, and the most influential. Unlike the short-lived Penikese laboratory, there was no teaching at the Stazione; the principal function of the laboratory was research. But Whitman was still attracted to the conceptual model of Agassiz's laboratory, where research and teaching were combined, where renowned scientists and inexperienced students met on an equal footing. Despite their disagreement on this point, Whitman and Dohrn became great friends, and in the summer of 1897 Dohrn paid a historic visit to the new laboratory at Woods Hole.[5]

From its inception the MBL was based on uniquely American principles of democracy and opportunity. The Woman's Education Association (founded in 1871) had initiated action to establish the coeducational laboratory as part of its effort to promote the education of women.[6] The first group of MBL trustees included women in its ranks. During the summer months, women investigators came from both large coeducational universities like Chicago and Pennsylvania and small women's colleges like Mount Holyoke and Agnes Scott. If men were not to be preferred to women, neither were prestigious universities to be favored over little-known colleges. Institutions like Chicago, Harvard, and Pennsylvania had large constituencies at the MBL, to be sure, even in the early days, but small colleges like Goucher and Amherst sent investigators too. In 1902, determined that no one institution would control operations, the trustees squelched a plan by William R. Harper, the first president of the University of Chicago, to place the MBL under the control of that institution.[7] The MBL was to remain free and open, a place for all who wanted to examine not only problems in marine biology but in general biology as well. Foreign investigators were welcome. Very early, the MBL attracted European scholars such as the internationally renowned physiologist Jacques Loeb; many Chinese and other Asian scholars were also regularly included on the laboratory's roster of investigators. Between 1888 and 1908 the democratic and international spirit of the MBL was firmly established under the leadership of Whitman, a spirit that was to be maintained in the administrations that followed.

During the 1910s, when the MBL was under its second director, Frank Lillie, Woods Hole consisted of a winding thoroughfare lined with laboratory buildings and small business establishments, and several back streets dotted with lovely old cottages, beautiful gardens, and old stone fences covered by wild honeysuckle. The town had the atmosphere of a Cape Cod fishing village. The population was divided into three categories: natives, summer people, and bug hunters. There were just a few summer homes occupied by senior investigators who returned summer after summer. These homes, sizable and sturdy, were located on the quiet back streets, within easy walking distance of the main MBL building.[8]

In 1909 the MBL itself consisted of old wooden buildings housing laboratories and classrooms and administrative offices. That same year plans were put together for construction of a fireproof modern brick laboratory. The building was completed in 1913. Still, the old wooden buildings were allowed to remain. Constructed in the mid-1890s, they gave the summer seaside laboratory a rustic feel that was enjoyed by many of the scientists who, like Lillie and Loeb, spent the winter months fighting the hurly-burly of urban Chicago and New York to get to their laboratories. The town's seclusion and its proximity to the sea gave scientists and students a unique opportunity to combine rest and research.

Woods Hole was an odd mixture of work and fun, professional diligence and social enjoyment. An unknown student was likely to meet up with a famous physiologist not only in the classroom or the laboratory but also in the Woods Hole ice cream parlor or the movie house in Falmouth, the next town over. Informal choral groups, Sunday teas, picnics, and clambakes brought people together as much as the specimen collecting and laboratory research did. It was a vibrant social and scientific atmosphere.

Frank Lillie assumed the title of director in 1908, though he had technically begun functioning as director as early as 1902, when Whitman became disenchanted with the MBL Corporation and the board of trustees.[9] Lillie had been properly groomed for the position; in 1899 he had received his first appointment to the board and had gone on to demonstrate a high level of diplomacy and administrative acumen. Further, his family connections were excellent: his wife's brother, the multimillionaire Charles R. Crane (of the Crane Plumbing Company), became the main benefactor of the MBL. Talented and wealthy, Lillie was the perfect man for the job, and it is difficult to imagine the MBL without the imprint of this scientist and administrator.

In 1909 Lillie brought Just to Woods Hole as his research assistant. There Just was introduced to the professional world of science, not as a mature scientist but as a student investigator with potential for development. He had entered a predominantly white social and scientific community. In a way, he was already familiar with that kind of community through his experiences at Kimball Union and Dartmouth, but at Woods Hole the social obstacles were larger and the professional stakes higher. The story of how this young aspiring black lived in the Woods Hole community, how he gained a foothold in the field of biology, and how he finally came to reject Woods Hole for other similar communities in Europe is one of the most fascinating turns of events in the social history of American science.

ℜ

Just was twenty-six years old when he started out at Woods Hole as a research assistant to Lillie and as a student in a number of zoology courses. He could not but have realized the opportunity he was being afforded: close and direct guidance from one of America's leading biologists. Woods Hole was *the* place for an aspiring marine scientist to find his bearings; other American biological laboratories, such as Cold Spring Harbor and Pacific Grove, did not offer as much exposure to the best minds in the field. To go to Woods Hole and work under a top biologist was an opportunity that many whites vied for, one that was almost unthinkable for a black.

Just shared Lillie's laboratory in the early days and was introduced to some of the most exciting problems in contemporary biology, particularly those pertaining to fertilization and cell cleavage. His duties as a research assistant were mostly routine, including maintenance of the laboratory equipment as well as collection of marine material and the setting up and monitoring of experiments. The two men were both unusually diligent and meticulous, quiet and reserved, so they got on well with each other. Their habit of working together in complete silence became a standard joke around the MBL.[10] Just wanted to please and excel, and he found this easy to do with Lillie, who was in many ways so much like him. Very early he became attached to Lillie, and formed deep loyalties to him over the course of time. Mutual admiration and a sense of kinship grew between the two.

While a student investigator, Just not only assisted Lillie in the laboratory but also, in order to defray his tuition expenses, waited tables in the MBL mess under Miss Belle. Although this job was a common way for white students in similar financial straits to earn money, it is probable that Just found his position as a kitchen helper demeaning. He went about his duties without complaint, but in the back of his mind he no doubt associated the job with the predominant role of blacks as domestic workers in the society at large. What most white students viewed as temporary, albeit unpleasant work could have caused him much concern about his social and professional position. The MBL job was not different in substance from the other low-status jobs he had held at Dartmouth and Kimball Union. But at school he had not been seeking to enter the professional world, as he was at

Woods Hole, and he had not yet proved himself a motivated academic of exceptional promise. As a matter of fact, even though he loved the science and the experiments he perhaps viewed Lillie's laboratory work much as he did his labors in Miss Belle's kitchen. One of his daily duties in the lab was to scrub the floors and wash the glassware. Of course, he was convinced that everything he did at Woods Hole, including the menial work assigned by Lillie and Miss Belle, would facilitate his career advancement. Never once did he voice any discontent, and it is not until twenty years later, in the mid-1930s, that we find his revealing assessment of his early work as a laboratory assistant.[11]

For the first few years at Woods Hole, Just kept mostly to himself and concentrated on his duties as student and research assistant. He does not appear to have made any attempt to mingle with fellow students and beginning investigators. In fact, he said not a word to anyone but Lillie during his first six weeks at Woods Hole in 1909.[12] He passed his spare time alone, in quiet reverie, watching "the swells from the Bay," sniffing the "salty wind," staring into the "clear sky" beyond Penzance Point.[13] At dawn on 22 June 1909 he spent an hour or two before work sitting on the wharf, gazing out over the water, and jotting down his impressions of the scene before him:

> The angels have hung out their robes, translucent, cerulean; on them heaven's dew has fallen—glittering beads.
> Diana's trains have scattered o'er the field of night thousands of (?).
> —drives his herd afield—how their eyes shine.
> This time-worn battered tent is rent in many places—we get glimpses of heaven.
> Once dire rebellion raged in heaven and the angels sharpened their swords. The sparks remain.
> The stars are shining—God with his thousands of eyes is looking down on his world.[14]

This was the kind of thing he did to avoid the "nervous strain" of coming in contact with "people, people, people."[15]

Solitude was a way of life for Just at Woods Hole in the early 1910s. Though he met Loeb in 1912, his feeling for him was more one of deference to a patron than of intimacy with a peer. The effort to get on at Woods Hole in a professional way was uppermost in his mind. Aside from assisting Lillie, he was trying to complete work for his doctorate from the University of Chicago. There was little, if any, time for socializing. Besides, Woods Hole was new to him and there were no other blacks there. He needed to proceed slowly, to carve out a place for himself before becoming friendly with fellow scientists, before beginning to take part in social life at the MBL.

In 1914 Just met L. V. Heilbrunn, his first real friend at Woods Hole. Nine years younger than Just, Heilbrunn was also completing work

under Lillie for his doctorate. The two men ran into each other one day
in the chemical supply room under the old laboratory. Just liked
Heilbrunn from the start, taken as he was by "the frankness" of
Heilbrunn's smile and "the honesty" of his eyes.[16] Rumors of Heilbrunn's difficult personality that had reached him, mainly from the
"Stone House" bunch, painted Heilbrunn as a terror. But these asper-
sions did not bother Just; in fact, they may have been a point of
attraction. As everybody was against Heilbrunn, so Just was for him.
Just took a vicarious pleasure in backing Heilbrunn. Racism and anti-
Semitism (Heilbrunn was Jewish) may not have been a topic of conversation between them, but Just nevertheless perceived many of Heilbrunn's problems as those of a member of an oppressed group. This is
not to say that their friendship was based mainly on the sympathy felt
by a member of one minority for a member of another, but it is true that
the similar problems they encountered at Woods Hole and in the larger
scientific community gave them good reason to be friends.

The friendship deepened during the 1915–16 academic year while
Just was completing his Ph.D. residency requirements at Chicago and
Heilbrunn was an assistant in the zoology department there. Just
planned to be at Woods Hole with Heilbrunn for the 1916 summer
session, but he was too tired after the rigorous course work at Chicago
to do anything but return to Washington and rest until September.[17]
The two scientists kept up their friendship by writing to each other
regularly over the next year.

Heilbrunn may have been Just's first friend at Woods Hole, but he
was by no means his only one. After receiving the Ph.D. in 1916 Just
opened up to other people there. A. H. Sturtevant became a close friend
during the summer of '17, when Heilbrunn was away at officers' camp
fulfilling his military service. Just arrived in Woods Hole on 9 June,
two weeks later than originally planned. Sturtevant arrived the same
day. Since the mess hall had not yet opened they immediately planned
to eat meals together, using T. H. Morgan's laboratory as a kind of
makeshift kitchen.[18]

Just and Sturtevant enjoyed themselves thoroughly. After a hard day
at the laboratory they would play a few games of cards, buy ice cream at
Eaton's, and talk at length about everything from the international
scene to university politics. We do not know what their ideas on the
war in Europe were, but we do know that their dream zoology department included Sturtevant himself, Heilbrunn, and H. B. Goodrich.[19]
Sturtevant was to handle the genetics, Heilbrunn the cell biology,
Goodrich the morphology and administrative "dirty work." E. Newton
Harvey was also to be included, though both Just and Sturtevant had
doubts about the quality of his teaching. This dream department had
all the characteristics of a mutual admiration society, and the late-
night planning sessions by Just and Sturtevant were not unlike those of

the captain and cocaptain of some varsity sports team. No mention was ever made, however, regarding what slot Just would fill, though he apparently wanted to include himself. The fact that there were no blacks teaching in white universities at the time must have affected their plans. For them to have talked about a position for Just would have meant bringing up complicated political and social issues in American academia that were beyond their control. The two aspiring scientists and new friends simply ignored the problem.

The summer of 1917 marked the first time Just showed real interest in MBL social activities and opened up to new acquaintances. The war was on, and as a result there was an abundance of women at Woods Hole.[20] The enrollment in the invertebrate course, for example, numbered fifty women to five or six men. Dances and parties were scheduled, in part to dispel the "lethargy" that had set in because of the war and the shortage of men. It is unclear whether Just himself took part in these affairs in 1917, but he did know about them in detail, and he was able to identify Harley N. Gould and I. J. Davies as "social arbiters" of the MBL community. His main interest was in getting to know a few fellow male scientists who, for one reason or another, were doing scientific work rather than military service that summer. To this end he joined the "stag table" at the mess, sitting with Goodrich, Sturtevant, H. H. Plough, Fernandus Payne, Davies, Gould, Mario Banus, William Perlzweig, Alexander Weinstein, Calvin Bridges, and W. T. M. Forbes. For the first time, Just was at Woods Hole as independent investigator, full professor, and Ph.D.—an equal in every sense.

The 1918 season was as slow as the one in 1917. As usual, Just worked hard collecting and sectioning and examining marine eggs. Though he had little time for extracurricular activities, he did manage to strike up a friendship with Franz Schrader, a young Ph.D. from Columbia. Schrader and Just liked each other for the same reasons. They were both reserved, some would say shy; both were refined in a way that Heilbrunn was not; both were connoisseurs of gourmet food and exquisite wines; each had a fear of life that sparked a sense of adventure, a touch of the risqué; each had a love of biology, an appreciation and perhaps even awe of the wonders of the animal kingdom; above all, both were up-and-coming in the professional world of biology.

When Schrader took his first job at the Bureau of Fisheries in Washington in 1919, the friendship deepened. Just visited Schrader's home and Schrader visited Just's. They often had lunch together at Howard or in town. From time to time Just would accompany Schrader on visits to Long Island, Schrader's childhood home. There he became close to Schrader's mother, an Old World matriarch with whom he often practiced his German and talked about Germany.[21]

In the fall of 1920 Schrader resigned his position at the Fisheries

Bureau to take up an appointment at Bryn Mawr College. Just encouraged the move.[22] The Fisheries Bureau was a fine place for a Ph.D. to relax between real jobs, but it was not the place an aspiring academic would want to stay for long. Just knew Schrader's departure from Washington would not affect their friendship. There would always be time during their summers at Woods Hole to socialize and catch up on news.

Right after leaving Washington Schrader married Sally Hughes, a beginning investigator at the MBL and a graduate student at Columbia. The newlyweds both took up appointments at Bryn Mawr, where they stayed for nearly ten years. Throughout this period they were part of Just's close circle of scientist friends at Woods Hole.

Just drew close to Sally; he had always felt more comfortable with women than with men.[23] They spent many hours together discussing recent novels. One of their favorite authors was D. H. Lawrence.[24] At the time, Lawrence's work attracted a readership composed mainly of avant-garde literati. That Sally and Just read and reread and talked of Lawrence during their evenings at Woods Hole suggests not only the breadth of their interests but also their bohemian spirit and love for the unconventional. Sally was Just's ideal of a woman—sophisticated and charming, witty and cynical, brilliant as a scientist, and above all loyal and helpful to her husband in a way that his own wife Ethel had not been. Ethel may have been sophisticated and handsome, but she was also imbued with the kind of moral and social stiffness that was characteristic of most upper-class blacks, especially women, at the time. She was not the type, for example, to enjoy the blues of Bessie Smith, so much a part of life in the Gay Twenties. Just was. He enjoyed jazz and burlesque, and even if Sally Hughes in fact did not, his image of her was that she did.

Among his other friends at Woods Hole in the 1920s were Donald and Rebecca Lancefield.[25] Donald was attached to Columbia University, Rebecca to the Rockefeller Institute for Medical Research. Their fields were different: Donald worked on the fruit fly *Drosophila*, Rebecca on bacteria. Unlike the Schraders, they never worked collaboratively. Every summer they came to Woods Hole together to take up their separate research problems.

Woods Hole was a place for husbands and wives and families. Frank Lillie had met his wife, Frances Crane, there in the 1890s; Franz had met Sally and Donald had met Rebecca there in the 1910s. The large enrollment of female biologists led to a number of romances and marriages among students and beginning investigators. The annual return of these couples made Woods Hole a special place. Everyone, including children and pets, joined in the summer fun at Woods Hole. Many husbands and wives formed research teams—the Schraders,

Harveys, and Brookses, for instance—that worked together during the day, socialized with other couples in the early evening, and often resumed work at night. Weekends were reserved for special family outings such as picnics, clambakes, movies in Falmouth, and cocktail hours at Dr. Warbasse's mansion out on Penzance Point.

This pattern of life affected Just in a curious way. He found himself in the peculiar position of being a single man among the married couples at Woods Hole, for there he was very much a bachelor. In the 1910s and early 1920s he never brought his wife and children to Woods Hole. While he welcomed the summer months as a relief from the day-to-day responsibilities of parenthood, there were moments when loneliness overcame him and he thought about how enjoyable it would be to escort his wife to Sunday tea at the Lillies'. The fact that he led two different lives—one at Woods Hole and one in Washington—proved to be not only an emotional burden but a financial one as well. He had to maintain two residences during the summer, a detail he often mentioned to foundations and the Howard administration as an argument for increased financial support.[26]

Living at Woods Hole was one concern, working another. His wife was no scientist and could not assist him in his laboratory work. He envied the Schraders and the Harveys; he could not help but notice how much they were able to accomplish through collaboration. Ethel would never be his assistant, much less his colleague. Besides, she seemed not the least bit interested in his work or his professional advancement. Occasionally, when she was downright hostile, he felt he had to give "kilos" of himself for "one ounce of freedom" to do scientific work at Woods Hole.[27] At the beginning of every summer there would be tension over his leaving for the laboratory—tension perhaps centering on Ethel's fear of the unfamiliar. Knowing no science, she could hardly appreciate the importance of his work. Also, the Woods Hole people were for the most part white, and she did not care to mingle with them, an attitude quite understandable given the predominance of jim crow in America in general and Washington in particular. Also, Ethel had come from a proud and snobbish background, and while she did not care for most blacks, she cared even less for whites. As far as she was concerned, Woods Hole was her husband's world and not hers.

After the war, when normal activity at the laboratory had resumed, Just tried to bring Ethel into the Woods Hole community. In 1920 he asked Lillie about the possibility of renting or leasing a house for the summer of 1921.[28] Lillie did not respond, but Heilbrunn urged Just to follow through with the plan.[29] Ethel rejected it, however.

Just continued to live in single quarters during the early twenties, but he never gave up the house-rental idea. Money was no problem: his grant from Rosenwald included summer expenses at Woods Hole. The real problems were finding a house big enough and convincing his wife that Woods Hole was a desirable place for summer living. He was unable even to approach solving these problems until the late 1920s.

Though he was socially somewhat out of place at Woods Hole, Just remained convinced that the MBL was of extraordinary help to him in his research efforts and his struggle for professional advancement. He regarded the summer of 1921 at Woods Hole as his "happiest" and most productive to date.[30] He wanted to stay through October and November, but that was not possible: there was no escaping the drudgery of teaching responsibilities at Howard. Sadly he returned to Washington, with the hope that things would be as good for him the following summer. Everything had been working in his favor; in particular, he was having remarkable success collecting and analyzing data on the eggs of the sea urchin *Arbacia* and the common sand dollar *Echinarachnius parma*.

By 1921 Just had carved out a professional place for himself at Woods Hole. His scientific output that summer was so impressive that Lillie had to say he had never had a student superior to Just in "general intelligence, devotion to scientific research, and loyalty to high standards."[31] Just's interpretation of the effect of hypertonic seawater on *Arbacia* eggs was an ambitious piece of work per se. Lillie was impressed by it, and even more impressed that it was one of *three* ambitious pieces done that summer. According to Lillie, Just's research on hypertonicity cleared up a difficulty that had troubled scientists for over twenty years, ever since Jacques Loeb's first experiments on artificial parthenogenesis. In fact, Just's work in this area was to bring into question many of Loeb's findings and lead to a bitter falling out with his one-time patron.

Just had come to be viewed as one of the most valuable men among the 180 engaged in research at Woods Hole. Lillie regarded his example of "steady hard successful work, carried on with enthusiasm and good judgment," as a real asset to the MBL.[32] By the early twenties he was one of the most popular MBL workers; people would go to him for advice more than to anyone else. He felt almost as proud of his popularity as of his productivity. Small wonder that, as he confessed, even the simple process of cutting sections seemed to go better at Woods Hole than at Howard![33]

As we have seen, Just's career as a professional scientist began in 1917, when, Ph.D. in hand, he took his place at Woods Hole with a clear eye to opportunity and advancement. By then he had already published several papers, though these, rather than distinguishing him as an independent investigator, had shown him to be one of Lillie's faithful student collaborators. Now seemed an opportune time to do something serious about establishing an independent reputation. The war was on and activity around the town of Woods Hole lagged;

perhaps Just was also at an advantage because colleagues like Heilbrunn had been called to military duty. Though he published nothing between 1917 and 1919, Just had never worked so hard on his research. The work for three major articles published in late 1919 and early 1920 was completed during the summers of 1917 and 1918. What kept him out of print, and relatively quiet and subdued, was the sense of "horror, uselessness and waste" that overcame him during the war years.[34] He was deeply affected by the "convulsive hysteria" of those "hyper-excitable times."[35] Even more worrying than the war itself was the "element of near tragedy" that it had inflicted on his scientific work and professional status.[36] Before the war his field in general and his own work in particular had been "going big" among European workers, particularly those on the Continent. Citations of his work had been more frequent in European journals than in American ones; workers in France and Germany had expressed interest in his work on the embryology of invertebrates, and had often referred to him as an important authority on the subject.[37] Then came the war. Mobilization took precedence over research, and European scientists had no more time for Just.

The disparity between Just's reputation in Europe and his relative obscurity in America had a scientific basis; simply put, more Europeans than Americans happened to be working on the problems Just was tackling. Just pointed out the difference to an official at one of his funding agencies in 1921. He did not mean to suggest that the difference was due to social factors or racial attitudes, though later in life he came to hold that view. At the time, he was merely expressing enthusiasm about his work in an effort to secure further support, and was anxious to make his work sound as influential as possible. These officials could neither confirm nor deny the veracity of his assertions.

During the war years Just stepped up his efforts to achieve professional status. In January 1917 he asked Loeb to recommend him for membership in the American Society of Physiologists.[38] Before complying, Loeb wanted to know whether Just's professorship was in the modern field of physiology or the traditional field of biology.[39] If the latter, he thought the application for membership should be made to the Naturalists rather than the Physiologists. He promised that on receiving a list of Just's publications he would send in the nomination to either society and "prepare everything" for the next election. Unfortunately, an illness in his family prevented him from keeping his promise. Disappointed but not discouraged, Just repeated his request two years later.[40] He noted that he held the position of professor of physiology at Howard, that his Ph.D. had been obtained under Lillie and Carlson at Chicago, and that his research had been in cellular physiology. Membership in the Society of Physiologists, he pleaded, would mean a great deal to him personally. He had been a little afraid to ask Loeb again for a recommendation, but he was "anxious" for membership and would be "ever grateful" should Loeb consider it "wise" and manage to arrange the matter.

Just also asked Lillie to nominate him for membership in either the American Society of Zoologists or the American Society of Naturalists. Lillie replied that he would be happy to send in a nomination to either or both societies.[41] He mailed Just an application for the Society of Zoologists, with instructions to fill it out and send it directly to Caswell Grave, the society's secretary, at Johns Hopkins. Nominations for 1919 had no doubt closed, Lillie feared, so Just would probably not come up for election until 1920. Another question was whether Just also wanted membership in the Society of Naturalists; Lillie decided to wait to hear from Just before proceeding with that nomination process. Within a few days Just responded that he would be "greatly pleased" if Lillie would give him a recommendation for membership in the Society of Naturalists as well.[42] This Lillie did promptly. He also sent Just an application, which Just in turn mailed to Bradley M. Davis, the Naturalists' Society's secretary, at Pennsylvania.[43] As with the other application, this one was too late for consideration in 1919; the final list of nominations had been compiled on 1 January. Just would have to wait until 1920.

It was around 1920 that Just began to enjoy a high level of professional recognition. He was elected to several scientific societies—the American Society of Naturalists, the American Society of Zoologists, the American Association for the Advancement of Science, the American Ecological Society, and the *Société nationale des sciences naturelles et mathématiques de Cherbourg.*[44] These were honors previously unheard of for any black man, honors which helped to give Just the self-confidence to challenge traditional scientific concepts in light of evidence of his own. Professional success encouraged his already independent approach to scientific research, even to the point of jeopardizing future opportunities for advancement. A case in point involves the bitter conflict between him and Jacques Loeb. Loeb was a venerable physiologist who was not used to being contradicted, especially not by a young and aspiring junior scientist who had just asked his support for membership in professional societies. It was indicative of Just's tenacity that he was able to pursue a penetrating line of scientific inquiry in the face of hostility from Loeb and his followers.

In the summer of 1917 Just began what later became pathbreaking research on *Echinarachnius parma*, investigating, among other things, the cortical response of the egg to insemination and the role of a substance called fertilizin in straight and cross-fertilization. His work on these problems during 1917, 1918, and 1919 provided the substance of four major articles published in *Biological Bulletin* in 1919 and 1920, articles which directly challenged Loeb's previous work on artificial parthenogenesis.[45]

In 1899, while Just was still a beginning student at Kimball Union Academy, Loeb had discovered that he could cause unfertilized sea urchin and frog eggs to undergo development either by pricking them with a needle or by changing the salt concentration of the seawater in which they had been cultured.[46] The "artificially induced" embryos of the sea urchin developed to the larval stage and sometimes even further, while some of the unfertilized frog eggs developed to sexual maturity. After bringing these things about, Loeb put forth theories to account for them. One of his theories, as outlined in *Artificial Parthenogenesis and Fertilization* (1913), was that two agents were needed in parthenogenesis: a cytolytic factor to start the breakdown cytolysis of the egg surface, and a corrective factor to prevent the cytolysis from going any further. Loeb would pour a little seawater containing butyric acid on unfertilized eggs, and they would then start to cytolyze. But the butyric acid could carry the cytolysis too far, which tended to be lethal unless corrected. Next, Loeb would either pour off the butyric seawater or treat the eggs with hypertonic seawater and a little magnesium. This supposedly stopped the cytolysis so that the eggs could proceed to develop.

Loeb soon turned his attention to the process of fertilization itself. Arguing by analogy, he arrived at his "lysin theory" of fertilization. He postulated that the sperm carried a molecular substance, a cytolyzer which he termed "lysin," and of course a corrective factor. Since lysin apparently activated the egg, he was convinced that he could recreate the fertilization process in the laboratory without the living sperm simply by providing a physical or chemical agent with the same molecular effect as lysin. In other words, for Loeb the role of the sperm in the fertilization process was not specific, since it could be imitated by physical and chemical agents, and fertilization could best be explained by his general theory of artificial parthenogenesis. No wonder phrases like "fertilizing agents" and "chemical fertilization" came to be the language of that theory.

Loeb's work on artificial parthenogenesis constitutes, as Just acknowledged in 1919, one of the "important chapters" in the history of modern biology, mainly for its stimulating effect on the new experimental zoology and, more generally, for its novel approach to understanding the molecular basis of life.[47] Known chemical reactions, as Loeb had shown, could recreate biological processes; these processes, he reasoned, therefore must be purely molecular in origin. It was not then too fanciful to imagine that soon life might be created in a test tube!

Though the theory captivated many biologists and stimulated research in biology at the molecular level, doubts soon emerged as to whether it might not be too simplistic an approach to the complex process of fertilization. Frank Lillie, for one, was undaunted by Loeb's work. Spurred on by his own observations of physiological changes in the eggs of the sea worm *Chaetopterus*, Lillie went on to examine what he perceived to be an underlying mechanism involved in the process of fertilization, a mechanism without which fertilization could not take place. This was around 1906, shortly before Just's first visit to Woods Hole. In that year Lillie published a paper setting out his thoughts, which he later incorporated into a more comprehensive conception known as the "fertilizin theory."[48]

Lillie's theory, propounded in a series of papers appearing between 1912 and 1921, asserts that fertilization involves the biochemical interaction of substances carried by the egg and sperm. The chief role in this interaction is played by a substance Lillie called "fertilizin." For Lillie, fertilization was a process in which three substances are essential: the sperm, the egg, and a third which is secreted by the egg and reacts with both egg and sperm. He argued that the third substance, fertilizin, causes agglutination in sperm suspensions of the same species. In other words, the spermatozoa become firmly stuck together and, unlike those that are merely aggregated, resist separation by shaking. According to Lillie, fertilizin has two active functions in fertilization: one with the sperm, the other with the egg, with both sperm and egg carrying receptors that react with fertilizin. Lillie's final conclusion was that fertilizin is an essential mid-substance in fertilization, and the linkage is not simply

$$\text{spermatozoon} + \text{egg}$$

but

$$(\text{spermatozoon} + \text{sperm receptor}) + \text{fertilizin} + (\text{egg receptor} + \text{egg}).$$

The latter formula indicates the importance of the cell surface in the fertilization process—a theme that was to dominate Just's later work.

One problem with the theory was that Lillie had adopted his terminology, for the most part, from Ehrlich's side-chain theory of immunity, which had been repeatedly criticized since it was published in 1900. Though Lillie urged scientists not to be prejudiced by the terminology but, rather, to be persuaded by the facts, his fertilizin theory fell into disrepute in 1914 as part of a general rejection of Ehrlich's work.

Just championed the cause of his mentor Lillie by accumulating a large body of evidence for the fertilizin theory. Beginning with the research for his first paper in 1912, "The Relation of the First Cleavage Plane to the Entrance Point of the Sperm," he assembled results that in the long run would bear directly on the fundamental questions of fertilization.[49] While working on his next two papers, studies of the breeding habits of *Nereis limbata* and *Platynereis megalops* published in 1913 and 1914 respectively,[50] he gained further experience in observing the fertilization process, which in marine invertebrates and amphibians is external but in mammals is internal. The *Nereis limbata* paper was coauthored with Lillie. Just was responsible for the observational work, Lillie for directing the experiments and writing the paper. The *Platynereis megalops* paper was completed by Just alone, at Lillie's suggestion and under his direction.

Just had compiled observations of the swarming habits of *Nereis limbata* and *Platynereis megalops* during the summers of 1911, 1912, and 1913. Much of his time was spent gathering specimens for his and Lillie's experiments. The worm *Nereis* is perfect for experimental work on fertilization because every swarming individual is always sexually active, with no immature sexual cells; the same is true of *Platynereis*. Almost every night Just would go out with his lantern and lie on the float stage in the Eel Pond behind the supply building. The best time for making a catch was between sunset and moonrise, during the swarming period that lasted for an hour to an hour and a half. First the tiny males would appear, darting rapidly in curved paths; then came the larger females, plowing laboriously through the water. At the height of the swarm there would be upwards of a hundred males, only ten or so females. It was a beautiful sight to behold—the twisting red males and sluggish yellow-green females coming together under the light of the lantern, shimmering on the surface of the water. Just never failed to be spellbound by the drama of the copulation ritual: the way the *Nereis* females sank out of sight to die after shedding their eggs in the water made milky by the sperm of the males, the way the *Platynereis* males grasped and lashed and finally thrust their tails down the "mouths" of the females.

These excursions put Just in a favored position at the laboratory; the other investigators came to admire him as a scientist who had had unusual firsthand experience with marine specimens. And it was not only shallow-water organisms that he was familiar with. Many a day he spent several hours at sea with Captain Robert Veeder, skipper of the *Cayadetta*, and his crew, dipping in nets and pulling them up again. Each year he would try and get to Woods Hole earlier than most of the others in order to take advantage of the relatively light demand on the specimen-collecting vessels, and often he remained late into the fall. In 1914, for instance, he stayed at Woods Hole collecting *Platynereis megalops* long after most of the investigators had left; he did not return to Howard until 17 September, a day or two before his classes were to begin. This sort of sustained effort, combined with his keen observational skills, soon made him an expert on the breeding habits of a variety of marine organisms.

Just first wrote on the subject of fertilization per se in 1915, a year before receiving his Ph.D. and the same year that he received the Spingarn Medal on Loeb's recommendation.[51] He treated the phenomenon from the point of view of initiation of development, including artificial parthenogenesis. In his experiments, he used various chemical agents to initiate development in the eggs of *Nereis limbata* and *Platynereis megalops*, and also observed insemination of these eggs by sperm. Comparing the two processes, he concluded that "whether by sperm . . . or by artificial agents, the initiation of development is fundamentally the same."[52] Without attacking Loeb directly in any way, Just simply added what he called "another link in the chain of evidence which supports the theory that fertilization is essentially a process of the egg."[53] In doing so, he utilized aspects of the theory of artificial parthenogenesis. He was by no means hostile to Loeb's theory, leaving any inherent conflict between it and Lillie's fertilizin theory unremarked. Such diplomacy was characteristic of most young scientists in the predoctoral stages of their careers. Both Lillie and Loeb had been helpful to Just; both were eminent scientists who could influence the career of an aspiring scientist. Just was careful not to become too controversial too early. Only after receiving his Ph.D. did he publish papers identifying the conflicts between Loeb's and Lillie's theories.

The four papers that Just published in 1919 and 1920 on the fertilization reaction in *Echinarachnius parma* mark the beginning of his direct assault on Loeb and firm defense of Lillie. The third paper in this group (the one treating the nature of the activation of the egg of *Echinarachnius parma* by butyric acid)[54] was especially devastating to Loeb's lysin theory.

Just argued that if a single factor, especially the corrective (anticytolytic) factor of hypertonic seawater, could be shown to induce development, then the lysin theory made no sense. Just designed experiments to prove his suspicion. He treated *Echinarachnius* eggs with carefully measured quantities of butyric acid and found that the ones which formed membranes were completely activated (fully developed), first because they responded to subsequent hypertonic seawater treatment with development that simulated the normal, and second because they failed to develop further after insemination, even when the membranes were immediately removed. Conversely, eggs that were under- or overexposed to butyric acid did not form membranes and did not develop normally. Just was trying to show that eggs treated with butyric acid parallel either fertilized eggs or the fragments of unfertilized eggs that fail to refertilize. This was in opposition to Loeb, who had found that sea urchin eggs were capable of fertilization by butyric acid even after membrane formation, provided the membranes were removed by shaking. Thus Loeb held that complete membrane formation involves incomplete activation. Just took the opposite view, referring for support to work by Lillie and C. R. Moore. He concluded that although the egg responds in a similar way to the artificial agent and to the sperm, observation of membrane formation does not by itself warrant subscription to Loeb's lysin theory. Furthermore, "to argue that the sperm carries a lysin because a host of agents activate the egg, is bad logic. It is far simpler to postulate that the egg contains the necessary mechanism for development."[55]

The egg, for Just, was not an inert cell unaffected by laboratory techniques. Half-jokingly, he often accused some investigators, such as Robert Chambers, of working with eggs that were not "normal" but "poached."[56] By 1919 he had become notorious for his rigorous standards of experimentation and was beginning to be viewed as the authority on the correct methods of handling *Nereis limbata*, *Platynereis megalops*, *Echinarachnius parma*, and most other marine invertebrates. His work on *Echinarachnius parma* required utmost sensitivity to laboratory technique. According to Just, many of the early experiments with artificial agents were open to suspicion because the investigators, including Loeb, had had no uniform and standard method for handling the eggs. He stressed the point that the response of eggs to agents that initiate development depends on the physiological condition of the eggs, which in turn depends on the presence of fertilizin.

Just continued his work on *Echinarachnius parma* in 1920, pursuing his attack on several aspects of Loeb's work. This time Loeb's theory of cytolysis, as outlined in his *Artificial Parthenogenesis and Fertilization*, became Just's primary focus.[57] Just did not believe Loeb's double theory of cytolysis and correction, and proceeded to argue against various aspects of it. He proved the theory wrong in an extremely ingenious manner. He first treated some *Echinarachnius parma* eggs with a hypertonic solution to initiate activation, and then added butyric acid to complete the job. But he also showed that with the right dosage of the two in the reverse order the eggs still developed normally. He had cornered Loeb. The older man had claimed that a cytolytic agent had to be followed by a corrective agent, but Just had produced the same effect the other way around.

Loeb made this mistake, Just demonstrated, because he had used sea urchin eggs that had been overexposed to the butyric acid treatment and had assumed for them the same development as for underexposed eggs, showing that Loeb apparently knew very little about the normal initiation of development. On several other counts Just was severely critical of Loeb's laboratory technique. To him, the cytolysis theory of development was unfortunate to say the least; it had blocked progress toward an understanding of the problem of fertilization. Just was doing something to clarify the question and to point toward its solution, and his work had influence on people studying parthenogenesis. It settled the fact that it is the reacting biological system that is important, not what is applied to it. Just went on to offer Lillie's fertilizin theory as "the best working hypothesis for the study of the fertilization-reaction."[58] For him, fertilization was not simply a type of artificial parthenogenesis: it was a process involving both egg and sperm, neither of which could be eliminated.

The work on the fertilization reaction in *Echinarachnius parma* established Just as an outstanding scientist. It was this work that brought him to the attention of foundation officials; that won financial support for him, his students, and Howard; and that prompted Lillie to ask him to collaborate on the fertilization section of E. V. Cowdry's comprehensive book, *General Cytology.*[59] It was this work that brought the attention of prominent scientists such as Morgan and Wilson to what he was doing and that, in the end, aroused Loeb's wrath against him.

༄

As a result of his publications and newly acquired authority in the field, Just was invited to participate in the symposium on fertilization at the annual meetings of the American Association for the Advancement of Science and the American Society of Zoologists in Chicago in December 1920. This was indeed a "rather unique invitation," one which he accepted not only as a personal honor but also "in the interest" of his race and Howard University.[60] Determined to make a good and lasting impression on his colleagues, he prepared his talk very carefully throughout the fall of 1920, checking with Lillie to avoid any overlap with others in the symposium. His sinuses flared up, in part because he was nervous, and the pain became so severe that he had to undergo an operation for removal of superfluous bone tissue from his nose and another to cauterize his tongue and tonsils.[61] Despite these problems, however, his spirits were running high, high enough for him to go out and shop for new clothes for his Chicago trip—"sox, shoes, hat, coat, shirt, collar."[62] In her typically proper manner, Ethel had insisted that he lacked "everything . . . the male person needs in order to come up to the legal standards set for appearance in public." The new wardrobe required a substantial outlay of money. Though the family was laboring under a financial strain, Just and Ethel decided together that the fertilization symposium was important enough to the advancement of his career to warrant some financial sacrifice. In addition to buying new clothes, he would register at the plush Congress Hotel in Chicago rather than accept the hospitality of his old friends the Bentleys, whose household he felt was not conducive to the intensive concentration desirable the last few hours before an important presentation.

Christmas festivities began early on Saturday morning in the Just home at 412 T Street. By late afternoon Just had finished preparing the Christmas dinner, complete with a fifteen-pound Smithfield ham and several fruit pies and pound cakes. It was something he enjoyed doing every year.[63] A better cook than Ethel, he always looked after the important dinners, while she attended to the domestic matters of cleaning and arranging. As evening drew on, Just found a chance to put some finishing touches on his symposium paper. In two days he would be caught up in the swirl of that busy period between Christmas and New Year's, the usual time for annual conferences in the academic community. His paper was uppermost in his mind. After last-minute preparation on Sunday, he got up early Monday morning, boarded the train at Union Station, and reached Chicago late Monday evening.

On arrival in Chicago, Just went straight to the Congress Hotel, registered for his reserved room, and settled in for a good night's sleep. On Tuesday he got up early and located Heilbrunn. They had breakfast together before attending the morning session on general physiology in Harper Library at the university. Sixteen papers were read, each lasting about ten minutes.[64] There was nothing of particular interest to Just in the morning session, and as a matter of fact the afternoon session on embryology, his own specialty, was equally boring. Luckily, however, the last quarter of an hour was livened up when a bat got into the conference room and distracted attention from the last two speakers.[65]

Just had dinner that evening with Lillie, and the two talked seriously about job possibilities for Franz Schrader. The rest of the evening was taken up with an address by William Morton Wheeler of Harvard, the retiring vice-president of Section F of the American Association for the Advancement of Science (to which Just had recently been elected). Wheeler's talk, entitled "The Organization of Research," was by and large well received. Articulate and witty, he chided the National Research Council for controlling the direction of scientific research and for allocating its funds mostly to the older, more established scientists in its own ranks. Scientists attached to the council, like C. E. McClung, were chagrined to say the least, while the younger scientists, Selig Hecht and Libbie Hyman for instance, were thoroughly pleased.[66] Just no doubt felt ambivalent toward Wheeler's speech: on the one hand he could sympathize with the plight of the younger scientists, but on the other he knew he had been receiving innumerable benefits from his own council fellowship. After Wheeler's speech there was a "Biological Smoker," where Just chatted with old teachers from Chicago, particularly Child and Newman, and fellow Woods Hole investigators, including Sturtevant and Lancefield.

Just's first day at the conference gave him, if not a deeper knowledge applicable to his own field, at least a sense of professionalism and camaraderie he could scarcely have found elsewhere. He felt as if he belonged to the larger scientific community. As he listened to scientists from all over the United States present results of months and sometimes years of hard labor, he was inspired with a sense of purpose. There had been no such comprehensive presentations at Woods Hole. The MBL was a place for laboratory work; reflection on data occurred elsewhere.

Just's friend Heilbrunn was scheduled to read a paper in the session on cytology on Wednesday morning, 29 December. His paper, "Protoplasmic Viscosity Changes in the Dividing Egg of *Cumingia*," was a success, at least in the eyes of Just and Libbie Hyman.[67] According to Just, it made "such a distinctly good impression" that Heilbrunn's senior colleagues at the University of Michigan could not help but promote him.

Symposium presentations could be a means of either acquiring new jobs or of advancing in current ones. Just could gain little, however, in the way of job placement; for a black, Howard University was the top of the ladder. However, he could increase his status within the profession so as to gain leverage in securing funds for his own research at Howard. With one eye on his presentation, scheduled for Thursday, and the other on his inadequate opportunities for research, he decided to skip the Wednesday afternoon session on evolution and genetics to spend part of the afternoon looking over his talk and the other part visiting the philanthropist Julius Rosenwald, who was supplying the funds for his council fellowship. The meeting with Rosenwald went well. Just was particularly grateful for the "kindly and sympathetic reception" he received from Graves, Rosenwald's private secretary.[68] Long afterwards, he continued his "enthusiastic ravings" about Graves, whose generous hospitality had given him the hope that he had at last found someone sincerely interested in his work and career.

On Thursday morning Just presented his work on the response of eggs to hypotonic seawater and tap water at various intervals after fertilization. He made quite an impression, even though Hyman thought he seemed "nervous."[69] His paper was scheduled to last twenty minutes. It was by far the longest, if the abstracts are any indication. It clearly outlined new results defending the fertilizin theory and questioning Loeb's work. The audience was divided into two camps: those who were open to criticism of Loeb and would entertain Just's arguments in support of the fertilizin theory, and those who were reverential toward Loeb and would fight any attempt to question his work. Libbie Hyman was typical of the former camp, Selig Hecht of the latter.

Hyman, who was in the Chicago department with Lillie and had been there as an instructor when Just was a student in 1916, agreed with many of Just's findings. She found his presentation "very interesting," and admired the gumption he showed in presenting it.[70] When she could not repeat some of Just's experiments with the same results, she was willing to give him the benefit of the doubt; Loeb, on the other hand, received no such courtesy. One of her big battles in the early twenties was to get brilliant young scientists to reexamine Loeb's work. She defended Just because she saw how critical he had been of Loeb. Whether her support affected him is a moot question; suffice it to say that she does not seem to have either increased or diminished his criticism of Loeb.

After the conference, Hyman defended Just's presentation in a letter to her friend William Crozier, a zoologist at Rutgers who later moved to Harvard. She stated her case with force:

Just is not prejudiced against Loeb. I have often heard him express admiration for Loeb's fertile brain and for the role which Loeb has played in the development of experimental biology. But like the rest of us he thinks that Loeb is frequently wrong and that in the name of scientific truth his mistakes ought not to go unquestioned just because of the prominent position he justly occupies. It also irritates Just as it irritates the rest of us to see people like yourself with highly critical minds refuse to apply that critical faculty to Loeb's statements just because they are Loeb's. Now Just believes that Loeb's theory of artificial parthenogenesis is erroneous and he is not the only person that thinks so. After receiving your letter I very carefully looked up in Loeb's book the quotations that Just gave, and I could not see that he had to the slightest degree perverted Loeb's meaning.[71]

Crozier was upset by this. Like Hecht, he was one of Loeb's special protégés. Loeb pushed with all his power and influence to secure prestigious positions for Crozier and Hecht. Out of loyalty to Loeb, Crozier must have responded to Hyman with something like the wrath of God. His letter has not survived, but her next few letters to him suggest considerable hostility on his part, and a strong feeling that Loeb was taboo as a subject of gossip. In a letter dated 26 February 1921, Hyman apologized for having hurt Crozier's feelings: "I have a bad habit of being very outspoken—and you have a right to be offended if you wish."[72] She protested, however, that Crozier was attributing to her a desire to be "nasty" or disparaging. Nothing could have been further from her mind, Hyman insisted. She had simply indulged in a little harmless pleasantry by cutting down others, among them Loeb. Her remarks were not meant to be taken too seriously: "I think it is psychologically impossible for one to be 'nasty' to one's friends. I should have to stop liking you before I could behave that way." She further assured Crozier that Just had not said anything about him, that she must have given "a wrong impression somewhere." As far as she could tell, Just was "merely irritated in general at the wide-spread acceptance of Loeb's theories (or is it wide-spread?)." His annoyance was with ideas, not personalities.

Crozier and Hecht were close friends, the two shining apples of Loeb's eyes. They were caustic about those who did not share their scientific theories or their way of approaching science. Hecht, for instance, was not amused by Just's support of Lillie's work over Loeb's at the December symposium:

> I pass the buck in re Just. He's a—what's the use of calling names? His paper at the Symposium on Fertilization of the Zoologists would have made you writhe. It is possible (?) that he might have amounted to something if he weren't so handicapped by his color. The more I come in contact with him, tho, the less "substance" do I find in him.[73]

The reference to Just's color is interesting. It could be interpreted as sincere, as positive rather than derogatory. Hecht, like his mentor Loeb, professed dedication to removing the social barriers created by oppression. His view of oppression, though, was narrowed by Semitic blinders.

Just assumed that Hecht, Crozier, and Loeb were still more or less friendly to him, if not on a professional basis then at least on a personal one. He was unaware that within Loeb's circle he had been stigmatized as "that poor fool Just!"[74] In his most intimate letters to Heilbrunn and Lillie, he never made disparaging remarks about Loeb or Loeb adherents such as Hecht, Crozier, W. J. V. Osterhout, and Otto Glaser; in fact, at one point he showed deep-felt concern about Hecht's job prospects.[75] Little, too, did he realize the intensity of Loeb's negative feelings about the direction and content of his research. When Loeb was informed by Hecht of the specifics of Just's talk at the Chicago symposium, he wrote to Osterhout sarcastically suggesting that it would be better for Just and Heilbrunn if they could live in "the exhilarating atmosphere of Cold Spring Harbor" rather than be "stifled" at Woods Hole by his, Loeb's, "proximity."[76]

Loeb had expressed serious doubts about Just and his work even before the Chicago symposium. In November 1920 he received a letter from Abraham Flexner reminding him that it was Loeb who had "originally called . . . attention" to Just's work, a specific reference to Loeb's nomination of Just for the Spingarn Medal.[77] Now that Julius Rosenwald had provided a fellowship for Just through the National Research Council, allowing him complete freedom from teaching duties at Howard during part of the year so that he could concentrate on his research, Flexner wanted Loeb's opinion on two matters: (1) whether Just's work could be carried on properly in his Washington laboratory, and (2) whether an opportunity to work in a place such as Jamaica, which Just had proposed to the council, was "indispensable or necessary or just advantageous." He requested a "cautious and conservative opinion." Loeb responded that since he was perhaps "indirectly responsible" for Just's fellowship award, he felt in fairness to Flexner an obligation to express himself "with complete frankness."[78] It was his opinion that Just would never be "a prominent investigator," but that he could be made into "a better type of scientific teacher through his research fellowship." Instead of going to Jamaica to continue what Loeb regarded as a deficient type of work (that is, research on *Echinarachnius parma*), Just should "stay in Washington and use the freedom granted him by his fellowship for developing along the lines of physical chemistry and organic chemistry." If he continued his present line of research after studying chemistry, his work would at least be "of a better character." The letter concluded with a request that Flexner not divulge the details of the consultation, as Just was in Loeb's opinion "hypersensitive."

Loeb's assessment is at variance with one that Lillie had already sent Flexner supporting Just's proposal for research in Jamaica.[79] Lillie explained that the problems on which Just and many other biologists were working were "seasonal," in the sense that material is "available in the Northern climate only during the summer months." In Jamaica, however, "the material is abundant, and is represented by several species" throughout the year. Convinced of the variety and abundance of material in Jamaica as compared to Woods Hole, Lillie stressed in no uncertain terms that one winter in the West Indies would enable Just to duplicate the results of an entire year in the northeastern United States. Lillie had himself gone to California the previous winter on an errand similar to that which Just was contemplating, and he had been abundantly compensated in results for the time and expense involved. He was positive, then, that Just's proposed expedition to Jamaica was "a wise plan."

The proposal was rejected. It is clear that Flexner was against it from the beginning, since he solicited Loeb's "cautious and conservative opinion" after receiving the strong statement of support from Lillie. In Flexner's view, help should be given to black professionals only very slowly and carefully, and only so long as they were "safe risks."[80] In Just's case the National Research Council and Rosenwald seemed to be "going a little too fast," Flexner warned. There was, too, "a little more danger of spoiling an exceptional colored worker than . . . of spoiling an exceptional white worker."

It is therefore understandable, if inexcusable, that Flexner blocked the proposal. The man who had almost single-handedly revolutionized medical education in the United States, who later was to engineer Einstein's emigration to the United States and to head the first major American research institute for scientists—the Institute for Advanced

Study at Princeton—had no trouble insuring that his opinion held sway in the matter of support for an aspiring black scientist.

Three years later, Loeb figured in a situation where the losses for Just involved much more than a trip to Jamaica. From all indications, Just was being seriously considered for a research position at the Rockefeller Institute for Medical Research in 1923. That year, in the first week of January, he met with Frederick P. Gay, chairman of the Medical Fellowship Board of the National Research Council, to discuss the problems of doing serious research at Howard. He explained that his teaching responsibilities at Howard were too great to permit him adequate time for research. Though "very sympathetic indeed," Gay was not sure how his division could take any steps toward a solution of the problem.[81] The two men conversed extensively on scientific matters. Just impressed Gay as "an extraordinary person . . . who should be aided . . . in any way possible in carrying out his desires."[82] A bacteriologist of some standing, he concluded that Just was so "good" that his contributions to the development of his race should be "indirect rather than direct"—that is, the opportunity for him to attain unique scientific eminence should be given priority over any benefits he could provide by raising general standards at Howard. Just's efforts at Howard were of some value, but they were being wasted on a "partial and perhaps temporary" endeavor. Gay suggested to Flexner that Just be provided with an opportunity for "pure research" over a period of years at the Rockefeller Institute, or possibly at some institute in France, "where, of course, his race would prove no barrier at all." He had heard from Wallace Buttrick, chairman of the General Education Board, that Flexner's older brother, Simon, was anxious to employ "a striking example of the negro race" at the Rockefeller Institute. Perhaps Simon would be interested in offering Just a permanent research post.

On hearing from his brother of Gay's high opinion of Just, Simon Flexner solicited the further opinion of Loeb.[83] He wanted to know, in particular, if Just was "the kind of man" the Rockefeller Institute for Medical Research should try to help. Loeb responded promptly, and scuttled any plans that may have been afoot to give Just a position at the Rockefeller. His response, a letter to Simon Flexner dated 16 January 1923, is revealing both in what it says and in what it suggests.[84] He wrote that he had known Just "for about ten years." Though he had tried "to help and encourage" Just from the start, he had come to the conclusion, "confirmed more and more during the last seven years, that the man is limited in intelligence, ignorant, incompetent, and conceited; in fact, his so-called research work is not only bad but a nuisance." Just was one of the men who were making Woods Hole "an impossible place for a decent scientist to live in." According to Loeb, Just should be persuaded to stop research and go into high school teaching, a move that would be "a good thing for science and in the long run for Just himself."

If there ever had been a chance of Just getting a position at the Rockefeller, Loeb squelched it. Such an appointment would have been a first—symbolic for the whole black race. A different letter could have changed the course of Just's career, and no doubt would have affected the role of blacks in American science generally. Could Loeb, the socialist, the NAACP supporter, have been unaware of the implications of his response? Could it be that he was in fact quite aware, and was striving to satisfy a different objective? Or could he have felt he had presented an accurate picture untainted by malice and his own political motivations? None of these questions can be easily answered, if at all. A further look into Loeb's personality should provide some insight into what turned out to be a crucial juncture in Just's life.

Tempestuous in mood, religious in politics, and catalytic in scientific influence, Loeb captivated the scientific world at home and abroad for over thirty years. His research fell into three broad areas: tropisms in animals, artificial parthenogenesis, and the colloidal behavior of proteins. His contributions in any one of these would have assured him a place in the history of modern biology, in spite of the fact that all three had one or another inaccurate conception embedded in them. Herbert Spencer Jennings, in his *Behavior of the Lower Animals*, disproved Loeb's doctrine of animal tropisms; the Lillie school, Just in particular, dismantled his conception of fertilization and artificial parthenogenesis; Heilbrunn and others modified his work on protein chemistry. And the ongoing controversy about Loeb's science spilled over into the professional and private lives of many scientists.[85]

There were no in-betweens for Loeb, only extremes. Those he liked, he loved; those he disliked, he hated. He loved those people who believed in either his science or his politics, preferably both, in a way that Just did not. He demanded that his friends be committed to the theories he put forward and modified at will, and in particular committed to his method of using physics, chemistry, and mathematics in biology. A kind of "self-constituted dictator of scientific progress,"[86] Loeb tolerated little deviance from his program and approach. In his eyes it was less important—though important nonetheless—to possess superb scientific ability than to promote the science of Jacques Loeb. For instance, Loeb resorted to clandestine tactics to block the publication of articles which offended him for one reason or another. Upset by the independence of the Lillie school, he founded the *Journal of*

General Physiology, partly in an ineffective attempt to keep Lillie's work from being published. With three (Loeb, Morgan, and Osterhout) on the journal's board refusing Lillie's work, there could be no accusation of autocracy.[87] Loeb's justification was that Lillie's work was inferior and, worse, that it often failed to refer to Loeb's previous work on the same subject.[88] Lillie's brother Ralph, a physiologist like Loeb, fared no better. Loeb thought Ralph Lillie's work on regeneration contained "worthless experiments" and was plagued by "hypothetical nonsense."[89] Ralph Lillie, Loeb charged, appeared to be engaged in "vague speculation of the Bergsonian type," which confirmed the fact that he was not "in his right mind" and "pathological . . . in some cases."

Loeb's dislike for the Lillies ran deep. Frank came in for special ridicule because he had married into a wealthy family, maintained his Protestant religious faith, and pursued a career as a professional administrator of science. Loeb ruthlessly satirized Lillie's science and religion:

> He is by temperament a mystic and his wife, who has recently joined the Catholic Church and suffers from a religious hyper-emotionalism, influences him in the wrong direction. She is devoting her time to reading the life of the saints and makes everyone of her friends, whom she can persuade, read that kind of stuff also. And since her husband's mind is given to nebulosity, I am afraid that the result is showing in his work. Wherever he works purely morphologically, I think he is all right. . . . But, whenever Lillie deals with phenomena which require knowledge of physics or chemistry his work is worthless. . . . I should not in the least be surprised to see him become a Vitalist.[90]

Loeb was an eminent scientist at Woods Hole, had special research quarters and facilities established there for him by the Rockefeller Institute, was well-respected by most investigators, and was afforded every courtesy by Lillie, as the director of the MBL and as a personal friend. Even so, Loeb could not refrain from making Lillie's scientific work the subject of crude personal remarks. He derived pleasure from a type of ridicule that could have been mildly amusing had he himself not taken it quite so seriously.

The Lillies were not the only ones to receive this kind of treatment. The fact that E. G. Conklin, an embryologist from Princeton, had begun as a Methodist minister was a point of special derision to Loeb, who remarked that he was "reverting to type" by working on a book which "might have been written by a liberal minister."[91] Though a "charming fellow" personally, he had "not made any progress in scientific work," and was without the training necessary to "leave the limits of descriptive morphology," so he could not be "critical in the field of experimental biology."

Junior professionals were also treated harshly by Loeb, even when their only crime was to pursue an independent line of research. Loeb thought Libbie Hyman was "a fanatical person," simply because she had been supporting the theories of C. M. Child.[92] He was of the firm conviction that the work of Child was "all based on error," and he saw no reason to perpetuate such work. He withheld from Hyman scientific information about the physical condition of some of her specimens. He apparently felt that it was better to let her make a fool of herself in print than to warn her of the facts.

Loeb was no more yielding in his politics than in his science. He demanded allegiance to radicalism. Conservatives like Lillie received little sympathy from him, while radicals like Svante Arrhenius were favored as much for their political views as for their scientific ones. Loeb was concerned with a diversity of political issues, all of which centered around the elimination of racism on an international scale and the uplift of oppressed people throughout the world. As a Jew, he was principally a fighter against anti-Semitism in America and abroad, but he never forsook his global perspective on racial problems. He refused, for example, to sign the Balfour Declaration in 1915, not because of any fundamental disagreement in principle, but because of the narrowly Zionist perspective which the program set forth.[93] He was not, however, against setting certain priorities in racial politics. He tried to explain to Just the rationale behind the General Education Board's decision to develop medical education in China rather than in black medical schools in America.[94] He saw no reason to be dissatisfied with the board's decision, since the Chinese perhaps needed assistance more urgently than American blacks did. To Just this must have seemed irrelevant, if not insensitive, despite its apparent logic and political sophistication. After all, Just was black, Loeb was not. As a professor in the Howard Medical School, Just had been confronted daily by the problems of black medical education. Loeb only knew about them secondhand through the pages of *The Crisis*.

Loeb was one of the most political-minded scientists in the Woods Hole community, in terms not only of world politics but university politics as well. Though he did not hold a university position between 1910 and 1924 (his tenure at the Rockefeller), he thought he wielded a considerable degree of influence in university circles, particularly concerning positions and promotions for aspiring young faculty members. He tried to guide the careers of his protégés with a zeal that did not always stick to fair play. He was also not above using his influence to promote the careers of his own children. He paved the way for his eldest son, Leonard, a physicist, to work in the famous laboratory of A. A. Michelson in Chicago and Sir Ernest Rutherford in Cambridge.[95] All he had to do was write and say what he wanted. He likewise helped arrange an appointment for his youngest son, Robert, at Physicians and Surgeons Hospital in New York.[96] In his mind, science would be better served if people whom he respected and trusted, including his sons, landed prestigious jobs. The particular means employed to accomplish this end were of little consequence to him. Any means were justified in promoting his brand of science, which he believed would be the savior of humanity. His protégés were to take the new biology, based on the quantification used in the physical sciences, to its limits and make it useful for a socialist society.

Loeb's relationship with one of his protégés, Selig Hecht, is revealing; examined in detail, it highlights those aspects of Loeb's character that were to have a strong impact on Just. The career of Hecht, whose work on color vision won him success and who went on to guide aspiring scientists such as the Nobel laureate George Wald, is the prime example of Loeb's efforts to channel young science professionals in the direction he thought they should take.

ℜ

Hecht earned his Ph.D. from Harvard under G. H. Parker in 1917. Eager to begin a career in science, he worked that summer at the MBL in Woods Hole. There he met Loeb, who was immediately attracted to him. Loeb was captivated by Hecht's work on the physiological effects of light, an extension of his own work on animal tropisms. He admired Hecht's brilliant scientific mind and was pleased, in particular, with the heavy use of physics and mathematics in Hecht's biology. Hecht approached biological problems as Loeb thought best: that is, he treated biological phenomena in a mathematical fashion, relying heavily on concepts from the physical sciences. Later he became one of the pioneers in the field of biophysics.

On leaving Harvard in 1917 Hecht took up a position at the medical school of Creighton University, a small Jesuit institution in Omaha. Almost as soon as he arrived he began to think about leaving for a more prestigious institution. He disliked Creighton from the beginning. While there was little in the way of research facilities or scientific ambience, there was a great deal in the way of teaching responsibilities. In many ways, Hecht's situation at Creighton was parallel to Just's at Howard. There was, however, one important difference: Hecht had an opportunity for professional mobility, Just did not.

As a graduate student Hecht had gained a reputation for being disruptive, and, true to form, at Creighton he became involved in a bitter controversy with another professor in the department, Sergius Morgulis. According to Morgulis, Hecht exerted "a destructive influence" on the department and was a "vicious thorn" in Morgulis's side.[97] As a striving junior faculty member, charged Morgulis, Hecht

had made "free use of underhand methods" and had become known
"more as a kind of politician than scientist." Aware that Hecht was
Loeb's protégé, Morgulis informed Loeb of the situation, insisting that
even scientific aspiration was no excuse for abandoning common
human decency.

Loeb sided with Hecht in the controversy. Though he had known
Morgulis a long time, much longer than he had known Hecht, he
refused to give any credence to what Morgulis had to say. Hecht was
"an excellent man," wrote Loeb, who had "the backing of the men at
Harvard"; Morgulis was unfair, had mistreated Hecht, and should go
penitently to him and offer to "bury the hatchet."[98]

Morgulis's charges against Hecht were grave, and Loeb knew this.
Whether or not the charges were justified is beside the point: in this
case veracity is less important than implication. Morgulis's charges, in
general outline and character, are not inconsistent with Hecht's behavior as displayed in his correspondence with Crozier. Hecht was both
politically astute and willing to be disparaging about colleagues and
friends. He indulged in petty characterizations of other scientists and
their work, as if he had little regard for professional protocol and none
for social graces. Not surprisingly, he considered Whistler's *Gentle Art
of Making Enemies* "a work of genius."[99]

In temperament and outlook, Loeb and Hecht had much in common.
Hecht's letters to Loeb are as vindictive as those to Crozier. Loeb
displayed similar habits in his own correspondence. It is no surprise,
then, to find Loeb anxious to help Hecht secure a more prestigious
position, to rescue him from "the wilds of Omaha."[100] What is remarkable, however, is that he carried his efforts as far as he did.

Loeb promised Hecht that he would handle things, that his future
was "absolutely guaranteed in this country."[101] Declaring his admiration for Hecht as "the most brilliant man in physiology in this
country," and "one of the most brilliant lecturers" in any field,[102] Loeb
wrote glowing letters of recommendation for him to the University of
Illinois Medical School, the University of Buffalo, Columbia University, the University of Michigan Medical School, the University of
Rochester School of Medicine, and, last but by no means least, Harvard
University. Loeb especially wanted to place Hecht at Harvard. This is
not to intimate, however, that he did not make every effort to find
Hecht less prestigious positions. He did, and in at least one case he
employed questionable methods in doing so.

In 1919 Loeb was asked for a professional evaluation of J. F.
McClendon, a young scientist who was applying for a post at the
Peking Union Medical College in China, with Frank Lillie's support.
Loeb's response was negative, based on what he termed McClendon's
"highly nervous temperament not infrequently bordering on the con-
dition of irresponsibility."[103] In the same letter, meanwhile, he described

Hecht as "a very brilliant . . . and a very reliable man" with "an unusual well-balanced mind." His personal criticism was not only of questionable propriety but unnecessary, since Loeb really wanted to place his own protégé on the faculty at Harvard, not send him to China.

In early 1920 Loeb wrote his old friend G. H. Parker, a biologist at Harvard, laudatory letters on Hecht's behalf.[104] When Parker mentioned "the objection" to Hecht's personality that had been raised in the Harvard biology department, Loeb admitted that he had already gathered from two of his Harvard protégés, Osterhout and Marian Irwin, that Hecht was perhaps not "of the most pleasant or meekest type." But the main issue, Loeb insisted, was that Hecht was "unquestionably the ablest of the younger men in biology and physiology." His scientific work on the effect of light was "the best work on this subject since the days of Helmholtz";[105] on the basis of this alone he merited a Harvard appointment.

The Harvard biology department decided otherwise. Along with the issue of personality, it seems likely that anti-Semitic feeling in the department was in part responsible for Hecht's rejection. Hecht was a Jewish immigrant of Austrian descent who had struggled through City College of New York and then through Harvard. Such a background was no advantage, as the famous mathematician Norbert Wiener revealed in his description of his own confrontation with anti-Semitism in the Harvard mathematics department around 1920.[106] For aspiring young Jewish scholars it was a very real problem. No doubt Hecht was referring to anti-Semitism in the biology department when he made cryptic remarks to Crozier about the "Hebrew question" at Harvard.[107]

Loeb certainly suspected that anti-Semitism was at the bottom of Hecht's rejection. He had always thought that too much stress was placed on social position at universities like Harvard, and the rejection of Hecht simply confirmed to him that the universities in the East were "run on the principle of a social club, scientific ability, knowledge, and production playing a very minor role."[108] Had achievement been more important than status, Loeb was convinced, Hecht would easily have found himself "a full professorship of physiology in one of the leading universities." Loeb had run into the problem himself a few years earlier, when he had been turned down for membership in the Century Club, in spite of his great scientific achievements, solely because of his Jewish heritage.[109] The answer lay in Jews looking after their own. Loeb comforted Hecht, advising him not to "despair of ultimate success."[110] If universities like Harvard would not "take care" of him, research institutes like the Rockefeller would. Hecht's work would ultimately "conquer" all.

Hecht left Creighton in 1921 and went to England on a National Research Council fellowship to work in the laboratory of E.C.L. Baly

—a few months, incidentally, after Loeb and Flexner had denied Just the opportunity for winter work in Jamaica. Hecht remained there for a little over a year. In 1922 he returned to the United States to work in the laboratories of William Duane and Lawrence J. Henderson at the Harvard Medical School. Loeb was still trying hard to get him a prestigious position, but these efforts went nowhere, partly, Loeb felt, because Henderson was traveling all over the United States "to stir up anti-semitism with the purpose of excluding Jewish students from American universities."[111]

Somewhere along the way the relationship began to sour. In the summer of 1922 two of Loeb's close friends, Osterhout and Parker, lectured in the physiology course at Woods Hole. To Crozier, Hecht described Osterhout's lecture as "the most rotten, *cheap*, irrelevant performance" that it had ever been his "misfortune" to hear; he labeled Parker's lecture "stupid."[112] This was the first in a series of attacks on Loeb's friends. And Hecht had also become disenchanted with his mentor's politics. He was beginning to wonder whether humanity was worth saving, through science or any other means.[113]

Loeb learned of Hecht's feelings, no doubt through Osterhout. Hecht did not care. He no longer felt obliged to please Loeb as he once had. He was still jobless; Loeb had not placed him at Harvard or any other leading university. Why, then, should he feel any loyalty, or take it upon himself to defend Loeb against personal and scientific adversaries?

In the summer of 1923 Hecht spoke to Robert Loeb concerning "derogatory remarks" someone had made about his father.[114] When Robert begged Hecht not to depress Loeb by telling him about the remarks, Hecht's response was that Loeb deserved any verbal abuse he received. Later, Loeb managed to "worm out" of Robert the details of this encounter with Hecht.

On 11 February 1924 Loeb died of angina pectoris while vacationing in Bermuda. Although his death was sudden, he had, in certain respects, put his affairs in order. The day before leaving New York he had sent Simon Flexner what was perhaps the most carefully thought-out letter he had ever written. It was a reflection on and summary of the professional and social conflicts that had marked his career.[115]

Loeb apologized for not saying good-bye in person, and thanked Flexner for having secured for Albert Einstein a two-year grant of $500 per year. Next he explained that he was being threatened by Arnold J. Gelarie with a suit for defamation of character. He had just received a letter from B. A. Younker, Gelarie's lawyer, accusing him of making "slanderous statements" to undermine Gelarie's "career and standing."[116] The lawyer quoted the damaging testimony of Reinhard Beutner, who had worked as an assistant in Loeb's laboratory and whom Loeb had helped to obtain a position elsewhere. It was a matter of grave concern to Loeb.

The Hecht matter was even more disturbing, and Loeb wanted to empty his heart about it. Hecht had been invited to the Rockefeller to give a lecture as part of a job interview. Intent on making a good impression, he had evidently overacted his role. The lecture was upsetting to John H. Northrop, Oskar Baudisch, and P. A. Levene, among others. Loeb found it "so painful" that he could not watch Hecht deliver it. He was appalled by Hecht's "excessive egotism . . . in disregarding the work of his predecessors," notably that of Northrop and Osterhout. Worst of all, Hecht seemed to talk like "an eastsider who wanted to sell some goods." It was one thing to be Jewish, polished and refined (as Loeb and Flexner were), and quite another to be Jewish, rough and aggressive (as Hecht and Heilbrunn appeared to be).

Osterhout had warned Loeb of Hecht's character. Loeb had disregarded Osterhout's warning. The whole situation, he realized, was his own fault. He had recommended Hecht strongly in spite of negative reactions not only from Osterhout but from others, going back to the early days at Creighton and the controversy with Morgulis. At the same time he had been persuaded by E. J. Cohn, a colleague at the Rockefeller Institute, to support Hecht, even though he knew Cohn to be a poor judge of human nature.

The Hecht incident was not an isolated one, but part of a clear pattern in Loeb's professional dealings. Loeb also regretted having written a recommendation for Leonor Michaelis to William H. Welch and Warfield T. Longcope of Johns Hopkins. He had written it without knowing anything about Michaelis's character, and was mortified when he later heard from Levene that Michaelis was "very unpleasant" and had been denied a position at the College of Physicians and Surgeons for that reason. The fact was that Loeb had "suffered" all his life from the mistake of accepting in his laboratory and recommending for university positions men whose "weakness of character and personal untrustworthiness . . . should have barred them from a position in a scientific laboratory." But now, as he realized, he was locking the barn door after the horses were out. He did not want to stand in the way of either Hecht or Michaelis, but he was anxious to undo his "excess of zeal" in getting those two into positions where they could possibly "do harm."

༄

Michaelis went on to a successful career in biological research, as resident lecturer at Johns Hopkins from 1926 to 1929 and then as a member of the Rockefeller Institute from 1929 until his death. Hecht's career was also successful, even distinguished. After having held the National Research Council fellowship in physical chemistry for five years, from 1921 to 1926, he received an appointment at Columbia in 1926 as an associate professor of biophysics. Morgan, Loeb's old friend, was responsible for securing Hecht this position. When he moved to the California Institute of Technology in 1929 to begin a new department, he offered Hecht a position there. Hecht decided instead to stay at Columbia, where he remained until his death in 1947.

Hecht's story suggests that it might not have been "safe" to be a Loeb disciple, to be dependent on Loeb for professional guidance and a job.[117] Loeb's temperament was too changeable, too volatile, too erratic. Often such qualities are regarded as an aspect of creativity, essential for anyone who strives to make grand and great innovations in science or art or politics. Osterhout, for example, described Loeb as "a scientist with an artist's soul,"[118] though it is true that this romanticized description appeared in a memorial tribute rather than a critical biography. Curiously, the destructive side of Loeb's character has never been portrayed. Loeb said what he wanted, when he wanted, and to whomever he wanted. His statement that Just was "limited in intelligence, ignorant, incompetent, and conceited" is simply one example of this impulse. At the time the comment was made, Just had been publishing significant articles in major scientific journals for over a decade. Loeb knew this. His statement to Flexner, then, was less an objective, accurate assessment of research than a muddleheaded defamation of character. Yet it weighed heavily against Just's career opportunities at the Rockefeller, and perhaps other places as well.

If Loeb was capable of reversing his opinion of Hecht, so was he with Just. When the two first met in 1912 he thought highly of Just. Just was the first black scientist he had ever encountered, a racial oddity with an inherited reason for political action. But Loeb never committed himself to helping Just fully, either scientifically or politically. Although he promised Just he would speak at Howard on a number of occasions, for example, he never was able to do so, for one reason or another.[119]

In 1914 there were reports that Just had been offered a position at the Rockefeller Institute for Medical Research. No record of this offer survives. Any such "appointment" must have been more or less a word-of-mouth promise from Loeb. Naively, Just construed this as an official offer.[120] It was an illusion that inflated his self-image and his career hopes. Loeb did nothing to dispel Just's misapprehensions; rather he "encouraged" them, and no doubt exaggerated the extent of his own influence. That he actually had little influence in American academia is illustrated perfectly by the Hecht story.

There are several reasons for this. First, Loeb was affiliated with a research institute, not a university. Since the Rockefeller Institute had no students and no curriculum, he could not offer the Ph.D., a qualification that was becoming more and more desirable for aspiring scientists. Second, he was too extreme in his negative characterizations of conservative academic types like Nicholas Murray Butler, president of Columbia, for his recommendations to be taken seriously.[121] And third, but by no means least important, he was a Jew facing a non-Jewish establishment.

As a black, Just had to do his utmost to maximize his career options. He realized the disadvantages of fully committing himself to Loeb. It was natural, given the choice, for him to select Lillie rather than Loeb as his mentor. Lillie was well entrenched in academia and had full authority to grant Ph.D.'s; he was highly respected, with a sobering and stable personality; he was white, Anglo-Saxon, and Protestant, and had married into the Crane fortune.

Though Loeb cared little about propriety, he cared a lot about socialism and anti-racism. The zeal with which he sought to end oppression is remarkable for both its sincerity and its intensity. Few men in any walk of life have possessed so much deep moral concern for the fate of mankind. But his opposition to racism was by and large theoretical. A man of ideas, he often was unable to act on his convictions in a practical manner. His temperament got in the way. Though a socialist, he could not overcome his personal animosity toward Just for the sake of the larger social good of the black race. No doubt he recognized that an appointment for Just at the Rockefeller would have been an important symbolic gesture. Unfortunately, his personality prevented him from acting on that recognition.

Loeb condoned intermarriage and fought the evils of eugenics and "racial biology,"[122] but he was not totally without prejudice in regard to blacks. His suggestion that Just should go into high school teaching is revealing on this point. The only other black attempting scientific research whom Loeb had ever heard of was Charles Henry Turner, a high school teacher in St. Louis, Missouri, and Loeb's views on the career potential of blacks engaged in scientific research were apparently based on Turner. It was one thing for blacks to publish articles as a part-time occupation, quite another for them to be committed to scientific research as a career. True, Loeb did consider medicine a career option for blacks; his politics gave him an awareness of the crying need for doctors in the black community. But doctors, in his mind, were practitioners, to be distinguished from scientists, who created ideas. Like teachers, doctors could only be part-time intellectuals. In his recommendation of Just for the Spingarn Medal Loeb had emphasized Just's contribution to medical education, not pure science. Later, when Just's concern for channeling funds into black medical schools had become a personal concern about finding support to insure his own survival as a research scientist, Loeb came to think of him as arrogant and apathetic. Loeb felt that no black should try to extricate himself from the commitment to his people or devote himself exclusively to the realm of ideas.

Loeb's hostility grew as Just became less and less concerned with medical education and more and more devoted to pure research. Loeb had done excellent work on artificial parthenogenesis. Who then was Just to be looking at fertilization as a separate and independent process? Why was he not attending to the problems of educating his own race at Howard? Crozier and Hecht, who were neither handicapped in intelligence nor tied to a racial cause, belonged in scientific research; Just did not. Also, Just was working with Lillie, whom Loeb viewed as a politically conservative and scientifically outdated plodder. Loeb disliked Lillie and took this feeling out on Just, Lillie's most ardent supporter. Lillie's aristocratic background, his religious faith, and the Crane fortune all irritated Loeb no end. Since Lillie was safe from Loeb's verbal daggers, Just was left to receive them.

He survived it all, but not without struggle and sacrifice. The Loeb episode marks an important fork in the long and twisting road of Just's life. He did not know of Loeb's letter to Flexner, but that letter nonetheless is testimony to the trials and tribulations of a black attempting to gain professional status in American science during the early twenties.

Just had begun to be recognized as a productive and brilliant zoologist in the MBL community some time before Loeb's death in 1924. In the spring of 1922 he was invited to give a series of guest lectures in the embryology course headed by H. B. Goodrich. He exaggerated the importance of the invitation, claiming it was an opportunity rarely afforded other scientists.[123] The fact is that brilliant young scientists, Hecht and Crozier for example, were often invited to lecture in the MBL courses. The real importance of Just's invitation lay less in its specialness than in the fact that it was general practice: that Just was invited to lecture indicates that he was receiving the same recognition as other scientists of his caliber. Of course, these invitations were an honor for anyone who received them, and no doubt they were a special honor for Just, the only black to receive them.

Just was apprehensive about lecturing. He felt it necessary to inquire from a number of white Southern men around the MBL whether the Southern women in the embryology course (students of Mary Stuart MacDougall from Agnes Scott College) would object to his lecturing to the class.[124] He was encouraged to lecture, and he did. His fears were unfounded; the lectures—a series entitled "Initiation of Development in the Egg of *Arbacia*"—were without incident, and he received high praise for them. The students were struck by how "well-organized," "concise," and "informative" the lectures were.[125] Just always possessed absolute command over his ideas and the details that backed them up, and at Woods Hole he had to be particularly well prepared to make the kind of impression he wanted to make.

By the early twenties, after the publication of his pathbreaking papers on *Echinarachnius parma*, and after acquiring a reputation as an expert on the habits of marine organisms, Just began teaching other investigators some of his skills and techniques. It was an unusual day that a student or investigator did not seek his advice; perhaps they felt more comfortable exposing their ignorance to a black like Just than to their white teachers and colleagues. One such investigator, Mary Stuart MacDougall, wandered into his laboratory for help one afternoon.[126] MacDougall was a Southerner born and bred, and had not forgotten the Civil War. But possessing the best qualities of a genteel Southern woman, she asked Just for advice rather timidly. He answered her question concisely and politely; he too was Southern. She asked another and then another. Before they knew it, a couple of hours had passed. MacDougall apologized for intruding on his time, but he insisted that he liked helping others. She took him at his word and left feeling good about going to him for help. Later that evening, she related the incident to a friend, recommending Just as a good person to consult on scientific matters. In her estimation, Just would not mind helping anybody.

Throughout the twenties Just helped MacDougall in one way or another, almost to the point of offending his old friend Sally Hughes-Schrader.[127] Hughes-Schrader sensed condescension in MacDougall's attitude towards Just. She did not like the way MacDougall used to wander into Just's laboratory and take up his time. On several occasions she scolded MacDougall for her attitude and Just for his submission to it. She could not understand why Just would want to help, as she put it, "such a wretch." Just tried to explain to her that MacDougall's attitude was not all that bad, given her background and upbringing.

MacDougall was by no means the only one to seek Just's advice. Many scientists consulted him. He is remembered as a technical expert, always willing to share his expertise. As a beginning investigator at the MBL in 1923, Herman Beerman was studying the cell permeability of hydrogen sulfide.[128] After thinking for some time about what would be a good biological indicator of permeability, he decided on *Arbacia* eggs, because they were white and would display a black intracellular reaction when treated with a lead salt solution. He performed the experiments and noticed that the eggs turned black even when permeability was not a factor. Puzzled, he approached Just, the acknowledged expert on *Echinodermata*, a phylum of marine animals consisting of starfishes, sea urchins, sea cucumbers, and other related forms. Just was kind to Beerman and listened patiently to the details of his experiment. Without reproaching Beerman for his ignorance of the habits of *Arbacia*, he simply explained that the idea would have worked beautifully—if developing echinoderm eggs did not turn dark normally. Beerman never, before or since, received "a better lesson in the need for good controls."

Throughout the twenties many embryologists availed themselves of Just's expert advice. He helped K. S. Cole obtain "a six to seven fold increase" in the fertilization of *Arbacia* eggs, and showed W. B. Baker "some . . . marvelous demonstrations of fertilization and early embryos."[129] His role at the MBL was so prominent that Lillie was entertaining notions of setting him up there in a permanent position, perhaps with an appointment combining the duties of librarian and assistant director. This would have meant virtually unlimited opportunities for research, not to mention a close to fifty percent increase in salary.[130] But Lillie, knowing the situation better than anyone, was afraid to act. Just was black, and this might cause problems. Besides, Lillie was beginning to wonder whether he and other concerned scientists had not done "more harm than good" by trying to develop Just as a research specialist.[131] They had brought him to Woods Hole and set him up with a laboratory and facilities, without paying enough attention to his working situation at Howard. Just insisted, of course, that his Woods Hole experiences had been "the finest of his life" and that he would never abandon Woods Hole, no matter what the professional consequences might be. But Lillie stood his ground; he was convinced that Just's "best outlook for happiness and usefulness" lay at Howard University, not Woods Hole. His opinion became more and more fixed as he sensed a growing disharmony between Just and the MBL community.

In the early twenties Just could often be seen laughing and talking in an animated fashion with Heilbrunn, the Lancefields, the Sturtevants, Calvin Bridges, and others.[132] After lunch he would join in a game of horseshoes, usually with Lancefield, Schrader, and Sturtevant. After a long afternoon at work in the laboratory he would play a game of tennis or chess, usually with Lancefield. Recreation always ended before twilight, for Just's evenings were spent collecting *Nereis* and other marine specimens in nearby Eel Pond or setting up and carrying out experiments in the laboratory. He worked hard, but he found some time for social activities. Though reserved and shy with investigators he did not know, he was comfortable and outgoing with his friends.[133] In fact, his intimate relationship with a few people gave the false impression that he was included in the social life of the MBL at large.

Just was vulnerable to racial prejudice at Woods Hole, as much as he would have been in any other American community at the time. That he was accepted by one group of friends in no way shielded him from the racial attitudes of other scientists around the MBL. Nor, for that matter, did it shield him from the racial attitudes of the friends of scientists who might visit Woods Hole for a weekend or so. Like any black anywhere in America, he was exposed to, and unprotected from, the racist whims of any white.

One distressing incident involved Ruth Hibbard, later the wife of Alfred S. Romer, one of Just's close friends and colleagues.[134] In the summer of 1923 Hibbard was in Woods Hole as a student and an assistant to her older sister, Hope, who later befriended Just in Europe during the thirties. The parents of the Hibbard girls visited Woods Hole briefly from their home in Missouri. After lunch one day they joined the usual group of spectators for the usual afternoon entertainment—a horseshoe tournament. As the pitching got underway, Mr. Hibbard, who was "very anti-Negro," began to make a scene. In Just's presence, he ordered Ruth to leave the area: he refused to have his daughter watch a black, scientist or no scientist, pitch horseshoes. Confused and embarrassed, Hibbard disobeyed her father, for the first time in her life. Although Just was never directly confronted by the man, the incident disturbed him a great deal. He did not take such things lightly.

An even more disturbing confrontation involved the MBL Club, an organization that planned social gatherings for the scientists and their families.[135] In the summer of 1924 the club purchased a small house on a small piece of property by the water. After the house had been enlarged and renovated, a celebration was planned. The organizers thought it would be nice to start with a reception and end with a dance. A few days before the celebration was to take place, the chairman of the planning committee, wife of one of the scientists from Virginia, informed Just that his presence at the dance would be an embarrassment. Though many people in the Woods Hole community urged Just to ignore the woman, nothing could undo what had been done, and nothing could persuade him to attend. The open house went ahead as planned, but the dance was canceled. That way close physical contact would be avoided altogether, while less intimate interaction could go on as planned. Needless to say, Just stayed away anyway.

How often such incidents occurred is difficult to determine, but each one had a lasting effect on Just, creating in him, according to one of his friends, a "bitterness of spirit."[136] Some scientists recall the incidents as a consistent part of the pattern of Just's life at Woods Hole. For instance, Curt Stern, the eminent geneticist, states without reservation that Just was not accepted by the larger MBL community for racial reasons and that a protective group made up of Schrader, Heilbrunn, Bowen, Lancefield, and Sturtevant seemed to have grown up around him.[137] A researcher from the Kaiser-Wilhelm-Institut in Berlin-Dahlem, Stern first met Just in the summer of 1925 and then again in 1926. The two men talked freely about racial conditions in American academia and American society at large. Stern was a foreigner, and Just found it easy to talk to him about problems in the Woods Hole community. Later, when they were together in Berlin during the 1930s, they exchanged views on the situation at even greater length.

❧

If Just met with bigoted opposition in some quarters, he was received as a charming, witty, sophisticated, and intelligent comrade in others. His close relationship with two couples, the Lancefields and the Schraders, gave him the strength to endure some of the more obvious expressions of prejudice in the Woods Hole community. In fact, he was part of a group that avoided activities sponsored by the MBL Club; ironically enough, Just's clique became snobbish in its own right. Its amusements were much more fascinating than any the MBL Club arranged. Its five members had a reputation for being avant-garde and bohemian, funloving and risk-taking, industrious and brilliant.

Whenever possible, the five would get away from the MBL for a day or so on the Schraders' Crossby catboat, the *Squidhound*.[138] Countless were the sailing junkets across Vineyard Sound and the lobster and steak and swordfish dinners on the Elizabeth Islands. Everyone ate and drank well on these picnics. Just would cook from time to time, and despite Prohibition he almost always brought along one of his well-loved bottles of whiskey. Where he got them, nobody knew. He and his friends were passionately fond of the islands, of the primeval forests and virgin beaches. Often they spent the afternoon reveling in the grandeur of the trees and kettleholes, in the beauty of the sand and the water. They shared many wonderful moments in the early evening, huddled together around a fire to escape the chill sea air. Their conversation covered a wide range of subjects, including literature, music, art, and science. Just, for one, talked with ease on topics as diverse as St. Augustine's *Confessions* and Picasso's Cubist paintings. Talk was not always cultured and refined; often it became vulgar and bawdy. Just was notorious for his store of dirty jokes, derived at times from his work on fertilization. He was, after all, a biologist.

Just felt comfortable with the Schraders and Lancefields, even though he could suffer some discomfiture on occasion. One day they were on the *Squidhound*, anchored off Kettle Cove. Always a meticulous dresser, Just decided to show off his fashionable new swimsuit. Everyone admired it before jumping off the boat, without suits, to swim ashore. Unfortunately, Just's elegant suit shrank to about half its original size shortly after he entered the water. The Lancefields and Schraders laughed for years afterward recalling the look of agonized embarrassment on his face!

Just's friendship with the Schraders and Lancefields extended beyond the confines of Woods Hole. He visited the Schraders at Bryn Mawr and later in Tenafly, New Jersey. Schrader's family on Staten Island had had the foresight to stock their cellar well with vintages of the Rhine and Mosel before Prohibition was imposed; and Just visited them on several occasions, along with Sally, Franz, Donald, and Rebecca. The five arranged many parties and managed to live through Prohibition without too much trouble.

Just's wife Ethel never became part of this intimate social circle. She simply was not as free and bohemian in spirit as her husband and his friends were. And how was she to know whether these two couples were really any different from the other whites she had encountered? Besides, Just perhaps did not want her to become a part of the group. He wanted her to go to Woods Hole, to be sure, but he may not have wanted to share with her the intense pleasures of "a friendship the gods might envy."[139]

As his friendship with the Lancefields and Schraders grew, Just tended to forget the racial slights he had suffered in 1923 and '24. In the mid-1920s he began to win respect not only from younger scientists like H. J. Muller, Stern, and H. H. Plough, but also, and more important for him, from the old guard at the MBL, like Wilson, Morgan, and Harrison. Wilson in *The Cell in Development and Heredity* (1925) and Morgan in *Experimental Embryology* (1927) cited Just's work heavily, showing particular interest in his articles on *Echinarachnius parma*.[140] These men saw Just around the MBL and occasionally mentioned to him their appreciation of his work. Just felt flattered by the attention. At last he could call himself a full-fledged member of America's most vital community of biologists; at last his research was being recognized alongside the breakthrough work of Whitman on sex, H. H. Newman on twinning and genetics, Loeb on parthenogenesis, tropisms, rejuvenation, and colloid behavior, and Ralph Lillie on protoplasmic and nervous action.

By the end of the summer of 1926 Just was participating fully in social life at the mess and enjoying afternoon recreation with the younger biologists. After lunch one day a group from the mess—including Calvin Bridges, Paul Reznikoff, and Just—was standing outside on the laboratory steps.[141] Soon the conversation wandered from the realm of biology to that of sex. Bowen, a Columbia scientist with rigid religious morals, walked by and overheard some of the stories. He was shocked. Turning to Just, he said, "Just, I'm not surprised at these other men telling stories of this type, but I am really surprised at you. I have always considered you a Christian gentleman." Just politely but pointedly responded, "I think you have made two mistakes there." No more was said. Bowen, embarrassed, scurried away, and Just, triumphant, rejoined the discussion. He belonged as never before. Perhaps next summer, he must have thought, he might bring his family to Woods Hole.

Throughout the early twenties, Just had inquired about the possibility of renting a house in Woods Hole for the summer, but Lillie never responded. Just's friends, the Lancefields and the Sturtevants, were building cottages, and the idea of making some kind of home away from home at Woods Hole appealed to him too. In the beginning of 1925, a year after the MBL Club incident, he began planning to bring his wife and three children to Woods Hole for the summer of 1926. He wanted to provide himself with a domestic environment he could call his own. But Lillie was against the plan. He told Heilbrunn that he doubted Just's family would "fit in at Woods Hole," and predicted that conditions for Just's wife and children would be "almost impossible to bear."[142] Heilbrunn agreed; he was convinced of "the impossibility of Just and his family living happily at Woods Hole."[143] More than anyone, Lillie had a good understanding of social attitudes around the MBL. Few people knew the MBL as intimately, and few were as objective and perceptive in their observations. Though inclined to understatement, he sensed a widespread unwillingness on the part of MBL scientists to accept Just and his family as full-fledged members of the Woods Hole community. Just did not bring his family to Woods Hole in 1926.

But the following summer, 1927, Just did. While his mother-in-law, affectionately called Bamba, remained in Washington to look after the house at 412 T Street, Ethel and the three children, Margaret (now fourteen years old), Highwarden (eleven), and Maribel (four), came to Woods Hole. Problems immediately arose. As Lillie had feared, Just's wife and children found conditions unbearable. Families refused to sit with them in the mess hall. The wives of the MBL scientists made it quite clear to Ethel that she was not welcome. Most people at Woods Hole had been able to overlook Just's race because of his scientific achievements and relatively light complexion. Aside from having no claim to an academic reputation, Ethel was "extremely dark in skin color."[144] At Woods Hole she was received with callous indifference, and even open insults. By the time the Schraders arrived (they were late that summer), Ethel was ready to leave.[145] Just had to interrupt his research and postpone a scheduled lecture for over a month while he got his family out of Woods Hole and back to Washington. Ethel and the children never went to Woods Hole again.

It was all so uncomfortable and controversial that it was whispered about for years, but never discussed with any real frankness.[146] But why should such an incident, in no way out of character for the time, have created any furor at the MBL? Surely most of the MBL investigators would not have been astonished if something similar occurred at the faculty club of a major white university. Perhaps they wanted to believe that the scientific community at Woods Hole was more enlightened in its racial attitudes than other American communities, academic and otherwise. Perhaps, but the fact is that Woods Hole was hardly unique. Race prejudice had come out into the open there in 1923, and had matured and festered by 1927.

Still, Woods Hole was the only place where Just could maintain a foothold in the professional world of biology. Racial incidents, however regrettable, could not be allowed to impede his career. Though hurt and frustrated, he returned the following summer.

There seemed to be no end to incidents. In the summer of 1928, a German biologist gave a seminar at the MBL.[147] His English was a problem; though he was fairly fluent, he was not always aware of the nuances of the words he chose.[148] At one point, trying to find a way to convey that something had not worked out well, he used the expression "nigger in a woodpile." Just got up and walked out. He did not appear at the mess for the next three days. He locked himself up in his laboratory; even Lillie (who was right next door) was not allowed in for a talk. It was Just's way of showing not only the German lecturer but the whole Woods Hole community that he abhorred racial slurs.

Samuel Milton Nabrit, a black biologist who is the source of this story, thought Just's response hypersensitive and exaggerated. He attended the lecture too and came away feeling that the German was simply trying to use American idiom. But perhaps Just was protesting precisely that fact—that the expression was part of the idiom in the first place. Lillie, Conklin, Heilbrunn, and others could listen impassively to such expressions if they liked, but they could not expect him to do likewise. He was continuously being scarred by what he increasingly perceived as racial hatred in the scientific community.

The hurt of racism did not leave Just. Nabrit remembers him as an embittered man.

The two first met during the summer of 1927. Nabrit, who had done his undergraduate work at Morehouse College and had received a master's degree at the University of Chicago, was doing doctoral work at Woods Hole under J. W. Wilson of Brown University. While at

Chicago, Nabrit had been advised by Carl Moore to plan a trip to Woods Hole and to ask Just for a letter of introduction. John Hope, the president of Morehouse, had written to Just on Nabrit's behalf. Just did not respond. Nabrit nonetheless made his way to Woods Hole. After a couple of weeks he met Just one rainy morning on the front porch of the mess hall. Just began to explain to the rather embarrassed young man why he had not written a letter on his behalf. He did not know Nabrit, Just said, and had not seen any of his work; he could not therefore jeopardize his own reputation by writing a recommendation. Just did not speak to Nabrit again that summer.

The following summer Just went to Wilson, a good friend of his, to inquire about Nabrit's scientific abilities and accomplishments. Wilson assured Just that Nabrit was at an advanced stage of research and likely to produce an excellent dissertation. After receiving these assurances, Just's attitude changed. He went to Nabrit's laboratory with some reprints of his own articles, sat down, and had a drink. The two men discussed the primordia of fish eggs for hours. At some point during the conversation, Just commented on the role of blacks in biology at Woods Hole. He believed, he informed Nabrit, that blacks should not go to Woods Hole until they had obtained (or were about to obtain) the doctorate. Like many blacks breaking into professions for the first time, he was convinced that blacks had to be better than whites if they were to have a chance to compete. His concern over his own reputation in the Woods Hole community was a factor in his attitude. He had wanted, worked hard for, and for the most part received respect from his colleagues. He had earned a name for himself despite adverse conditions at Howard and racial prejudice in the society at large. He was winning out in what was sure to be an ongoing struggle.

In the summer of 1927 Just brought his assistant, a young black woman named Roger Arliner Young, to Woods Hole for the first time. Her situation was different from Nabrit's. Just knew her personally, had observed her work, and was willing to support her. As his student at Howard she had shown great promise. As a junior faculty member there she had been helpful to him both in his teaching program and his new area of research, the effect of ultraviolet light on marine eggs. At Woods Hole she was closely supervised by Just, working beside him in the same laboratory and on the same problems. From all indications, she had the potential to be, like him, a leader in the field of biology.

The two shared many social activities, especially in 1928 when Ethel did not go to Woods Hole. Often they went out with the Schraders and Lancefields. When she was not with Just and his friends, Young would go on picnics and canoe trips and drives up the Cape with the younger scientists.[148] Although she was as dark as Ethel, she was accepted more readily at Woods Hole, perhaps because her purpose for being there was scientific rather than social. The MBL community could deal with her in a professional or social way, whichever happened to be more convenient. The option did not exist with Ethel.

☙

Just wanted, demanded, and received respect for his scientific accomplishments. By the late twenties, in fact, he was viewed with deference. The Woods Hole community considered him a first-rate experimentalist, "a genius in the design of his experiments."[149] The beauty of his cytological preparations was generally admired, and his contribution in the field of experimental embryology was seen as fundamental not only at Woods Hole but throughout the country, and abroad as well. He knew more than anyone else about the natural history of the life forms the investigators were using at Woods Hole. He was usually there in the spring, before anyone else, sometimes as early as mid-March. By the time other scientists started arriving in mid-June, he could tell them whether *Chaetopterus* was going to be normal and whether *Nereis* was going to be in abundance that summer.[150] He knew more than anyone else about what constituted normal conditions for marine eggs, and his concern for the proper handling of eggs had a great influence on experienced scientists and beginning investigators alike. Though he never had graduate students of his own to carry on a scientific tradition, he was, in the words of Sally Hughes-Schrader, "responsible for training a whole goddamn generation of embryologists at the MBL."[151]

Just was meticulous in the way he carried out his experiments.[152] Knowing that it was impossible to obtain decent cytological results from dead or dying material, he invariably handled his material with precision and care. Unlike Loeb, he could never have been criticized for carelessness. One of the things he stressed, for example, was the importance of keeping organisms at the temperature to which they are accustomed. After years of observation, he established that sea urchin eggs do not develop normally above 25°C, sand dollar eggs not above 23°C. In the laboratory he always kept the material under study on a table fitted with apparatus for simulating normal seawater circulation during experiments. Other scientists used to put material in dishes under a microscope, often near windows with the sun blazing in. Just's seawater apparatus maintained a constant temperature of 21°C, whereas the method used by other scientists resulted in temperatures as high as 28°C.

Cleanliness was another point stressed by Just. While some other scientists did not thoroughly clean their glassware, Just washed and scrubbed and sterilized every dish and test tube. He insisted on *Bon Ami* as the best cleanser. If he wiped his glassware with cheesecloth from a package, that cheesecloth was sterilized beforehand to take out any factory impurities. After cleaning, each piece of glassware was placed face down, to avoid dust accumulation. Just had known the importance of treating material with deliberate care from the time he did his earliest experiments on artificial parthenogenesis, a process that requires the investigator to eliminate every trace of spermatogenic material. He reproached careless scientists, often in a cutting, even devastating manner. His approach was, after all, essential for scientific accuracy.

Once Harold H. Plough, a geneticist from Amherst, sucked some sea urchin eggs through a pipette to strip off the external membranes. When the eggs developed abnormally, he eagerly reported the phenomenon at one of the informal Tuesday evening gatherings in the MBL lecture hall, talking to the crowd of scientists for over an hour with the confidence of a discoverer of some important theory. Just stood up, and said that he would stake his reputation on the fact that Plough had damaged the cell surface of the eggs and had not simply removed the external membranes. The crowd was stunned. Just did not retract his statement; he knew that his own method of removing the membranes—by putting the eggs through bolting silk—was the only safe one. Nobody in the Woods Hole community could argue with him on questions of method. Nobody dared.[153]

However, reputations are rarely built on pinpointing mistakes in the work of others, but rather on helping others rectify those mistakes. Just's reputation, by and large, was the result of the years of good advice and help he gave numerous investigators. In the summer of 1928 Paul Reznikoff and his research group were studying fertilization in starfish and sea urchins. Only about 50 to 60 percent of their specimens were showing cell division after fertilization, whereas 90 to 95 percent of Just's were dividing. They asked Just to troubleshoot and he agreed. He asked to see how they collected their specimens, so they took him over to the traps on the dock, put the specimens in the pails, carried them back across the street to the laboratory, and placed them in a tank of seawater. Just spotted the problem at once. He told them that when he collected specimens he always made certain to cover the pails. Otherwise the short trip from the dock to the laboratory subjected the animals to the heat of the sun, enough to inhibit the multiplication of eggs. The group adopted the method suggested by Just and improved the results significantly.[154]

MBL scientists continued to seek Just's advice and finally prevailed on him to publish his methods in the MBL weekly paper, the *Collecting Net*, beginning in 1928:

The Collecting Net considers itself most fortunate in obtaining a series of articles on Methods for Experimental Embryology. Our contributor is Dr. E. E. Just, professor of zoology at Howard University. He is the acknowledged authority on the field which will be covered by this treatise. There is no one, we are confident, who is more competent to write concerning the technique of handling the various marine embryological material. The information presented in these articles has been accumulated by Dr. Just over a period of twenty years in his work at the Marine Biological Laboratory.[155]

The purpose of the articles was to teach biologists some basic biology. Just loved to tell the story of the eminent physiologist who once showed him two dishes of sexual excreta and asked which were the eggs and which the sperm.[156] The *Collecting Net* provided a perfect forum for doing something about this kind of ignorance. Scientists young and old benefited from Just's articles, the consensus being that the *Collecting Net* had "never printed anything so valuable."[157] Almost immediately two presses, Blakiston's and Darwin, were competing for the opportunity to publish the articles in the form of a laboratory manual or handbook. Blakiston's brought out the manual, *Basic Methods for Experiments on Eggs of Marine Animals*, in 1939.[158]

One usually pays a price for a reputation, and the price of Just's was high. Ironically, it was Just's role as an authority on marine methods that clinched his already growing disenchantment with life in the MBL community. Many feelings came together in him during the summer of 1928. He became disheartened by the "really *bad* biology—physicochemical stuff" that Lillie and others were allowing to go on at the MBL.[159] Biology was changing, becoming more heavily allied with physics, chemistry, and mathematics, and Just was beginning to feel passé. He, for one, was ready to stand aside and give way to younger workers who, despite their biological ignorance, might be able to break new paths. Nevertheless, his classical techniques of observation were still producing results and uncovering new leads. He got a "genuine thrill" out of his work and would not, under any circumstances, forsake "old fashioned embryology." There was a problem, however. As the expert on the habits of marine animals, he was sought after almost daily by younger scientists whose interests lay in advanced physics and chemistry but whose needs included help with basic biology. He began to feel that he was giving too much of his time to helping others for nothing. He had written the articles on methods for handling eggs and sperm in part to reduce interference with his own work. The plan did not succeed; after publication he was interrupted perhaps more than ever. He was interested in doing his own work, not just in helping others do theirs. He wanted to avoid placing himself in "the position of a mid-wife to the investigations of others,"[160] which had proved to be a rather thankless task. No, he would not help Gregory Pincus, a young Harvard scientist, with his work on the fertilizin theory. In fact, he was more and more loath to help anybody.

As he observed Just's uneasiness, Lillie sensed the approach of the end of an era. How long could Woods Hole be a source of attraction and stimulation for him? Just was losing the eagerness of young adulthood and gaining the sobriety of middle age. His social life at the MBL was odd, consisting of intense relationships with a few close friends who protected him from the larger community. Lillie was approaching retirement as director (he retired in 1930), and the help and guidance he had so far given Just might not be forthcoming from Lillie's successor. The technical expertise that had established Just's reputation was now becoming a detriment to the progress of his research. Was there really any reason for Just to stay on?

Lillie tried to find an answer. There was little he could do about the social situation at the MBL, but he could attempt to ameliorate the conditions under which Just conducted his research. He was anxious for Just to feel accepted and wanted at the MBL. With this in mind, he arranged for Just to be appointed a member of the editorial board of the *Biological Bulletin*, the official journal of the MBL. Alfred C. Redfield had just taken over the general editorship, and he was counting, he told Just, on the latter's help as editor and contributor. Just would no doubt encourage "work on the appearance and habits of the organisms in the Woods Hole region."[161] The value of this kind of work lay less in its use as "a handmaid to physiological and biochemical investigation" than in its "own intrinsic interest." The *Bulletin* would not view zoology as a source of material for physiological study; rather, it would treat physiology as a point of view from which zoological problems would be attacked. Such a perspective was appealing to Just, but it was not enough to keep him at Woods Hole.

Woods Hole had in many ways provided Just with "a haven of refuge,"[162] but in 1928 he was up against the problem of how to push forward. Difficulties at Howard and Woods Hole were on the increase; effective teaching and productive research seemed impossible. Lillie could do little about the Howard situation, but he did not want Just to feel "forced out" of Woods Hole on account of "the numerous interruptions" experienced there.[163] He advised Just to hang a "no interruption" sign outside his door. He even offered to institute a "no interruption" policy throughout the laboratory; signs would be distributed to every room, "so that their use might become general and not reflect on the individual using them." Any effort to improve working conditions for Just would be well worthwhile.

Lillie never had a chance to institute his plan, however. In December 1928 Just made a special trip to Woods Hole to collect his notes and check on some specimens. He was about to make his first transatlantic voyage. A dream had come true: he was going to work for six months at the Stazione Zoologica in Naples.

Woods Hole seemed quiet and serene. Snow began to fall as Just moved about his laboratory. Looking through the window, he saw the moon shining dimly through the clouds. He thought back over his Woods Hole experience. This was the last time for a while that he would be able to enjoy the brick building; it was the best time too, because the noise and fuss of the summer workers had been carried back to the big cities, out to sea, wherever. He could think alone. He could mull over his past experiences, his present feelings, his future expectations. Nostalgically, he wandered next door into Lillie's dark, empty laboratory. All the time that had passed since the first day with Lillie in 1909 flashed through his mind. He would be eternally grateful to Lillie; his life belonged to Lillie; his one purpose was "to lay . . . something of worth" at Lillie's feet.[164]

Standing there in the laboratory Just had an overwhelming experience, a moment of quasi-religious ecstasy. He felt overcome by "the glow of a sort of transfiguration"; he seemed to be "an earnest and devout disciple" praying at the shrine of "his master." It dawned on him then that "nearness" to Lillie was the most worthwhile thing in his life; inspiration had flowed into him through Lillie. He had done much; now it was time to do even more. He was off to Naples, having finally realized his ambition to go there for work on the complicated life history of *Nereis dumerilii*. He would try and prove his hypothesis that Woods Hole *Platynereis megalops* and Naples *Nereis dumerilii* were not the same. He would study cortical change in *Amphioxus* eggs, the behavior of the middle piece in *Cerebratulus* eggs, and cortical change in sea urchin eggs. He would be away about six months, but he would return to his mentor Lillie and his "Mecca, Woods Hole!"

CHAPTER 4

The Role of Foundation Support, 1920-29

Money has always been crucial to a career in science. Even in the early part of this century, the initial investment in undergraduate and graduate education was costly, as were the ongoing expenditures for laboratories and equipment. The bills were often footed by a university, a scientific society, a philanthropic foundation, or a private benefactor— but often also by the scientists themselves. Most scientists, including the independently wealthy, sought and received some outside support. Frank Lillie, for instance, obtained funds from various sources throughout his career, even though he was married to the Crane fortune.[1]

As a result, scientists frequently presented themselves as geniuses with colossal talents, an image that helped them in their pursuit of financial backing. Benefactors in both the public and private sectors were suitably impressed. Nonscientists themselves for the most part, they did not understand atoms and gases and enzyme functions, but they were fascinated by the mystery of what went on in the laboratory and especially by the idea that science held the key to knowledge and power in a flourishing industrial state. They gave liberally to scientists, convinced that the returns would more than justify the expenditures.

Blacks did not benefit from all this. The extensive education and rigorous training required of anyone who chose a career in science could be found only in white universities, where blacks were at best admitted only in exceptional cases. Even if a black was lucky enough to get the training, there was no hope of long-term support for his work. He had no chance of a post in a white institution, so he was obliged to seek employment as a teacher in one of the black colleges. And the philanthropic Establishment, on advice from the white scientific community, did not think of black colleges as a good investment for the training of scientists and the production of high-quality research. Only the most niggardly support was extended to blacks and their institutions, and since science required such a great deal of money, few black colleges had viable science departments and few blacks pursued science as a career.

ℛ

Contrary to popular belief, Howard University was not set up for blacks alone. Founded on 1 May 1867, it first opened its doors to four white girls. The institution, situated in the nation's capital, was meant to provide an educational experience for people of all races, to attract and accommodate students from throughout the world, to create a "cosmopolitan" student body drawn from "all classes, conditions, and nationalities."[2] This goal was never achieved. Over the next twenty-five years the white enrollment gradually dropped off, and around the time of the "separate but equal" decision in the case of *Plessy* v. *Ferguson* in 1896 the student body became, and has since remained, virtually all black.

Howard was a poor institution. It started out with funds from the Bureau of Refugees, Freedmen and Abandoned Lands set up by Congress after the Civil War to aid in educating the freed slaves and other destitute persons. At first it was merely a high school, but its administration strove hard to make it more, by developing programs as quickly as possible in medicine, law, education, theology, and agriculture. Financial support flowed in swiftly and the prospects for the future seemed encouraging. But in the early 1870s funds began to dwindle. The Freedmen's Bureau closed, and private donations fell off during the panic that crippled the nation's economy in 1873. Howard suffered enormous financial strains, but it kept afloat with the help of dedicated faculty and staff. Run mostly by whites in the service of blacks, the institution attracted leaders imbued with missionary zeal. Its first twelve presidents, in fact, were all white Northerners connected to the church in one way or another. Much of the faculty was white too, dedicated to the cause of black education. The black professors were the cream of the black intelligentsia, highly educated men and women often trained at white Northern schools. Barred from jobs in white colleges, they found their niche at Howard, the only black institution with a wide, growing range of collegiate and professional programs.

In 1879 Congress made its first appropriation to Howard, in the sum of $10,000, very little compared to the institution's total debt of over $100,000. The wording was simply "Howard University for maintenance."[3] It was a small gift, but one that Congress continued on an annual basis, and gradually increased. The Howard trustees, mindful of the size of the appropriation, used the money only for necessities. Much of it was spent on repairs to the school's old buildings. Aside from the president's house and the chapel for the University Church, no new buildings were erected between 1872 and the turn of the century. A new building was finally put up in 1909, when Wilbur P. Thirkield, Howard's hardworking ninth president, persuaded the federal government to fund the construction of a science center. Thirkield had begun an ambitious campaign to channel money into science and to recruit scientists for the faculty. Under his influence, Howard was to become *the* school for educated blacks migrating to the cities, paralleling the role already filled by Tuskegee for blacks remaining in the countryside. Thirkield's vision led him to solicit big money for the institution from government sources and private philanthropists. There could not have been a better time for Just to enter the Howard scene.

In 1910 two events important to science at Howard took place. The first was the publication of the Flexner report, commissioned by the Carnegie Foundation for the Advancement of Teaching and compiled by Abraham Flexner. A layman with no medical training, Flexner had taken a strong interest in medical education in North America and set out to discover the truth about it. Immediately after joining the staff of the Carnegie Foundation for the Advancement of Teaching in 1908, he visited almost every medical school and took detailed notes on his findings. Two years later he brought out *Medical Education in the United States and Canada*. It was a dramatic report, with startling anecdotes about the primitive conditions in just about every medical school: minimal entrance requirements, uninterested faculty, and inadequate space—dissecting rooms crowded with clucking hens and putrid cadavers, laboratory tables cluttered with dirty test tubes standing in pans and cigar boxes.[4] Only Johns Hopkins was above serious reproach. Of the black medical schools, Flexner reported, Howard and Meharry were the only two worth maintaining. But even there conditions had to be improved if black doctors were to be trained to keep their people healthy. In Flexner's view, such a goal was important not only in itself but also as a way to minimize the problem of blacks passing on hookworm and tuberculosis to whites.[5] The evaluation was rather backhanded, but at least it put Howard and Meharry in a favorable position to receive grants throughout the 1910s.

Flexner felt that one of the best ways to upgrade standards in medical schools, black and white, would be to hire fewer clinicians as part-time instructors and more research men as full-time professors. And Just was exactly the kind of medical school teacher that the Flexner report called for. He had recently joined the faculty, and under Thirkield's influence was moving into science teaching and research. Indirectly, the Flexner report increased his chances for support.

The dedication of the science building, Thirkield Hall, was the second important occasion for science at Howard in 1910. Leaders in the world of government and education attended and delivered speeches.

The politicians spoke in general terms. President William H. Taft urged blacks to achieve success through education and promised to do his utmost to help Howard secure support; James Bryce, British ambassador to the United States, commented on the all-important role of the teacher in black schools and colleges; former president Theodore Roosevelt remarked that the responsibility for leading the race out of misery lay largely with Howard professors and their students. The educators, on the other hand, focused on science. Henry S. Pritchett, president of the Carnegie Foundation, expressed pleasure that blacks were fast waking up to the idea that "progress . . . lies in the adoption of the scientific attitude of mind and of scientific methods"; Booker T. Washington, national black leader and head of Tuskegee Institute, stressed the practical need for developing medical education in black institutions; William H. Welch, professor of medicine at Johns Hopkins, reiterated the importance of scientific training and praised Howard as "an exceptional institution . . . with the greatest promise of future usefulness."[6] The ceremony was inspiring to a Howard newcomer and recent science convert like Just. Though the university was poor and offered no opportunities in science beyond basic course work, at least it had managed to attract national attention and a great deal of moral support. Progress seemed certain.

Thirkield retired from the presidency in 1912, and the energy he had put into advancing science was not exhibited by the next administration. S. M. Newman took over as president in 1913, and science was not a priority for him, although he more or less carried on the existing programs begun by the Thirkield administration. He spent his waking hours writing poetry and was "more successful as a poet . . . than as an administrator."[7] When discussing labor relations with Just, for instance, he wistfully and poetically alluded to the breezes playing and the sun shining.[8] During Newman's tenure as president, the Howard Medical School continued to hold the attention of funding institutions such as the General Education Board, however. The dean, W. C. McNeill, who handled budgetary matters, composed and presented requests for support to the foundations, while Newman simply observed the process from afar.

Because money was scarce at the medical school, great caution had to be exercised in fiscal affairs. The administration tried, for the most part successfully, to keep faculty salaries close to par with white universities. As professor of physiology at the medical school, Just made anywhere between $800 and $1,200 a year in addition to his regular Howard salary, depending on his course load and student enrollment. The extra money was welcome: it made summer research at Woods Hole possible, and it financed a sabbatical year at Chicago. It also gave him a way to defray the growing costs of research. He was well aware that if the medical school were to sink into serious financial problems, so would he. Consequently, he used what influence he could with Lillie and other science administrators to find funds for Howard, but only partly out of "selfish" motives.[9]

When the United States entered the First World War, Just was thirty-four years old, with a wife and child to support. He had just completed his Ph.D. and was pushing himself toward a career in science. At this crucial stage in his life he wanted no part of active military duty. No doubt there was some pressure on him to get involved. Though blacks in general found it difficult to enlist when they tried to offer their services to the war effort, Howard was a prime source of recruits and commissioned officers. The campus was converted to a makeshift camp and the faculty helped in on-site training. Aside from the time factor, for Just there was also a question of principle. Woodrow Wilson had instituted jim crow in regiments throughout the country, and even training camps on New England college campuses like Amherst and Bates were segregated. Just could not have tolerated such conditions, nor could he have endured taking orders from white officers who were higher-ranking but less educated than himself. Besides, war and the military went against his natural grain; he was "horrified" by violence of any sort.[10] But rather than declare himself a pacifist, he increased his teaching duties at the medical school and began some clinical experiments in electrocardiography so as to avoid active military service. The problem was that the extra work at Howard—"two full days . . . every day"—was making it difficult for him to carry on his own research to the extent that he wanted.[11] He knew he would have a better future if he could get time away from the drudgery of teaching. What he needed more than anything was a reasonably undemanding classroom schedule and a chance to begin his summer research at Woods Hole in early May rather than in mid-June.[12]

As the war drew to a close in 1918, the Howard administration again changed hands. James Stanley Durkee took over the presidency, serving until 1926. One of his first policy changes was to centralize the authority of the university in his office. Since 1880 power had been distributed among the offices of the various university deans, but all that was now changed. The medical school, once autonomous, became directly responsible to Durkee. The special relationship that Just had established with Dean McNeill and other administrative officials under Newman's loose administration was no longer of any use. Durkee ruled with an iron hand, ordering the faculty around like schoolchildren; once he went so far as to give Just a verbal knuckle-rapping for not attending morning prayers on a regular basis.[13] Worst of all, the prospects for science in general at Howard were threatened, for Durkee seemed not to be sympathetic to its development.

Just disliked Durkee's methods from the start, particularly the "unfair distribution of teaching burdens and of compensation."[14] He foresaw trouble in his arrangement at the medical school, and wanted out. Besides, he had been teaching too long and too hard. His research was sure to suffer unless he changed his schedule and modified his duties at Howard. One solution, he decided, was to look for support outside the university.

It was a pioneering move. No other black had ever received foundation support for research in pure science, let alone substantial funding. As a result, no black had achieved an outstanding career in science. Edward Bouchet, the first black Ph.D. from Yale in 1870, remained a high school physics teacher in New Haven for his entire career; Charles Henry Turner, a Ph.D. in biology from Chicago in 1907, taught high school in St. Louis until his premature death in 1923. Turner's case was particularly tragic. The black community knew him as a struggling and frustrated scientist who died of "neglect and overwork," who had once been denied a university position because "the head professor . . . would not have a 'Nigger'."[15] Without money, support, and position, men like Bouchet and Turner were unable to enter, let alone compete in, the world of American science. So it might have been with Just.

But as we shall see, Just turned out to be adept at winning support. He acquired the art of getting money, studied and learned the ins and outs of presenting persuasive proposals to philanthropists and foundations. He convinced donors of the worth of his research—for himself, Howard, science, the black race, and mankind—and people backed his work. Just managed to keep their enthusiasm and support, but he also kept his dignity at a time when charity and philanthropy were often confused, especially for blacks.

In the fall of 1919 Just pursued a casual reference that Lillie had made that summer to the possibility of his getting a fellowship from a newly established organization, the National Research Council. With strong support from Heilbrunn, he urged Lillie to give him a recommendation. The idea was that Lillie's friend and colleague C. E. McClung, one of the council's administrators, could handle the matter at the council headquarters in Washington.[16]

The council had been set up in April 1916 under the charter and constitution of the National Academy of Sciences to coordinate all nongovernmental scientific and technical resources with those of the Army and Navy for use in the First World War.[17] After the Germans sank the *Lusitania* in 1915, the academy formulated possible lines of action in the event the United States entered the war. When the U.S.S. *Sussex* was attacked in early 1916 and America officially joined the war, the academy moved quickly to mobilize science in the interest of national defense. President Woodrow Wilson accepted the academy's offer to organize the country's scientific agencies, and the National Research Council was born. It began operating immediately and carried on its duties throughout the war.

When the war ended in 1918 the council quickly sought a broader mission for itself in times of peace. No longer would it remain under the control of the government; now it could adopt civilian status and help to evolve long-range goals for American science. The agency would seek to develop and utilize the scientific and technical resources of the country, to formulate comprehensive projects for research at home and abroad, to encourage and nurture the initiatives of individual scientists. One of its prime aims was to support the work of men and women who, like Just, showed great potential for advancing the frontiers of science.

The council was a clearinghouse for scientific projects. Its administrative members, often older and highly respected scientists, would meet from time to time and discuss the merits of supporting this or that promising young scientist. If the decision was favorable, efforts were then made to match the young scientist with a source of financial support, either a foundation or a private philanthropist. Often this was difficult because donors preferred to support established scientists whose work had been recognized as high quality for decades. For blacks it was especially difficult. They were new to science and to the sources of support. Their professional opportunities were limited: few jobs, fewer colleagues, and no advanced students. Most donors were unwilling to gamble against such odds.

Lillie was aware of the problem, but he had a source of money in mind before nominating Just for a council fellowship. The source was Julius Rosenwald, a philanthropist with a strong and long-standing commitment to improving the lot of blacks. Born of Jewish parents in Springfield, Illinois, in 1862, Rosenwald had spent his early childhood in a house at Seventh and Jackson Streets, one block west of the home of Abraham Lincoln.[18] The family identified with the causes espoused by the president, particularly the abolition of slavery, and young Rosenwald grew up with the hope that some time in the future blacks would have a chance for a better life. He very much wanted to help them himself, and later he had the opportunity. After working hard at several business ventures he bought Sears, Roebuck and Company and laid the basis of a great personal fortune. Sears was a mail-order business that took advantage of the railway system developing throughout the country. Its customers were mainly rural people who read the company catalog and ordered goods sight unseen. Mail-ordering was an especially attractive innovation for blacks, because it precluded discrimination on the basis of race, color, or social position. Much of

Sears's profit came from black customers. No doubt business interest was one reason for Rosenwald's generous contributions to black causes; he was making big money from blacks, so he felt he owed them some in return.

Uppermost, however, were his genuine humanitarian instincts. He loved people and wished that they could find ways to live together more amicably. There were two men—one white, one black—he idolized for their sense of cooperation, their understanding that everyone should learn the art of living together with decency and forbearance. The white was William Henry Baldwin, Jr., the black, Booker T. Washington. Rosenwald was struck with admiration for them when he read John Graham Brooks's poignant biography of Baldwin, *An American Citizen* (1910), and Washington's deeply moving autobiography, *Up From Slavery* (1900). And among other things it was because these books were life narratives rather than political essays that he was so captivated.

Baldwin was manager of the Southern Railway at the turn of the century, and spent his spare time promoting liberal causes, such as civil rights for Jews in Russia and education for blacks in America. As a trustee of Tuskegee Institute he became a lifelong friend of Booker T. Washington, founder and principal of the institute. Washington was a former slave who devoted his life to proving that social and economic success for blacks could best be achieved through accommodation with the white status quo. Together the two men worked hard for social reform and justice based on their idea of man's fellowship with man. In Rosenwald's mind, they stood as a living example of interracial harmony.

Rosenwald met Washington in 1911, six years after Baldwin's death. Immediately appointed a trustee at Tuskegee, he began donating thousands of dollars to Washington's pet cause, industrial education for blacks. These gifts were a follow-up to the General Education Board's appropriations, begun in 1902, for educational programs at Tuskegee and the construction of schoolhouses throughout the rural South. Rosenwald's other contributions to black causes ranged from funds for black YMCAs to support for medical education at Meharry, the black medical school in Nashville. In 1917 he became a member of the board of trustees of the Rockefeller Foundation, developed and broadened his philosophy of philanthropy, and set up the Julius Rosenwald Fund in Chicago.

The purpose of the Rosenwald Fund was to promote "the well-being of mankind."[19] Its early activities focused almost exclusively on the rural school program, because Rosenwald felt that basic education— something that the state and federal governments should have been but were not providing—was one of the most crying needs in the black communities of the South. But other kinds of help were not ruled out. In 1919 the fund offered six fellowships for blacks entering medical research—the first Rosenwald Fellowships. A committee on awards was set up under the auspices of the General Education Board. Its members were Abraham Flexner; William H. Welch; David L. Edsall, dean of the Harvard Medical School; and Victor C. Vaughn, dean of medicine and surgery at the University of Michigan. These men, who would be responsible for shepherding the top black medical students along the path to a career in the black community, were all white. It had been suggested that a black be included, but Edsall thought the idea was fraught with "danger."[20] Suppose the committee did not find the "right kind of man" with the right "point of view"? A black member would enjoy special communication with other blacks and might convey confidential information to them. Edsall knew no black man of the "right type." The committee came up with no candidates, let the matter drop, and remained all white. It went about its duties cautiously, selecting only four Rosenwald Fellows in Medicine for 1920: W. S. Quinland, Theodore K. Lawless, Carrie J. Sutton, and George W. Adams. Each fellow received $1,200.[21]

Lillie heard about the fellowships and was struck by the possibility that Rosenwald might be a source of funds for Just, even though Just was not engaged in medical research. He knew Rosenwald personally, but did not want to make initial contact in the matter. He thought it would be best if McClung and the council took the lead. In this way Just would gain by being the concern of a national agency rather than the protégé of a particular individual; also, the larger issue of blacks in science would be brought directly to the attention of powerful scientists and science administrators. But Lillie intended to use whatever influence he could with Rosenwald once the proper initial steps had been taken and the council had made the first move.[22]

While Lillie was working along these lines, Just had already gone about doing something on his own. First he arranged an interview with R. S. Woodward to request support from the Carnegie Institution of Washington. Woodward appeared to be sympathetic, but turned Just down outright. While the Carnegie did support pure research in various scientific fields, it had committed itself heavily to the war effort, or so Woodward said, and funds for research had been depleted. Just sent Woodward's letter of refusal to Lillie, and urged that he be kept in mind for "*anything* that may turn up."[23]

The only thing to do was to keep trying; there was little point in feeling despondent about the Carnegie rejection. Next, Just went straight to the source of power in American medicine and medical education: William Henry Welch. He knew medicine was a sure route to money and realized it might be wise to feel out some of the sources of support for medical teaching and research.

In 1919 Welch was a venerable scientist.[24] He had made a formidable reputation for himself by almost single-handedly establishing Johns Hopkins Medical School as the best in the country. Medical people everywhere, especially the two Flexner brothers, revered his accomplishments. In their view, the example of his life was as solid a guide for young academics and physicians as "the Northern Star" had been for sailors prior to the invention of the compass.[25] He sat on many boards of trustees, including that of the Carnegie Coporation, and consulted with medical administrators on the most important matters affecting the field. His favorite student had been Simon Flexner, whom he had handpicked to be director of the Rockefeller Institute for Medical Research in 1902. He had also arranged Abraham's appointment at the Carnegie Foundation for the Advancement of Teaching.

Just had heard Welch speak at the dedication of Thirkield Hall in 1910, as well as at the commencement exercises at Chicago in 1916, and had subsequently kept his eye on him as the most important and influential man in the field. It was a shrewd move on Just's part to write Welch in 1919 and introduce himself as a "colored man" and former student of Lillie's.[26] Just was careful not to approach Welch in his role as a trustee of the Carnegie, but he did mention Woodward's rejection. Along with a copy of Woodward's negative though not unsympathetic letter, he sent a glowing recommendation from the geneticist T. H. Morgan. Then he went on to present himself as a research scientist, prudently stressing, of course, the importance of his work for black medical education. He outlined his scientific research and his medical and dental teaching, placing them both in the context of his solid liberal education at Dartmouth. His urgent hope, he said, was to advance the "cause of medical education for negroes" and to improve "the health of the country, the South in particular." For this he would need some research funding and some relief from teaching duties.

Welch was impressed. In recent years his attention had been drawn more and more away from medical research and into public health policymaking; he had appreciated the Flexner report, with its point about blacks learning disease control as a means of protecting both themselves and their white employers. Just's proposal seemed relevant to all this. Welch brought it to the attention of Abraham Flexner at the General Education Board, commenting how "distressing" it was that an exceptional scientist like Just "should have to carry such a tremendous burden of teaching, and have so little time for research."[27] Flexner got busy and consulted his brother Simon about possible funds at the Rockefeller Institute for Medical Research. Simon agreed with Welch: Just needed and deserved support, and "something should be done for him at once if at all possible."[28] Simon's idea was that Just should be provided with an assistant to relieve him of routine experimental and administrative work, to allow him more time to "improve medical education . . . and produce the kind of men to profit by . . . fellowships for colored students." But there were no funds available at the institute to help in the case. Simon joined Welch in urging Abraham to look elsewhere.

Again Rosenwald came to mind. He had been on the board of trustees at the Rockefeller Foundation since 1912, and Abraham Flexner had gotten to know him. The only problem was that Just's case would not come within the current Rosenwald Fund scheme, which was "meant for students rather than teachers engaged in research."[29] Nevertheless, Flexner thought Rosenwald might do something on a personal basis. He wrote him in January 1920. At about the same time, on Lillie's suggestion, McClung sent off a letter from the headquarters of the National Research Council. Just had thus set in motion two lines of concern that led to the same place: Julius Rosenwald's office in Chicago.

For years Rosenwald had received hundreds of applications for financial assistance, so many, in fact, that as early as 1912 he had found it necessary to hire a special assistant, W. C. Graves, to deal with such matters. Graves had had wide experience as an administrator with charitable organizations in Illionis. Under Rosenwald, his job was to put together summaries of each case that came through the office. Just's case was handled no differently. Graves prepared it by summarizing the letters of recommendation and highlighting the purpose of the support Just sought. In his dossier, McClung's emotional portrayal of Just as wearing himself out in "a hopeless struggle" to continue his research was complemented by Flexner's measured advice that "service would be rendered to humanity through giving a fitting opportunity and support to a really able scientist of the Negro race."[30] It was a solid case, and before the end of February Rosenwald had decided to help Just for three years at $1,500 per year. A little later he increased the amount to $2,000 per year, adding $500 per year for summers at Woods Hole. The decision was straightforward, but the ramifications were not.

The grant went into effect on 1 July 1920. Rosenwald would supply the money; the National Research Council would administer it and send Rosenwald progress reports on Just's work; Abraham Flexner would be the liaison, the middleman and pivotal force of the whole arrangement. The importance of Flexner's role became clear very early. President Durkee agreed to accept the gift and alter Just's teaching arrangements at Howard only on condition that the General Education Board make a contribution to the Howard Medical School.[31] His condition was accepted: Flexner gave Howard a large grant, $250,000 out of the $50 million Rockefeller fund he administered for the reorganization of American medical education. The way was clear for Just's fellowship, and plans could be made to reduce his teaching load. Apparently, he had more clout than he knew. Behind the scenes high-level officials were using his situation to haggle over huge sums.

Setting up the new arrangement was problematic, however. Lillie and McClung wanted Just to have a whole semester off from teaching, so that he could devote a period of uninterrupted time to his research. In their view, teaching and research could not be mixed; half days devoted to each would cause one or the other to suffer, most likely the research. It would be difficult for Just to go into Howard without being disturbed at all hours, they argued. Students from his afternoon classes would inevitably find their way to his laboratory in the morning, and such interruptions would hamper his high-powered, nerve-straining microscopic work. The National Research Council had a responsibility to maximize the advantages of the fellowship, to insure that Just was allowed "sufficient time free from teaching in order to continue his important investigations."[32] The entire fellowship program was in fact being taken to task. On the other hand, if Just succeeded with the Rosenwald grant, finding money for other scientists would become easier.

But Durkee did not like the idea of giving Just a semester off. He thought it would be best if Just spent the year doing research in the mornings and holding classes in the afternoons.[33] He was adamant on this point. Perhaps he feared that other faculty members would also want leaves of absence, and hoped that if Just could be kept busy throughout the year, less attention would be called to his privileged status. Flexner vacillated. When he was with McClung or Lillie he would agree with them; but with Durkee, he would note that the "free morning" arrangement Durkee proposed was more liberal than that under which his own brother, Simon, had been doing research. Rather unfairly, he likened the circumstances of a fledgling scientist and those of a seasoned administrator, assuming that what was good for the one was good for the other.

The controversy went on, with Lillie and McClung, Flexner and Durkee each setting forth his viewpoint. Even Rosenwald stepped in on one occasion, asking Flexner for an opinion on Lillie's suggestion that Just teach half a year and do research the other half, instead of dividing the day between the two.[34] For his own part, Just agreed with Lillie and McClung, and he told Heilbrunn privately that there had been "much damn foolishness" in the arrangement of research conditions under the grant.[35] But he was so delighted to have the fellowship at all that he was content to let Flexner and Durkee win out, at least for a while.

Except for minor routine mix-ups, the first year of the fellowship (1920–21) went well. As usual, Just spent the summer at Woods Hole. He was "sour as a goat" because he could not get away from Howard until mid-June, but once at the MBL his research went better than ever.[36] He found himself "going strong" on the fertilization work, and felt particularly pleased that his starfish slides were "looking good" and promising to help advance the general theory of the role of the cortex in fertilization.[37] He spent day and night in his laboratory observing the fertilization process, which sometimes took as much as thirty hours from the moment of insemination to complete development.

Back at Howard in mid-September, he began a regimen of four hours in the morning in the laboratory and four hours in the afternoon in the classroom. Even though he managed to do more research in a single quarter than in "all ten years previously" at Howard,[38] his responsibilities to the university were not substantially lessened. He taught at least 750 students, and was responsible for counseling some 300 prospective medical and dental students in the college and junior college. Part of his time was spent thinking about "pedagogical" questions, such as the need to maintain "the personal element" in the classroom.[39] Also, he became "one of the most careful and painstaking" advisors to the administration on matters relating to science, making an effort to bring to the university the special knowledge he had gained from his outside experiences concerning the place of science in American education and life, and its directions for the future.[40] The university would have been unable to meet the larger responsibilities of the science curriculum without Just's "wise counsel" and "eager and hearty cooperation." His fellowship had put him in a very special position. He was also in demand to do committee work, which turned out to be as time-consuming, if less exhausting, than teaching. He contributed so much, in fact, that Durkee relaxed restrictions and allowed him time off in the spring before the close of the academic year.

In early May he went to Woods Hole again and worked hard at his research for four months. Progress was slow at first because unusually bad weather had depleted the fauna along the Massachusetts coastline. But things picked up by mid-June, and the summer turned out to be one of Just's most productive to date. Glowing reports went to Rosenwald from all concerned: Just, the National Research Council, Lillie, and Durkee.[41] There was little chance that Rosenwald would discontinue his three-year commitment. Just seemed set.

One problem, though, was that he and Flexner got off on the wrong foot. Whenever they met anywhere, at Woods Hole in particular, Flexner would remind Just that the grant was "tentative and conditional," and that its continuation depended on satisfactory report cards from Lillie and McClung.[42] He seemed to harbor, or so Just thought, a deep-seated fear that Just's head would become "swollen" and that he would quit work, especially on his "sudden accession of wealth." This whole attitude was "a bit childish" and "the whole business . . .

irksome," as far as Just was concerned. Flexner was, however, the line
to sources of money, and too powerful a man to be crossed directly. Just
had to be careful lest he jeopardize future help for himself and other
blacks. But he also had his pride, and he was not going to stand for
paternalism. He was an established scientist, not an inexperienced
student, so why should others be "responsible" for him and his work?[43]
He let Flexner know his feelings obliquely, through McClung and
Lillie. He could not change what Flexner thought privately, but he
could at least try to change what he said and did publicly.

As Just must have expected, Flexner was intelligent enough not to
continue in the same vein. He more or less called a truce, stressing at
the same time that he did not want Just to read into his statements
meanings that were not there.[44] Over the course of the next two years
they worked out a mutually respectful relationship, though Flexner
never truly became the "friend and advisory sponsor" Just had hoped
for.[45] Flexner expressed concern, offered advice, and promised to do
everything in the world to promote Just's welfare and progress. But he
never seemed completely sincere.

Flexner was a Southerner, albeit a Jewish one.[46] Born and bred in
Louisville, Kentucky, he had the outlook of a typical educated Southern
white. Though he acknowledged that blacks deserved some opportunity
to better themselves, he was a paternalistic segregationist. He believed
that segregation would work if blacks were given a chance to develop
their own institutions. Just would work at Howard, Rosenwald Fund
recipient W. S. Quinland at Meharry—and that was that.[47] It did not
matter what black professionals wanted for themselves. Flexner had
decided that everyone, black and white, would be better off if blacks
kept to themselves.

Flexner's goal was constructive in a way. At least he was attempting
to help blacks and to solicit support for Howard, which was more than
most whites were doing at the time. He had a point when he once
linked himself and Rosenwald as "two Jews of liberal spirit and lofty
purposes" who might be able to "do something to encourage tolerance
in this country—and perhaps in other countries."[48] Without doubt he
was one of the best allies a black could find among whites in that era.
As a result, there was hardly any black other than Just who insisted on
being treated with real respect by Flexner—for most blacks it was
satisfying enough to find a white man who was even cautiously
progressive.

Still, Flexner's attitude left much to be desired. Despite his importance as an administrator, he spent much time talking about blacks
in a small-minded manner. Even in his professional correspondence
with Rosenwald he did not refrain from passing on racial slurs. A case
in point involves W. S. Quinland, the black doctor from Meharry.
Wallace T. Buttrick, president of the General Education Board, had
interviewed Quinland concerning fellowship support. These are his

comments, as reported to Rosenwald by Flexner: "I believe he is a real find. To begin with—he is about my height, not yet having attained to my equational dimensions. He is as black as Major Moton. Close your eyes and you'd think you were talking with or rather listening to a refined country gentleman."[49] For Buttrick and Flexner, Quinland was the day's comic relief. True, Quinland, a Jamaican, spoke with a British accent and had a complexion as dark as that of Major Moton, president of Tuskegee Institute. But what did these facts have to do with awarding him a fellowship? Such petty considerations, deeply ingrained in the American mentality, were part and parcel of the larger problem of support for blacks. That Flexner quoted these remarks without condemning them is revealing; that he might have thought them complimentary is appalling. Blacks knew remarks like these were being made everywhere, though they did not usually know the details. The best way for them to deal with such things, most felt, was to ignore them if at all possible or, like Just, complain through third parties. Most often they suffered silently, in the interest of winning support for themselves and their race.

Just's problems were not limited to the slurs and paternalism of white philanthropists and donors. The low-keyed intellectual life at Howard also weighed heavily on his mind. He was an oddity in that environment. Few Howard faculty members were actively pursuing research in any field, scientific or otherwise; even fewer were receiving fellowship support. Most, including Durkee at first, believed research was "merely time off to do nothing."[50] And those who did consider research worthwhile employment were jealous of anyone who had time off to do it. Evidently there was no pleasing anybody.

Just had difficulty pursuing his work under these conditions. He felt guilty whenever he spent time in his laboratory rather than in the classroom, and ill at ease when he saw how much newspaper publicity his grant was receiving. Therefore, during the first three years of his fellowship he found himself giving in more and more to requests that he teach large classes and serve on numerous committees. Even though the restrictive "free morning" arrangement was abandoned after the first year in favor of leave for the entire spring term, Durkee was still making extra demands on Just's time and energy during the fall and winter terms, involving him in work far removed from biology.[51] Just complied willingly. In part, his intensive devotion to university affairs was a means of offsetting adverse criticism at Howard and the only way he could feel free to go to Woods Hole early in the spring; in part, it was a response to the serious needs of his students.[52] He had no obligation to squeeze in more work during the fall and winter quarters; the official arrangement now simply called for him to do a *normal* amount of work during the fall and winter, and to use the spring for research. No one intended him to cram three quarters of work into two. The burden was of his own making—with a little pressure from Durkee.

Just felt as strong a responsibility to the National Research Council as he did to Howard. He knew that he and two other scientists—Albert Mann, a specialist on unicellular algae, and Carl H. Eigenmann, an ichthyologist—were test cases for the council's Division of Biology and Agriculture, that they were the first three biologists to receive grants from the council, and that the success or failure of their work would be a crucial factor in determining whether the grant program was continued.[53] The pressure was on, and Just had to produce. Unfortunately, the gimmicks he devised to keep his time for himself were often not effective. During the 1921-22 academic year, for example, he scheduled his larger courses at eight o'clock in the morning to try and cut down on the enrollment.[54] Fewer students, less work. The scheme had only limited effect, however, and he continued to work on university matters throughout the day and evening, able to snatch only bits and pieces of time here and there for science. He was unable to find time for in-depth research. All he could do was perform mechanical chores such as slide sectioning; real experiments and serious writing had to be left for Woods Hole. Even publication was becoming a problem. In the early 1920s Just and most other scientists had to worry about getting their work into print: American journals were cutting back their expenditures drastically, and the *Biological Bulletin* in particular had instituted a frustrating policy of "cold storage for a year before publication."[55] A scientist either had to canvass far and wide for a place to publish or give up in despair. As might have been expected, Just opted to make the extra effort needed to find journals with space available. At one point he considered publishing privately, but the cost was clearly prohibitive.

Just's regimen brought on periods of great mental and physical fatigue. As early as the fall and winter of 1920 he was having such "a hard time" with his health that Lillie became quite concerned.[56] To make matters worse, Just's home life was in confusion. Some time in the spring of 1921, Ethel learned she was pregnant, but she did not want another child. She and the two children, Margaret and Highwarden, spent the summer on Martha's Vineyard, an island five miles south of Woods Hole. Just visited them on weekends and tried to placate Ethel, but he was unable to relieve her distress. Throughout the fall he worried about her so much that teaching became a painful routine and research impossible. His low output was making him feel guilty about the Rosenwald stipend.

At two o'clock in the morning on 9 December Just had to rush Ethel to a private hospital.[57] Her two previous children had been born at home without complication, but this time she went through a difficult ten-hour labor before giving birth to a daughter, Maribel. It took her three weeks in the hospital to recuperate. Just, unable to find domestic help—his maid had quit and Ethel's mother, Bamba, could not take a leave of absence from her post at the State College in Orangeburg, South Carolina—had to "look after the kids, . . . cook, wash, do every damn thing."[58] In addition, he held classes daily from eight to noon. There was no escaping the pressures, and Lillie voiced concern that Just would "burn himself out" unless he found a way to relax his pace.[59] Still, Just kept hoping that he would be able to "buckle down to good work for the rest of the year."[60]

The spring of 1922 brought a further reason for anxiety. Just was about to put ten papers into press: one in *American Naturalist*, one in *Science*, six in the *Biological Bulletin*, and two in the *American Journal of Physiology*. He also had four others in the final stages of preparation. To top it off, he was trying to make up for "time lost" from research owing to duties at school and pressures at home.[61] His experiments were under way again. Results were coming in, but only after "extremely hard" work which could not be kept up "indefinitely."[62] In three months he cut, mounted, stained, and examined 635 sections of twelve stages of *Arbacia* eggs. His hectic research schedule was complicated by his personal sense of duty "to his University and the people of his race." He was still carrying out heavy committee assignments, teaching as much as thirty hours a week, and carrying the "onerous burden" of responsibility for the study programs of over five hundred premedical students. Many people, Lillie included, did not think he should relinquish the commitment to the education of his people. Others disagreed, arguing that a sense of racial duty "should not mean either suicide or abandonment of research."[63]

Just was simply working too hard—going to Woods Hole in mid-April, rushing back to Washington for end-of-term duties at the end of May, returning to Woods Hole in early June. In 1922 he had to leave Woods Hole for a rest around mid-August, before the end of the MBL session. He had worked every day, including Sundays, for four months, and wound up the experiments for no less than sixteen separate research problems.[64] He was badly worn out and on the verge of a nervous breakdown, causing much worry among the administrators in charge of his grant. They admired productivity and had let Just know this, but they did not want a breakdown on their consciences. At one point Graves urgently asked an official at the National Research Council whether Just was "undertaking too much," whether his health might "break under the strain."[65] What they feared materialized in the summer of '24, when Just had to take two months of complete relaxation in order to recover his health.[66]

᠁

By the early 1920s Just had begun thinking about leaving Howard. The institution was demanding too much of his time, and under Durkee it had come to be plagued by internal struggles. Just's short leaves of absence made him dream about what he might be able to accomplish under a less stringent regimen and in a more congenial atmosphere. It was a dream shared by a growing number of Howard faculty members.

Many of Howard's faculty were black, members of the group W. E. B. Du Bois had termed "the Talented Tenth," the intellectual elite who would be role models for blacks in all aspects of social reform. Among them were Alain L. Locke, a philosopher, Harvard graduate, and Rhodes Scholar; Benjamin Brawley, writer and professor of English; Dwight O. W. Holmes, author and administrator; Charles H. Houston, civil rights lawyer; Lorenzo Dow Turner, linguist and sociologist; and Carter G. Woodson, historian and administrator. They were part of the Negro Renaissance that had come to dominate the collective consciousness of a new breed of black American writers, artists, and intellectuals.[67] Claude McKay, Langston Hughes, and James Weldon Johnson had shaped the Renaissance movement in the 1910s and '20s; Johnson had even written the lyrics for a song, "Lift Every Voice," which later became known as the Negro national anthem. Howard University was a part of the new sense of identity these men had generated. So was Just. Although a scientist, he came to be identified with a movement centered primarily on arts and letters in the black community of Harlem. A prominent member of the Howard faculty, he was engaged in high-powered teaching and research and his achievements received extensive national news coverage.

But intellectual distinction was no solace in the face of Durkee's increasingly crude tyranny; in fact, it may have aggravated the problem. The new sense of self-confidence among black scholars, the feeling that they should no longer have to submit like children to white authority, caused Durkee some alarm. Confrontations increased during the early and middle 1920s. Professors were fired, transferred, and even personally harassed by the tyrannical president. Carter G. Woodson lost his post as dean of liberal arts in 1920. Kelly Miller, who had brought Just to Howard in 1907, was fired and then rehired in 1925 after a series of head-to-head disputes. On one occasion he was verbally abused by Durkee, who cursed him as a "contemptible puppy."[68] Just's close friend Alain Locke was "kicked out bag and baggage" after a run-in with the president.[69] Thomas W. Turner, a botanist attached primarily to the School of Education, was once forcibly ejected from a conference by Durkee, who "pounced" on him and "grabbed him around the shoulders, pushed him over chairs, and around the room like a mad man."[70] These incidents became an issue in the black community nationally and ultimately forced Durkee's resignation. The new black consciousness was gaining momentum at Howard.

Just seems not to have participated in the various fights and counterattacks, the "perennial scraps" as he called them.[71] Unlike most of his friends and colleagues, he had worked out a relatively satisfactory relationship with Durkee and was trying to concentrate on his work.

Even so, Just was warned of rumors to the effect that Durkee was "not satisfied" with Just's work and was getting ready to "dispense" with his services.[72] The environment was hardly conducive to a healthy state of mind. He wanted out, if not altogether, then at least to the extent that he could spend part of his time during the academic year somewhere else.

In the fall of 1924 Just again contacted the Carnegie Institution for assistance. Rather than going through the main office this time, he wrote directly to the head of a specific branch—the Department of Embryology in Baltimore.[73] His appeal was a cautious one. He expressed a desire to be a "sort of non-resident worker" in the Carnegie department. Such an arrangement would give him a link with an institution other than Howard, and a place where he could occasionally go to work and think. He was not asking to be hired for a permanent, resident position with the Carnegie; he knew no white establishment would want him on that basis. His primary affiliation would continue to be with Howard. But perhaps the Carnegie could see its way clear to giving him a makeshift position with a nominal salary. What he needed most was a little extra money to carry on his research, now at a stage demanding more material and equipment. In fact he should really do some traveling, which would require funds for "work at Naples or some other laboratory with rich echinoderm fauna." If only he had some affiliation other than Howard.

A response came from George Streeter, head of the Carnegie Department of Embryology.[74] Streeter informed Just that the department subsisted on "a fixed budget . . . entirely taken up by the staff now at work"; there were no openings at present and probably not for the foreseeable future. He advised Just to contact John C. Merriam at the head office in Washington, but admitted that even then the possibilities were "not very promising." Though the Carnegie had some fellowships, the general policy was "to help programs . . . rather than to assist individuals." Just might have a chance if he made his appeal not for himself but for his institution. Otherwise there was little likelihood that he would find support at the Carnegie.

Just did not contact Merriam, but he thought seriously about what Streeter had said about foundation policy. The foundations, it seemed, were beginning to view themselves as philanthropies rather than charities, to treat their work as dedicated to larger causes such as educational institutions rather than to the particular troubles of individual researchers. If that were indeed the case, a change of strategy was in order. Perhaps the way to get money was for him to take a more positive attitude toward Howard.

With two rejections from the Carnegie on his hands, and the Rosenwald fellowship scheduled to expire in a few months, Just applied directly to Flexner for help from the General Education Board. They were more or less on friendly terms, and Just knew that Flexner had control over large sums of money and was concerned about the direction in which Howard was going. In any case, there was nowhere else to turn.

More boldly than with Streeter, Just characterized his research program as "definite" and "clear-cut," asserting the need for "an adequate grant of a more permanent nature, . . . a grant of a sufficient sum to extend through a sufficient term of years to guarantee . . . working to the best possible advantage."[75] Also, keeping foundation policy and Flexner's concerns carefully in mind, he linked his research to Howard University and education for blacks in general. He stressed that any research he did would have "educational value," in fact that any attempt at scientific work such as he was engaged in would be of "inestimable value for the cause of Negro education." What blacks needed most of all was a "dispassionate scientific attitude," and Just made it clear that he could contribute toward that goal if only he could find some support for his research program.

Flexner offered no hope that the board would step in for Rosenwald, but he was interested enough in Just to pursue the matter. The first order of business, he thought, was to put on record the opinions of Carnegie scientists and administrators regarding Just. Merriam agreed to gather the information. Letters requesting comment on the "value and soundness" of Just's work were sent out to several Carnegie people by W. M. Gilbert, administrative secretary of the Carnegie Institution. If the responses did not lead the Carnegie to reconsider its earlier decision, Flexner thought they might at least open up other "possibilities."[76]

Streeter was one of the first to reply.[77] In his opinion, Just was a "capable and industrious investigator," though the connection to Frank Lillie made his "originality" questionable. He had been "given every opportunity" at Woods Hole and "estimated . . . entirely on the basis of his work." The question of color was "neither here nor there." Just's difficulty in gaining recognition should not be attributed to his race so much as to the "difficulty that all new workers have" establishing a reputation. Admittedly color had limited his "opportunities in the matter of obtaining a teaching position," but this was to be expected. On the other hand, there were no grounds for Just's exaggerated view, expressed in conversation and correspondence, of the effect racial discrimination had had on him in the allotment of research funds.

Another Carnegie scientist, C. B. Davenport, thought Just was doing "sound" work, but did not consider him to be a man of "any exceptional brilliancy or initiative."[78] Though willing to concede Just's standing as "the most prominent (if not the only) colored man" in zoology, he hardly meant it as a compliment. An ardent eugenicist,

Davenport had spent a good part of his scientific career analyzing mental differences between races and warning of the horrors of universal mongrelism.[79] According to him, blacks were incapable of producing brilliant results in an intellectual area such as science.

The longest letter came from Otto Glaser, a Woods Hole biologist and scientific disciple of Loeb.[80] According to him, scientists had "leaned over backwards in order to be fair," and Just had received "more consideration than a white man of the same mentality." There was no prejudice toward Just in the scientific community, Glaser insisted. Just associated "freely" at Woods Hole and could be found "eating at the mess and sitting in at little parties as an equal." His chance to secure a first-class teaching appointment was "undoubtedly limited by racial prejudices," but his opportunities "as an investigator and as a human being among scientists" had been "unusually good." Besides, most scientists did not know the papers he had written were "by a mulatto," so his reputation as a biologist was not affected on that count. The real question, Glaser suggested, was whether Just's admittedly high scientific reputation was warranted. Glaser thought not. As he bluntly put it, "Just begins and ends with Frank Lillie." No matter how true it was that Just conceived and planned out his own experiments, the fact remained that the thrust of his work was to "justify and defend Lillie's theory." He did not appear to have much originality. His papers suggested a lack of "independence," and at times they took on "the coloring of prejudice." The loyalty he showed to Lillie was extreme. What remained to be seen was whether some time in the future he could "break through to a new standpoint of his own." No doubt the potential was there, Glaser conceded. Just had "a better mind than most of the white men of his age as they exhibit themselves at Woods Hole," was "extraordinarily well-read in the literature of his field, knowing it critically and in detail," and always carried out his experiments with "unusual care." But at the moment, despite his diligence, he seemed unable to "shake off the influence of his teacher." The reputation he had acquired was the result of a free ride on Lillie's coattails. As an afterthought, Glaser said that he liked Just personally and that there was no reason for not wanting to spend the summer with him in a research laboratory. He would not object to being with Just at Dry Tortugas in the Florida Keys, any more than at Woods Hole in Massachusetts. Just was in no way an unpleasant person; he "swims regularly, is interesting to talk with and has the reserve and self-consciousness that prevents him from being omnipresent." Glaser's back-handed personal recommendation was hardly more complimentary than his negative professional evaluation.

Small wonder that Just received no offer from the Carnegie. The scientists there were unimpressed with his work, or so they said, and they were unsympathetic to his peculiar problems. If Flexner wanted to help him, the place to go was *not* the Carnegie. The alternative, though, was not clear.

Just realized that any attempt he made to expand his horizons would involve serious effort, even bitter struggle. Moving permanently into a better position, perhaps at a white university, had crossed his mind, but he knew it was wishful thinking. There was no point expecting the impossible. White universities simply were not ready to hire black scholars at all, let alone allow them to teach key courses, contribute to important departmental decisions, and socialize at the faculty clubs. Much to the relief of everyone, Just did not complain about the situation. He acquiesced to the opinion held by Flexner and the Woods Hole scientists that "Just at Howard would be less of a problem than Just at some other institution."[81] As an internationally recognized biologist, of course, he should have every right to expect a good position in a top university. In compensation for not demanding such a position, he hoped he could get white scientists and administrators to help make his search for outside funding successful. Though the Carnegie experience did not offer much hope, Just was not discouraged. What was required on his part was persistence. There was even a realistic chance, he believed, that with the aid of people like Flexner he would be able to secure not only occasional grants for summer work but also some kind of permanent endowment that would make conditions at Howard bearable for him.

Just had always sensed the importance of keeping long-term needs in mind. Early on when he tried to get close to Rosenwald it was in the hope of establishing a relationship that might continue for a number of years. He first saw Rosenwald in March 1920, to discuss research arrangements under the grant Just was to receive from the Rosenwald Fund. The meeting took place at the Shoreham Hotel in Washington and was also attended by Flexner and Howard's treasurer, Emmett J. Scott. Everything went well. Just was in awe of Rosenwald, but summoned up the courage to send him a letter of thanks that August, a month after Rosenwald's grant went into effect.[82] For some time afterwards he could not bring himself to write to Rosenwald directly; instead he addressed his letters to W. C. Graves, Rosenwald's secretary. He shared personal and professional concerns with Graves, asked his advice on numerous matters, and tried to make him aware of the unusual problems faced by a black scientist. For the most part Graves kept aloof, responding in a perfunctory way.[83] But Just was not put off, and continued to write detailed letters about his professional life, still addressed to Graves but no doubt intended for Rosenwald's eyes.

At first Rosenwald did not think of Just as unique, and regarded the grant to him as simply another of his charitable gifts to black causes and individuals.[84] But Just soon changed that. By 1922 he was beginning to make direct contact with Rosenwald, to try and make his benefactor more aware that the work he was doing had important ramifications for science and the black race. He wrote long thank-you letters filled with intimate details about his childhood and education, how he had come to decide on a career in science, what he hoped to contribute to mankind. Rosenwald treasured these letters as mementos of the help he was giving to a hardworking, unusually talented black. Though not a trained intellectual himself, his instincts told him that Just's deep dedication to science might be important in the effort to show that blacks could make worthwhile contributions in academic fields as well as in athletics and entertainment. A bond of genuine affection grew between the two men, a kind of father-son relationship at a distance.

The bond was apparent early. In 1922 Just asked permission to list his title officially as "Julius Rosenwald Fellow in Biology, National Research Council." The request was bold, since except in one case Rosenwald had invariably refused to allow his name to be used "in connection with any publicity."[85] But after consultation with Flexner and council officials, Rosenwald agreed to Just's request. It was a unique honor.[86] For the next ten years Just published under the title "Julius Rosenwald Fellow in Biology." The Rosenwald name became identified with the fate of E. E. Just.

Just kept Rosenwald abreast of all his scientific honors and distinctions. In 1922 he received a notice from his first alma mater, South Carolina State College, that he was to be awarded an honorary doctorate. He did not attend the ceremony, but he brought the letter of announcement to Rosenwald's immediate attention.[87] In 1924 he received and accepted an invitation to cooperate with a group of European biologists, mostly Germans, in compiling a comprehensive scientific treatise on fertilization. His contribution was to be a monograph on the physiology of fertilization, his magnum opus as Rosenwald Fellow. He lost no time in sending Rosenwald a copy of this invitation, "the first of its kind" ever extended to an American black.[88] In 1926 he told Rosenwald of still another "flattering mark of appreciation" from the Germans, this time an invitation to act as associate editor of the *Internationale Zeitschrift für physikalische Chemie des Protoplasmas* ("International Journal for the Physical Chemistry of Protoplasm").[89] The purpose of all this was to show Rosenwald that Just had established a considerable reputation for himself on the international scene. He assumed that Rosenwald knew about his national stature, but he did not let evidence of it go unmentioned either. He stressed that his work had been cited about eighteen times, "with full acceptance of the value and import," in E. B. Wilson's classic 1925 book, *The Cell in Development and Heredity*.[90] He constantly reminded Rosenwald that he was one of the few select contributors to E. V. Cowdry's 1924 *General Cytology*, perhaps the most important text in the field.[91] In 1928 he proudly announced that he had finally been listed in *American Men of Science* (the fourth edition, published in 1927).[92] His name was starred, no small matter, indicating that he took a place among the top thirty-eight zoologists in America. He was the only black "so honored"; five or six others had been included in the general list of scientists, but without stars. His peers had voted him this distinction, and the results had been compiled by J. McKeen Cattell, an old friend of Jacques Loeb. But the crowning glory, he thought, was his election to the editorial board of *Physiological Zoology*, "the first time in history" that a black had been honored in America for "pure scholarship, wholly apart from work . . . *on* the Negro or *with* Negroes."[93]

Though proud of his accomplishments, Just felt "at a disadvantage" when telling Rosenwald about them.[94] There was a tension between wanting to show Rosenwald the value of continuing the fellowship and at the same time not wanting to seem too obvious about "importuning" for renewed support. The problem lay in the short-term nature of the grant. During the seven years he received the Rosenwald grant it had to be renewed four times, with terms ranging in length from one year to three years, and at the end of each term he had to wait, hopeful and anxious, while Rosenwald considered renewal. A more permanent kind of support was necessary to make him feel less like a charity case and more like a bona fide scientist with something to offer the world. All the worrying was making him less productive than he would otherwise have been, and he knew it. The question was how to approach Rosenwald with a proposal for long-term support without appearing presumptuous, and perhaps even ungrateful for the earlier grants.

The National Research Council served as a kind of buffer. Early in 1925, when Rosenwald was again considering renewal of Just's grant for another term, the council decided to do something about the uncertainty of Just's situation. Many of the scientists and administrators there were convinced that Just should have a career in research, that it was unfair for him not to know from one year to another if he would be able to continue his experiments. A plan was developed to seek out sources other than Rosenwald, who, it was thought, had done more than his share by supporting Just for five years. Three foundations were considered: The Anson Phelps-Stokes Fund, which gave grants to blacks and Indians for research and also to Negro schools and colleges for the improvement of teaching and facilities; the Ella Sachs Plotz Foundation, which financed research by individuals up to a couple of thousand dollars at a time; and the John F. Slater Fund, which offered a subsidy of salary to educators who trained young blacks in industrial fields.[95] These foundations did not really offer the possibility of long-term support, but one of them might at least be able to tide Just over while the council sought out better alternatives and in the event Rosenwald decided not to renew his commitment.

Rosenwald did come through again in 1925, but this time with a grant for only one year, the shortest term yet. The urgency of the problem seemed to be growing. Maynard M. Metcalf, chairman of the council's Division of Biology and Agriculture, put the first step of the council's plan into effect. He steered through the division a formal resolution recognizing Just's research ability and recommending the indefinite continuation of his fellowship. The resolution was a major event, especially because "no such individual endorsement" had ever before been carried through the council.[96] More importantly for Just, it meant that the problem of status and funding might be nearing a solution. Even if nothing immediate came of it, the resolution was at least "an indication of the genuine esteem" in which he was held by council scientists. The council was now committed on record to finding a means for him to continue his "valuable research," whatever Rosenwald might decide about the fellowship in the years to come.

Metcalf also raised the larger question of establishing long-term support for science at black institutions. To him, Just's situation seemed to highlight the problems encountered by the growing number of black scientists in America, all of whom spent their time teaching and had little opportunity for research. No one else in the division seemed particularly interested, so Metcalf passed his idea on to the council's executive board, which turned it over to the Division of Educational Relations, which handed it back to Metcalf—a classic case of buck-passing. Metcalf next made an effort to include his plan for aiding institutions such as Howard and Fisk under a broader council scheme to promote research in American colleges. He wanted to invite Just to one of the council's conferences to speak on "the possible relation" between the "scheme and racial schools."[97] But some members of the organizing committee protested, saying that while they were "very cordial" in their "appreciation" of Just and the black schools, they thought there would be "but little opportunity . . . to be of actual assistance" in developing these institutions. Metcalf was embarrassed and went out of his way to tell Just that there was "no suggestion of race prejudice" in the committee's decision. He later consulted with Just privately and managed in the end to get his plan officially included in the council's scheme. It was a token inclusion and nothing came of it, but at least he had planted in everyone's mind the idea of long-term research arrangements for blacks.

Metcalf's positive attitude encouraged Just to ask Flexner outright whether or not the Rockefeller or some other foundation would consider endowing research work at Howard. He envisioned a research program "of modest size" devoting itself to the analysis of biological problems, "not only with the aim of advancing pure science but also with that of giving a scientific background to such work as sociology, psychology, and education as a means of helping to adjust the Negro to life in this country."[98] Improvement in science research at black institutions would facilitate improvement in the quality of teaching done in black schools throughout the country. According to Just, "lack of preparation" and the "old-fashioned point of view" of black teachers had kept the race from making educational advancements. His plan was to increase social awareness in blacks by training them in scientific methods at a science center affiliated with Howard. Such a center would, as one of its first orders of business, develop "the basis of a new approach to the race problem."

The plausibility of this argument was not enough for Flexner, who, while claiming that he did not wish to "throw cold water" on Just's hopes and aspirations, stood firmly opposed to any kind of permanent support.[99] Endowment of research at a black institution, he felt, was "a pretty serious undertaking" requiring a steady expenditure of money which would "go on forever." It was preferable to help blacks in research by finding one or two "really promising young men" and developing them individually, as in the case of Just. Large-scale funding for ambitious institution-wide programs was out of the question, at least for the time being.

This attitude was a reflection of the general policy of foundations in America, particularly in regard to support for black institutions. There would be no endowments for the institutions themselves, only a few grants here and there for talented individuals. Curiously, at the same time that foundation officials rejected any plans for building up black institutions through endowments, they insisted that individual researchers must hold institutional affiliations to be considered for grant support. It was a way of keeping control in their own hands. Flexner, for instance, wanted to use Howard to guide Just's activities and career choices, knowing that if the foundation committed itself to supporting science per se at the institution it would no longer have much of a hand in determining what went on, with either Just or the university. After all, an endowment for Howard would inevitably be controlled in the main by Howard administrators.

A further barrier to support for Howard was its elitist image, which the university had gained because many of its graduates refused to work in the deep South. This was particularly true of the medical graduates, in marked contrast with those of Meharry Medical College. Doctors trained at Howard sometimes accepted secure positions at black hospitals in the North and most often developed lucrative private practices in large urban centers there. Foundation officials, who wanted black doctors to work among the rural black poor, were disturbed by this trend, partly because it indicated social apathy, but also because it meant they had little control over the career choices of Howard students. As a result, few funds were channeled into Howard, and those that were came under careful scrutiny.

While they may have appreciated Just's potential as a scientist and wanted to help him further his career, Flexner and other foundation officials apparently did not appreciate that it was difficult, even impossible, for an individual, scientist or nonscientist, to flourish in an academic community that had no broad base of moral and financial support. It was a frustrating problem. The extent and success of Just's work depended on the general standard of education in all departments at Howard, on the general quality of intellectual interaction among colleagues, and on the equipment and laboratories allocated to science. Without full-fledged backing from funding agencies, there could be no hope of getting these conditions up to par. Howard's only reasonably dependable source of support continued to be the Congress, and even then the annual appropriations had to be fought for year in and year out. There were simply no funds available to be devoted to developing the science program into something larger and more worthwhile. For Just the future at Howard seemed dim.

The black community was not blind to the problem of funding black education. W. E. B. Du Bois, editor of *The Crisis*, wrote many articles on the subject, noting that blacks had little say about philanthropic decisions relating to them. Blacks could not "tell the truth or disclose incompetency or rebel at injustice."[100] If they did, a "Sh!" would arise from the whole black race: "'Sh!' You're opposing the General Education Board! 'Hush!' You're making enemies in the Rockefeller Foundation! 'Keep still!' or the Phelps-Stokes Fund will get you." A black had to go with "his hat in his hand"; he had to "flatter," he had to "cajole" in the hope of being awarded a pittance for his trouble.[101] If he showed the slightest independence, he was summarily dismissed. Du Bois provided a running commentary on foundations throughout the 1910s and '20s, opposing most of them with caustic verve. The General Education Board, in which Flexner was a powerful figure, came up for special criticism in the columns of *The Crisis*.

The board had been set up by the Rockefellers in 1902 to promote education "within the United States of America, without distinction of race, sex, or creed."[102] John D. Rockefeller, Sr. had made millions of dollars in the oil industry, and the Rockefeller name had come to mean big money and unrestrained greed. In order to improve the family's public standing and to fulfill his own personal interest in bettering the condition of blacks, John D. Rockefeller, Jr. decided to set up a foundation for philanthropic purposes. He pledged a million dollars to the venture, with the stipulation that the money be spent over a period of ten years. At first he had hoped to create a subsidiary Negro Education Board to upgrade black educational institutions throughout the South. But he compromised his plans in the face of pressure from Southern whites, especially Henry St. George Tucker, president of Washington and Lee University, and agreed to support instead education for all in the South, white and black. To that end he helped organize the Southern Education Board, which launched a widespread, popular campaign for tax-supported elementary schools. Most of its trustees were also trustees of the General Education Board. It was phased out in 1914, and the General Education Board took over most of its work.

Frederick T. Gates and Wallace Buttrick, both former Baptist ministers, planned out the program of the General Education Board. Gates served as chairman, Buttrick as secretary. In 1912 Flexner joined the board as assistant secretary, and when Gates resigned shortly afterwards, Buttrick became chairman and Flexner rose to the position of secretary. By this time the major thrust of the board had been set. Most of the money for Southern education was being channeled into the education of whites, and what little did go to blacks was put into the so-called industrial plan, not into higher education.

The issue was complicated. Should blacks be trained in the manual arts or should they be educated to become intellectuals? There was no easy answer. The problem of progress for blacks was socially and politically complex. Two opposing schools of thought grew up around the question and two blacks, Booker T. Washington and W. E. B. Du Bois, came to represent the split between the two.

Washington, a mulatto born into slavery in Virginia, showed real respect for the manual arts.[103] At an early age he developed a determination to lift himself out of oppression, to better his lot in the world. It was a difficult struggle, but he found what he considered to be the answer in his own diligence and in the cultivation of a submissive attitude toward whites. Later he broadened this into a kind of philosophy of racial behavior. Blacks, he asserted, should learn to improve their plight through self-help programs, through industrial and manual training, and through service to the black and white races. Washington had found success in this way, bettering himself by using his own hands and by avoiding racial conflict. Industrial education was the line he pushed.

Du Bois, also a mulatto but born a free black in Great Barrington, Massachusetts, was committed to intellectual pursuits. He had attended Fisk and Harvard, worked under leading American thinkers such as William James and George Santayana, written numerous sociological studies, and taught at black colleges in the South. In 1910 he became editor of *The Crisis*, a position he used to promote his ideas about how to improve the quality of black life. In his view, blacks could gain dignity only if they demanded the same opportunities, educational and professional, that whites enjoyed as a matter of course. There could be no compromises, no resignation to long-term, low-status situations such as those Washington was advocating. Intellectual development was the line he pushed.

Washington had far more followers than Du Bois, even though both men had access to large public presses. Washington's conciliatory attitude did not threaten the structure and stability of American society, so white Northerners and Southerners found it easier to accept. Philanthropists like Rosenwald and John D. Rockefeller, Jr. found his arguments persuasive, and his ideas and plans, for the most part, feasible. Blacks also tended to follow Washington rather than Du Bois, perhaps because the Tuskegee Institute head's approach seemed to promise more practical and immediate benefit. Most blacks were brought up on the Washington philosophy, hoping only for a decent job at a decent wage.

Just was no exception. His father and grandfather had been wharf-builders; his sister Inez became a nurse-practitioner, and his brother Hunter went "into service" as a chauffeur and part-time cook. His mother, a prime source of inspiration, had placed practical arts at the core of her school's curriculum. After giving him a start there, she had sent him on to South Carolina State College, where young blacks were trained in "the arts and sciences, . . . mechanism, agriculture, domestic economy, business, frugality" that would "fit them for the battle of everyday life."[104] Though his mother later encouraged him to continue his studies at Kimball Union Academy and then in college, and in so doing helped him become part of the "Talented Tenth" described by Du Bois, she saw to it that his earliest education was practical. Indeed, he kept up his cooking and cleaning skills; he was always careful to maintain his laboratory in perfect housekeeping order.

Washington would have been pleased by the appearance, if not by the purpose, of the zoology department at Howard. His idea of the biological sciences followed the "industrial plan." In 1896, when he was looking for practical men to help him build up Tuskegee, he managed to persuade George Washington Carver to come to Alabama from Iowa. Also a former slave, Carver had been trained in botany and agriculture at the Iowa State College of Agricultural and Mechanical Arts, where, as a faculty member in the mid-1890s, he had built greenhouses and developed soil programs. He committed himself to the Washington philosophy. At Tuskegee he achieved renown by creating hundreds of new food products from the peanut and the sweet potato. Newspapers nationwide nicknamed him "the peanut wizard," and in 1923 he was awarded the Spingarn Medal. The Spingarn award seems to have been the only connection between Carver and Just. They never met and never talked about each other. They were not linked as black scientists, except once, and then only very loosely, on a Pittsburgh radio program devoted to "notable colored men." Their work and outlook differed. Carver had a higher visibility than Just—partly because of the practical nature of his discoveries, partly because he was willing to conform to the image that was held of blacks. The white public at large felt more comfortable reading about a black scientist wearing dirty aprons and doing manual labor than about one dressed in elegant suits and preparing high-powered articles for serious scientific journals.[105]

Like most foundations and private philanthropists, the General Education Board supported Washington's industrial plan, as carried ᴐut at black institutions such as Hampton and Tuskegee. Blacks were to be trained to serve whites: black farmers for white plantations, black artisans for white factories, black cooks for white homes. The plan was to create a labor force, nonunion and only minimally educated, "never to vote nor strike and always to be happy."[106] Only in the late 1910s, and then gradually, did the board begin to shift its focus from the black artisan to the black scholar. Washington died in 1915, but his influence lingered and it took time before more liberal whites could exert direct influence on the board. Not until several old trustees died did the board begin to channel funds into higher Negro education —and then only very slowly and very carefully. For years the board was too cautious and skeptical about a place like Howard to establish a large endowment there, but by the mid-1920s it had become an important source of funds for Negro education, if not in the form of endowments then at least as help given on an individual basis. For instance, when the Rosenwald medical fellowships were discontinued in 1922, only two years after being set up, the board stepped in with $100,000 to keep the program running.[107] Blacks looked to Rosenwald first, the General Education Board second. It was natural for Just to turn his attention to the board when doubts arose about continued funding from Rosenwald. Perhaps, he thought, the board might offer a more permanent arrangement. There was no question that it was a bigger operation, with more money and a more experienced staff. If only he could get the staff to see reason, to set him up on a long-term basis; if only he could make them feel comfortable in doing so. It was worth a try, even though Abraham Flexner's uncompromising opinion that the board should not support "pieces of research or research workers"[108] had a dampening effect on Just's hopes.

❧

By 1925 Frank Lillie was as anxious as Just to arrangé some kind of long-term support for his protégé. He thought about setting Just up at Woods Hole, with Rosenwald footing the bill. But arranging something at Howard seemed a more attractive prospect, since racial tensions were building at Woods Hole and Just seemed none too comfortable there. Some decision on Just's future had to be made. Unless a plan of reasonable permanence could be arranged, Lillie was of the opinion that it would be better for Just "to make the transition at once to regular teaching duties" and forget about research, rather than have to do so later on, when conditions would be "even less favorable."[109] He was concerned that the attempt to "develop" Just only part of the way might not be right. Perhaps it would be better to forget the whole race than not to run the full distance.

Just was forty-two years old, an age when most scientists were long since settled. But now he was faced with a crucial turning point in his career. Should he continue? Would he get adequate support? He was grateful for Lillie's help in finding money, but he did not like being such "a nuisance."[110] Besides, Lillie could neither "do impossible stunts" nor "buck a machine." He was already doing more than anyone could expect. Heilbrunn and others interested in the situation should therefore "lay off" him, Just told Heilbrunn. Rather than pressure Lillie, they should join in and help. And, since home is where charity begins, Just decided to devise schemes of his own for securing more permanent support.

In August 1926 he handed in a report to the National Research Council on experiments for treating wood against shipworm.[111] It was an important practical problem. Even as a child he had known only too well that the damage done to wharves and wooden vessels each year cost millions of dollars. Now he hoped he had made a discovery that would protect wood from the shipworm *Teredo*. At Woods Hole he had treated wooden planks with copper sulphate and potassium ferrocyanide solutions to form a colloid gelatin coat or membrane of copper ferrocyanide. The ferrocyanide coat should, he hypothesized, withstand the attack of *Teredo*. At the beginning of the summer he put the planks in Eel Pond. After a couple of months the treated wood was free of the mollusc and the untreated pieces were infested. Just was careful not to push his conclusions too far, since two fundamental questions remained: first, would the method be effective on bigger pieces of wood? and second, would it be of large-scale commercial value? Whether he was convinced that he had actually come up with a money-making solution is uncertain, but he definitely wanted the council to look into the matter. Further, he assured them that he had no intention of profiting on the deal by taking the plan to a shipping corporation; rather, he wanted to give the council first chance. He had, after all, worked under its aegis, and his work was the council's property and, by extension, Rosenwald's. Just wanted the council to pursue his experiments by testing them on a larger scale to determine the commercial feasibility of the venture. After setting this straight, he saw his way clear to asking the council and Rosenwald, almost in the same breath, for permanent support. Nothing came of it, and the *Teredo* plan did not catch on either.

Just thought hard. Perhaps an abrupt change in career was in order; perhaps it would be better to "go in for something else" than to "prostitute" the deep love he felt for his work.[112] To him, research was "very dear, . . . a sort of pastime," but, unlike his white counterparts, he did not have to produce quantities and quantities of results in order to stay ahead of what was expected of him—from family, friends, and colleagues. His commitment was self-imposed. He could get out of it, and he might have to do so if he wanted to keep his place at Howard. Research at Howard was the exception rather than the rule, and it had put him "between the devil and the deep blue sea." Administration, on the other hand, offered a real opportunity to move onward and upward. He had been at Howard for nearly twenty years; he would have no problem getting a deanship and climbing the political ladder. In 1925 rumors that the school might choose a black as its next president filled the air, and the thought of filling that role himself must have crossed his mind. In any case, he could not leave Howard: "Just at Howard is not a problem, but Just wanting to get out from under might be an irritation." The only answer was to "adapt" himself as well as possible, one way or another, to the Howard environment. There seemed to be no point in rebelling against what people were constantly telling him: "You can never hope to get anything outside of what you have at Howard. No institution would have you, etc." There was no use being "sore" about the inevitable and unchangeable.

Just loved his research too much to drop it, however. He clung to the hope that there was some way for him to remain at Howard and continue the work to which he had dedicated himself for nearly twenty years. The problem was that he needed money, equipment, and staff— resources which scientists at white institutions had as a matter of course. For years there had been no assistants in his laboratory, and he had served as janitor, technician, typist, and experimenter all rolled into one. In 1920 there had been talk about hiring an assistant for him with funds from the General Education Board's appropriation to the medical school, but nothing came of it; in 1922 he had waited in vain for an assistant the administration had promised him a year earlier.[113] To make matters worse, he had never had any graduate students to help him carry on his work, none he could set to work on a particular problem as Lillie and Patten had done with him. The whole situation was discouraging, and he seemed not to be having any luck solving the financial aspects of the problem. It struck him, however, that there might be something he could do on a limited basis about the lack of support services and colleague relationships within the Howard community. To this end he began to devote some attention to training and encouraging a former undergraduate student, a young woman named Roger Arliner Young, in the field of biology.

℆

Born in 1899 in Clifton Forge, Virginia, Young grew up in Burgettstown, Pennsylvania, and entered Howard to begin an undergraduate course in music in the fall of 1916. Not until the spring of 1921 did she take her first science course. It was in general zoology under Just, and she received a C. In 1923, after six years of study, she finally obtained her bachelor's degree, with an undistinguished record overall and no strong preparation for work in the sciences. Aside from general zoology she had taken only vertebrate and invertebrate embryology. Her grade was B in both.[114]

It is not clear what led Just to encourage Young to pursue a career in science. There were rumors that they were involved in a love affair, but there is no evidence to support those early suspicions. Only later, around 1929, is there evidence of such a relationship. Just probably had another motive, a practical one, for bringing Young into the field. He often lamented the fact that men competent in zoology pursued medicine as a career rather than academic zoology. No one was to blame for this: the simple fact was that a man with scientific ability would have to be slightly off his head to battle for position and eke out a meager existence in the academic world when jobs and money were to be had so readily in medicine. Just was such a man, and his friends often referred to him as a "benighted idiot" who had thrown away the chance to earn some real money.[115] But Young was a woman and had little hope of a medical career, since woman's place, it was thought, was in the classroom if not in the home, and definitely not in the operating theater. Her one chance was to teach. Still, it was unusual for a woman, especially a black woman, to be involved in the teaching of science at the university level, instead of music or English or home economics. Young was in fact one of the first black women to go into science seriously.

After graduating from Howard, Young began teaching there as an instructor in zoology. She and Just worked well together, and he was so impressed that he tried, unsuccessfully as it turned out, to help her locate money to begin graduate studies. Somehow she saved the money out of her small salary to enter the University of Chicago on a part-time basis in 1924. For three summers she took courses there and in 1926 she obtained the master's degree. At Chicago her grades were so excellent that she was elected to Sigma Xi, the national science honors society— an unusual achievement indeed for a nondoctoral candidate. During the academic year she continued to work in the Howard zoology department as a faculty member and research assistant to Just. She was becoming more and more helpful to him, and her name began to turn up regularly in his funding proposals.

In 1927 Just and Young went to Woods Hole together, and she accompanied him there the following summer. He continued his research on fertilization, studying various marine forms with special reference to cortical changes as the basis of an approach to the kinetics of fertilization; she helped him perform experiments and examine cytological preparations. Together they studied the effects of ultraviolet rays on the eggs of *Nereis*, *Platynereis*, *Arbacia*, and *Chaetopterus*, a subject which Young was later to follow up in articles coauthored with L. V. Heilbrunn and Donald P. Costello. Also, it was about this time that Just was beginning to work on the problem of hydration and dehydration in living cells. He had a hunch that studying the mechanism of water loss and water retention in cells would result in an important breakthrough in cancer research. Together he and Young made a detailed analysis of the role of electrolytes in hydration and dehydration.[116]

Young helped Just along with the mechanical and some of the theoretical aspects of his work. She gained his respect for being "a skillful worker of merit," one who had shown "real genius in zoology" and whose work surpassed his own in "technical excellence."[117] Young was becoming a scientist in her own right. Her work had been "favorably commented upon" by well-known biologists, including L. L. Woodruff at Yale and C. M. Child at Chicago.[118] At one point she made quite an impact on American scientists when she brought out the results of a study two months before Dmitriy Nasonov, the eminent Russian cell physiologist who had been chosen to visit the United States as a Rockefeller International Fellow, published similar research. Young's study reexamined the accepted theory on the role of the contractile vacuole in *Paramecium* excretion. She had worked on it early in 1924 and, according to Just, had come up with "some interesting conclusions . . . and . . . hitherto undescribed results."[119] Her study received notice alongside Nasonov's in several European journals.[120]

At last Just had found an assistant who would not only help him but perhaps in time become a collaborator. The future seemed a little brighter, though the problem of long-term financial support still loomed large.

At least people were beginning to think about the problem. In 1925, when Rosenwald heard about the National Research Council's "permanency" resolution, he began right away to look into the question of setting Just up—at Howard—on a permanent basis. He thought of endowing a professorial chair that would allow Just unlimited tenure and free him from worries about salary and research money, and he even instructed Graves to ask officials at Howard and the University of Chicago about possible pension plans.[121] But before any such arrangement could be put into effect, serious fundamental issues had to be clarified.

Rosenwald wanted a brief résumé of Just's past achievements, a statement on his future work, and an evaluation of his research in terms of its "value to humanity"—all three documents to be written in layman's English.[122] He also wanted to determine whether Just's service to his race would be "greater in research work or as a full professor at Howard University." Perhaps Just would leave Howard altogether if he had a chance to develop his research plans, a risk that might not be worth taking if it was clear that the benefits of his teaching were as important to the black community as his research results. And finally, Rosenwald wanted to find out whether Just's attitude toward other blacks was "one of helpful association or aloofness." The question of Just's racial commitment would be crucial in the final decision on increasing support for Just's work.

The inquiries went to Lillie. Just was in an equally good position to answer most of them and was no doubt in a better position to answer those concerning his own attitude toward Howard and his people, but he could not, in Rosenwald's view, be relied on to respond candidly or objectively. After all, Just had a vested interest in painting a certain picture; Lillie presumably did not. Even though Lillie's response might not be as accurate as Just's, Rosenwald assumed that what it lacked in accuracy it would gain in detachment.

Lillie agreed to respond, but insisted that his testimony in this complex matter not be relied on exclusively. He took the questions in order. First he offered a "nontechnical" description and evaluation of Just's work—past, present, and future.[123] He described Just's work as being "largely on the subject of fertilization of the egg in animals, that is, the process of union of the maternal and paternal elements with which every individual life begins." The subject was "one of the most fundamental problems of biology" and an essential key to understanding "problems of heredity and eugenics"; it was "so complicated" that it was sure to be "a theme for study for many years to come." Jacques Loeb had established his "great reputation" by demonstrating "the possibility of substituting for one of the processes in fertilization the action of chemical substances," but Just's work extended and modified Loeb's results "in important ways." Just was in fact *the* current authority on fertilization. He and Lillie had coauthored a chapter on fertilization in E. V. Cowdry's *General Cytology,* and contrary to general opinion Just had done most of the work. At the moment he was in the process of writing a long general article for Buchner's monumental series on cell biology. In effect, he had "established a national and international reputation in the short period of five years," the period of his National Research Council fellowship. There was ample reason to think that, with support, this reputation would develop even further.

Lillie went on to analyze Just's work in terms of its "value to humanity." One of the greatest of all human problems, he said, was "how to deal with the quality and quantity of human population," and the research Just was doing lay "at the foundation of conscious efforts" that would have to be made in the future to control the problem. Just's work in fertilization, heredity, and developmental physiology was "as fundamental for social problems" as work on anatomy, physiology, and pathology was for medical problems. Even though Just had not pursued the applications of his work in allied fields, there was no doubt that in time others would.

In Lillie's mind, Just could contribute as much to his race through research as through teaching. While his research, unlike his teaching, did not involve direct contact with young, aspiring blacks, it provided an example for them of diligence, perseverance, and achievement. In any case, the dichotomy was false. Lillie pointed out that in no instance had Just asked to give up teaching entirely. The choice was not between research on the one hand and teaching on the other. Just wanted to *combine* research and teaching, to devote half of his time to each. He was a dedicated scientist and educator who entertained "no thought of leaving Howard permanently," who planned to continue "part-time teaching there as in the past five years." Any attempt on his part to secure research support, as with the Carnegie, did not imply a withdrawal from his teaching commitment to Howard.

The inquiry from Rosenwald that most surprised Lillie was no doubt the one about Just's racial attitudes. Lillie kept his response to this deliberately brief and pointed, stating that Just was "one of the most loyal men of any race," that his attitude toward blacks was "quite perfect," that he displayed no "aloofness" and often spoke of "his strong desire to continue to work for his own people." What more could be said?

Though the question might have appeared insensitive, Rosenwald had had to raise it to help himself in making the final decision on whether to support Just in a more substantial way. There were demands on his resources from all quarters, and he needed to make sure his donations reaped maximum benefits. If the donations were small he could make a decision on the basis of intuition or personal taste, but if they were large he had to think about the situation in greater depth. When it came to supporting individuals, in particular, he had to be careful to find out their plans, motives, and attitudes, to insure that they fitted into his overall scheme of social improvement.

But there was an inconsistency in all of this, one that was typical of funding agencies generally. The foundations were all too ready to suspect blacks of being elitist, separationist, disloyal to their people, but no such doubts arose when it came to supporting whites. Invariably, more delicate questions were asked about blacks than about whites—and the right answers were expected, indeed demanded. The Rockefeller Foundation gave large grants to the eugenicist Raymond Pearl, but no one expected him to identify with poor whites in Appalachia. Rosenwald supported the work of the philosopher Morris Cohen, but he did not inquire about Cohen's ethnic loyalty or suggest that he had to identify with unfortunate Jews living in East Side ghettos. The inconsistency angered commentators like Du Bois, who went so far as to suggest that it was a strategy used by funding agencies to prevent black scholars from getting substantial support and doing anything worthwhile.[124]

The fact is that intellectuals of any race must undergo a degree of separation from the masses; their work demands solitude and reflection, hence a certain withdrawal from active life. Black scholars were under added social and professional pressure to be different from their people. In most cases they had to deny their background and history. A good education, at a black or white institution, meant emulating the white value system. Kimball Union, Dartmouth, and even Howard were bastions of white America. Blacks who studied there tended to lose touch with their people, yet at the same time they did not gain acceptance by the white world. Isolation was thus their key to survival. They carried on their work in a vacuum, far from the black reality and outside the white mainstream. The frustrations of such a life were enormous, and there was no way to deal with them except to retreat even further into the ivory tower. In the end, many black scholars and other educated professionals became unproductive and apathetic, interested only in constructing an affluent little world for themselves, a phenomenon that the black sociologist E. Franklin Frazier was later to analyze in his 1957 work *The Black Bourgeoisie*.

Only in a certain sense can Just be labeled a "black bourgeois" in Frazier's use of the term. He was a member of the Episcopal Church; he enjoyed whist and chess, swimming, tennis, and golf; he employed domestic help; the Just family ate well and lived in comfort. But he was distinguishable from Frazier's clique in several fundamental ways. He worked hard, produced in his chosen field, and never overlooked the condition of blacks as he observed it in Washington and as he remembered it from the Carolina lowlands. Though these factors might seem independent of each other, Just consciously linked them. Science, he felt, had much to offer the black community. It could help instill in blacks an "objective and cold-blooded" attitude, an effective complement to blacks' robust *"joie de vivre."*[125] Like Loeb, Just thought science could be a savior of black people, their road to an improved life.

Contrary to what Rosenwald seems to have feared, Just was interested in black people, their heritage and history. Originally he had gone into biology as a way to enter anthropology, back in the Dartmouth days when he combined the natural and social sciences in his degree program. Though in the end he decided to remain in biology, he never lost interest in broad scientific questions such as racial origin and development.

A good example of this interest was his connection with Melville J. Herskovits. In the early 1920s, during a visit to Columbia University for informal research discussions with the T. H. Morgan group, he met Herskovits and became interested in his anthropological studies of American blacks. Herskovits was the star student of Franz Boas, the foremost anthropologist of his day. Just was not long in extending him an invitation to come to Howard as a visiting scholar, to observe the student body, interact with the black faculty, and in general experience life in a black community. Herskovits took advantage of the offer in the fall of 1924. Though he stayed at Howard for only one year, he developed a close friendship with Just and other Howard scholars, especially Alain Leroy Locke.[126] Several articles resulted from Herskovits's work at Howard and his famous book, *The Myth of the Negro Past* (1941), owes much to his time there. While Just may not, as he later boasted, have given Herskovits his "major thesis,"[127] the fact remains that he provided invaluable help and guidance to a young scholar whose work on black culture went on to achieve great renown.

By the mid-1920s Just had begun to be openly assertive about the problems and needs of blacks, especially those at Howard. His fate was intertwined with that of his colleagues, and he had nothing to gain by being, in Rosenwald's words, "aloof." Not that he was intimate with every faculty member, but he did have a few close friends and he kept up his ties with the ever-expanding Omega Psi Phi fraternity. Also, his relationship with Alain Locke was deepening. Locke had edited a book, *The New Negro* (1925), which was an important analysis of the Negro Renaissance and the new consciousness of black intellectuals. Harvard-educated and the first black American Rhodes Scholar, he was the kind of cultivated black man with whom Just felt comfortable. Locke's knowledge of black culture and history, his enthusiasm about the way blacks were shaking off "the psychology of imitation and implied inferiority" and achieving "spiritual emancipation," made an impression on Just.[128] Both men wanted a new leadership at Howard; they were looking forward to the time when a black man of strength and vision would be at the helm. Perhaps, they thought, the university could pull out of its moral decay under the direction of a black—one who would provide solid leadership and serve as an inspiring symbol, who would do something about the "debauchery," the "hopeless mess" that Durkee had made of Howard.[129] Their hopes seemed fulfilled when Mordecai Wyatt Johnson took over the reins of the presidency in the fall of 1926.

Johnson came to Howard from his post as minister of the First Baptist Church of Charleston, West Virginia. He was not the trustees' first choice. His reputation for being "too outspoken in his condemnation of social wrongs" made certain white members of the board apprehensive about the political consequences of electing him president.[130] The board's first choice was Bishop John Gregg, a devoted black missionary in charge of Congregationalist activities on the west coast of Africa. Gregg decided to remain in Africa, however, and there was no choice but to elect Johnson. The news was greeted with overwhelming joy by many blacks. Howard had at last found a black man to assume its highest position of leadership.

At first Just harbored reservations about Johnson, most likely because Johnson turned down some of Just's early recommendations for the zoology department. But Just changed his mind when he learned that Johnson had also turned down recommendations from other departments, in order to "study the situation impartially."[131] Just came to see Johnson's administrative decisions as sound and fair, his leadership ability strong. The Howard presidency was "a big job," given the "delicate situation" of black education at the college and professional school levels, but Johnson appeared to be equal to the task. He was, in Just's words, "a simple, straightforward man of tremendous sincerity and honesty," and there was not another "Negro in this land so free of pose and frill." If given a chance he would "do for Negro education" what Booker T. Washington had done earlier. His electrifying presence gave promise of an almost messianic ability to stop blacks from chasing "the false gods of show and extravagance," to make them "realize . . . opportunities," perhaps even to work out "the spiritual destiny of the Negro race."

Once he shed his initial skepticism, Just idolized Johnson for the next year or two. He became "utterly and passionately devoted" to the new president and felt as if he had been, in a sense, "redeemed."[132] Before 1926, Just claimed, he had lacked self-confidence, but he experienced a kind of rebirth when Johnson arrived in the fall of that year. Now Just felt himself to be "a purposeful man, inspired with an ideal"; his work took on "new color and light"; he had been transformed from "a groping scientific clerk and day laborer" into a researcher "more sure" of himself and "keener for the implications" of his work. It was almost as mystical as the James Island experiences of his childhood. His instincts told him that Johnson would prove to be a driving and dynamic leader. There was an element of egotism in his appreciation of the new president (as the outstanding scholar in the Howard community he received a certain amount of flattering attention from Johnson), but the primary reason for his glowing optimism was the sense that something new and wonderful was afoot for Howard and, by extension, for the black community.

Hopes ran high for Howard's first black president, and the black press expressed many of the same sentiments as Just. Johnson was seen as a kind of Moses, a leader who would deliver his people from the oppression of the Durkee regime. He had a heavy responsibility on his shoulders. If he was to pull Howard out of its financial crisis and end the period of low student and faculty morale under Durkee, he would need material support as well as hero worship. A relatively unknown administrator, he would need to be introduced to the numerous funding agencies before he could win their confidence and secure funds from them for the university. Just tried to help by bringing Johnson's name up to officials at national agencies, particularly the Rosenwald Fund and the General Education Board. He arranged a visit to Howard by the top board administrators, all of whom were impressed by Johnson's campaign (imminently successful) to secure a sizable annual appropriation from Congress.[133] Here at last, they must have thought, was a Howard president with the financial know-how to make good use of federal and private funding. For his own part, Just had something of a personal stake in helping the new administration. His career in science, after all, might depend on how successful Johnson was in solving Howard's fiscal and other problems.

There would be no overnight improvements at Howard, and Just knew it. The main problem for him was to maintain at least enough outside support to insure the continuance of his salary at the same rate; he had worked too long and too hard to give that up. In 1927 Rosenwald decided not to continue the original grant, much less to support Just on a long-term basis, and Just was in jeopardy of losing half of his income. Fortunately, the General Education Board took over the commitment, voting to give Just a five-year fellowship of $2,500 per year. H. J. Thorkelson, a high-ranking board official who had heard Just's presentations on ultraviolet and water at the Society of Zoologists' meeting in Philadelphia in December of 1926, pushed the measure through and set the stage for a more positive approach to Just and his problems. This was the first time that the board had given financial support to Just's work. Flexner was now essentially a lame duck director, only one year away from retirement, and new personnel at the New York office were beginning to take a fresh look at the situation. Before long, the board decided to give Just an additional appropriation of $1,700 for the purchase of apparatus to be used in studying the effect of ultraviolet rays on egg cells.[134]

Just was in no way put off when he heard about the termination of the Rosenwald grant. He wrote and thanked Rosenwald profusely for the seven years of support—support that had given him a chance to become a real scientist. He said what he felt and also what he thought Rosenwald wanted to hear:

No matter how important scientific work may be, we must evaluate work in terms of significance to human life and the deeper aspirations of the

human soul. I have wanted to stay on at Howard despite hardships
peculiar to this situation which are doubly hard on my kind of work
because I have felt that the little that I do might have some human
value. . . . For me, teaching is more than a profession—it is a sacred
heritage. . . . You will never know how much you have meant to me and
what an abiding light you and your life have meant to me. I do not want
to seem to be sentimental and I have no need to be. But as I write these
lines my heart is full. I do so want in some way to measure up to
something for what you have done not for me alone but for the Negro
race.

I hope that the fact that the General Education Board has taken over my
fellowship will not mean that we shall cut off this wonderful relationship.
It has meant more than money or chance to do scientific work. It has
meant an almost holy alliance—a thing of the spirit which I shall always
remember.[135]

His tone was mystical, his purpose strategic. Though deeply grateful
for Rosenwald's past support, Just was by no means abandoning the
hope that Rosenwald might provide more in the future. Before long, in
fact, he asked Rosenwald for a five-year grant of $1,000 per year to pay a
research assistant. Rosenwald committed himself to a three-year grant,
promising $500 for the first year, $650 the second, $750 the third.[136] The
plan was for Roger Arliner Young to use this money to support herself
at Woods Hole during the summer.

Around this time there was yet another change in Just's life. New
horizons were opening up for him on the international scene and he
was anxious to go abroad and meet his European colleagues. In Europe
his work was read with "interest" and considered to have "far-reaching
consequences."[137] Recently he had been elected to honorary societies in
France, appointed to the editorial boards of German journals, and
invited to conferences everywhere on the Continent. In 1927 he had a
chance to speak at international zoologists' meetings in Berlin, Budapest, and Rome. But he could not attend—there was no money. Rosenwald looked into the cost of the trip but decided not to foot the
bill.[138] And it was hard for Just to forget that earlier he had had to
decline, also for financial reasons, an invitation to spend a month or
two at the seaside laboratory of F. A. Potts, the Cambridge zoologist.[139]
Still, the desire to travel remained with Just, who kept on dreaming
about how his experimental work would take on new dimensions in a
place like Naples, Italy, or Plymouth, England. Woods Hole had
become a less desirable place to work, and the thing to do now was to
find another niche. But to travel abroad he would need much more
backing than ever before. Could he find that support anywhere?

In December 1927 Just was invited by officials of the Laura Spelman Rockefeller Memorial to attend an informal conference on "problems of contemporary negro life and of the field of interracial relations" in the Zeta Psi House at Yale University.[140] The conference was intended to be exclusive and confidential, with only a few high-level participants, such as "heads of negro educational institutions, leaders of negro national organizations, and white administrators of organizations working in the interracial field." On its roster were Edwin R. Embree, vice-president of the Rockefeller Foundation; John Hope, president of Morehouse College; James Weldon Johnson, black poet and critic; Mordecai Johnson, the new president of Howard; Major R. R. Moton, president of Tuskegee Institute; Arthur B. Spingarn, brother of Joel Spingarn and a long-time official of the NAACP; Anson Phelps Stokes, founder of the Phelps-Stokes Fund; James R. Angell, president of Yale; and Leonard Outhwaite, director of the Laura Spelman Rockefeller Memorial and organizer of the conference. It was definitely an elite affair, but its goals were kept modest. The aim was not to formulate "a unifed program, . . . resolutions or public pronouncements, or even . . . a plan for research," but simply to "take a look ahead at the probable and possible developments in the field of race relations over the next ten or twenty years."

Just attended the conference, his first dealing with these kinds of problems.[141] The experience meant a great deal to him, for it allowed him the chance to spend some time thinking deeply and talking about broad issues concerning blacks. He chaired some sessions and took part in others. At times he was noticeably silent, at times quite talkative.[142] There were certain subjects—economics, industry, business, legal status and citizenship, for instance—he stayed away from, but he had a great deal to say about black education. He regarded himself as an expert on the subject. His familiar attack on "sentimentality" and advocacy of the "cold-blooded scientific method" were heard over and over again. He expressed concern about getting people to understand the advantages of the scientific method, and he put himself forward as one of the few blacks who had become a master of that method. His long-term educational aim was to make other blacks see science as a means of advancing on merit alone, as a field unencumbered with subjective evaluations and therefore potentially more amenable than the humanities to equality of opportunity. First and foremost, Just felt that blacks could learn much about their own situation if they laid out facts as if on a dissecting table and examined them clinically, in microscopic detail. Then, and only then, could blacks hope to determine their future. Just took for granted the nonintellectual aspects of black life— particularly the family and the church. As far as he was concerned, the time had come to move beyond them.

Just mingled with the national leaders at the conference, discussing the broad issues and also hoping to arouse interest in his own work. He made a conscious effort to seek out the people who might be able to help him. In particular, he spent a great deal of time with Edwin R. Embree.

Making friends with Embree was a smart move on Just's part. Embree had close connections with many major philanthropists and foundations. When Just met him he was in the process of leaving the vice-presidency of the Rockefeller Foundation to take on the presidency of the Julius Rosenwald Fund. The fund was expanding operations and its assets had been dramatically increased by a gift from Rosenwald of twenty thousand shares of Sears, Roebuck stock, to be spent within twenty-five years of his death. Embree appeared to Rosenwald to be the kind of energetic and experienced administrator needed to head the revamped fund. He had run the Rockefeller Foundation's Division of Studies with "vision, persistence and interest" and had provided "very great service" as a senior administrator for the entire foundation.[143] Besides, by his own admission he was becoming a little tired of the foundation. He wanted to do something less "ultra-conventional," go somewhere that offered the chance to cover "a larger field of service."[144] The presidency of the Rosenwald Fund was the perfect opportunity for him.

Just probably knew all about this before attending the conference, or learned of it very soon after his arrival. Conscious of Embree's power and influence, he drew Embree into conversations that often lasted into the wee hours of the morning, telling the full story of his life and work, his experiences at Woods Hole and Howard, his current needs and future prospects. Embree seemed interested and sympathetic, and when the conference ended shortly before Christmas Just went away facing the New Year in a state of euphoria. There was now hope for his cause. The conference had provided him with a chance for a major breakthrough.

Just did not let up. Letters flowed to Embree reminding him of Just's predicament. In addition to requesting material support, Just asserted his need to be appreciated "*not* as a Negro who works in biology but as a biologist who is working in biology for his satisfaction and joy."[145] Because he was black, however, his work so far had been "looked upon by non-biologists as just the same sort of thing done by any other Negro in a Negro school." He could no longer stand for this. After years of silence he was now summoning up the courage to demand his due, "no more and no less than that which would be given any other worker on the same line." He was a "scientist," not simply a "Negro scientist." To be considered the greatest living black scientist was in his view not much of a compliment. He took no pride in being, in his own

words, "the only Negro doing work of preeminence in science to-
day."[146]

It was not long before Embree was echoing these sentiments. Even in
public speeches, before the NAACP for instance, he proclaimed that
"Dr. E. E. Just of Howard University is not a great Negro biologist, he
is a great biologist."[147] Embree felt nothing but respect for this black
man who had somehow defied the odds and built a name for himself in
the white-dominated world of science.

Early on, Embree advised Just that one of the best ways to further his
reputation would be to develop a strong set of disciples at Howard. Just
replied that he was not so sure, and that there were serious obstacles to
starting a graduate program at Howard. Embree thought this a selfish
attitude, and even accused Just of diverting blacks away from zoology
into allied fields.[148] That suspicion was not uncommon. Black zoologists were positive that Just discouraged blacks from pursuing graduate
work in zoology so that he alone could occupy the field.[149] Lillie and
Thorkelson did not attribute such base motives to Just, but they too
were concerned that he had not developed a graduate program to bring
blacks into the field to carry on his tradition.

Just tried to explain. He had indeed devoted his time to teaching
"the most elementary courses" in zoology, but always with the hope of
building "a really sound undergraduate department," never because he
wanted "to keep other Negroes from going into zoology."[150] There
was plenty of room for more black biologists. In his classes at Howard
he had taught men with high academic ability and research potential,
men who might have made first-rate biologists. The real problem was
that they could not find positions in biology and had gone instead into
medicine. Just furthered the medical careers of these men by helping
them obtain National Research Council fellowships, not because he
wanted to keep them out of biology but because he was genuinely
interested in their future as professionals.

A plan for a graduate program at Howard would be "premature,"
Just warned. Such a program would require well-trained faculty members and well-equipped laboratories and classrooms, none of which
was available at Howard. And then there was the money problem.
People did not seem to appreciate fully that the success of an academic
program, especially a serious one on the graduate level, depended on
the level of funding involved. In general Just felt "a little hurt,"
and annoyed perhaps, that black colleges and universities were expected
to produce wonders on tiny budgets. In 1925 he had asked Flexner
about the possibility of setting up a Rosenwald Institute of Zoology at
Howard with an accompanying endowed professorial chair, but was
turned down because Flexner did not consider substantial funding for
Howard a prudent investment. For years Just had tried hard, without

success, to win interest in and support for his work and for Howard, and he therefore found it unfair of Embree and others to accuse him of "lacking imagination" about how to upgrade Howard's science programs. His real trouble was that he "imagined" too much and got too little.

But despite his reservations, Just was willing to reconsider his position against graduate work now that people were showing more interest and seemed ready to provide some funds. A limited program would be good to start with. There was a great need for black teachers with at least the master's degree in zoology—teachers who would not become outstanding researchers, advancing the frontiers of scientific knowledge, but who would contribute in an important way to the development of black education. Just was willing, indeed happy, to attempt to satisfy that need. But without funding he could do nothing. Now was the perfect time for Embree to give Howard more money, to set up a substantial endowment for science in a black institution.

Just revived the idea he had mentioned to Flexner and worked tirelessly for three months to develop detailed plans for setting up a Rosenwald Institute of Zoology at Howard. He sent the proposal to Embree by way of Howard's President Johnson.[151] It was an elaborate eight-page report covering the categories of building, equipment, program, personnel, and funds. The overall cost was projected to be a minimum of $600,000, half for building construction, purchase and upkeep of equipment, and so on, and half to subsidize the research program and provide for emergencies, publications, trips to Naples and Woods Hole, and attendance at scientific conferences. Just stressed that there was no point in putting up a building and purchasing equipment without due consideration for long-term operational expenses.

The proposal called attention to the need for basic instructional equipment such as laboratory tables and chairs, lockers, demonstration tables, aquaria, animal rooms, microscopes, projection apparatus including a motion picture machine, incubation baths, paraffin embedding, model charts, museum specimens, life history mounts, microscope lamps, and full electrical outlets for heating devices. Also deemed necessary were facilities and apparatus for researchers and advanced graduate students: a library housing leading current biological journals in English, French, and German, and a laboratory equipped with microscopes, temperature-regulating devices, cold and constant temperature chambers, and various instruments for precise zoological measurement.

It was a practical proposal, requesting essentials routinely found at other major universities, but it also reflected Just's dreams about what kind of atmosphere a scientific organization should have. The building was to be designed and constructed in the manner "so characteristic of German universities," with a common room full of easy chairs, one or two lounges for light luncheons and teas, and a large fireplace around which the graduate students would gather daily for "cozy" chats about their work. Everyone would be part of one happy family, bound together into a working unit unlike anything that Just had ever experienced before, even at Woods Hole. They would come together in "perfect comradeship," not as "teachers and students" but as mutual "seekers after truth." The idea was to simulate "all the fine features of a monastery life or the life of explorers thrown together away from the world unto themselves." Just's own scientific career had demanded this kind of separateness, and he was of the opinion that results could only be achieved "by apostacy [sic] of the things of the outside world, by genuine feeling of kinship, by real blood brotherhood democracy." His scientific institute was utopian in concept, much like Francis Bacon's *New Atlantis*. He was searching for an intellectual comradeship within the black community, a sort of idealistic fraternity to be located on the little island of Howard University. He was not, however, so carried away as to forget the immediate fundamental needs of his department and laboratory.

Embree was not taken by Just's plan. Though very interested in Just's "brilliant work," he was not at all interested in building an institute at Howard.[152] This he told Rosenwald in no uncertain terms. In his view, the Rosenwald Fund would be ill-advised to erect "an elaborate institute" for Just's particular program. The men who followed Just as department heads might be committed to very different aspects of zoology, and besides, there was no assurance that they would even approach Just in ability. Both arguments implied an objection to the whole concept of endowment. Embree had a definite point of view, formed from many years of experience at the Rockefeller Foundation. The wisest procedure, he had always thought, was to make a great amount of work possible during the active and creative periods of a scientist's life, rather than to try to build a permanent institution around what might or might not turn out to be a long-term research program. He could see no reason to modify his position in Just's particular case, or to refine his assessment by further consultation with science adminstration experts such as Frank Lillie.

Embree felt that Just's proposal to the Rosenwald Fund should be handled in much the same way that the Rockefeller Foundation had dealt earlier with a similar proposal from Raymond Pearl, the Johns Hopkins eugenicist. At first foundation officials had discussed the possibility of creating a large permanent institute at Johns Hopkins, with Pearl as director. But in the end it was decided to furnish increased resources for Pearl and his department on the basis of a five-year appropriation, with the understanding that before the end of that period the foundation would assess the program and perhaps appropriate funds for another five-year term. The advantages here were twofold. On the one hand, the foundation would not have to tie up "large sums of money," and on the other, Pearl would not be encumbered with the "administrative routine that inevitably follows a special building and a permanent endowment."[153] While there could be no arguing with this in general terms, it was not quite fair of Embree to apply his principles rigidly in Just's case and make a parallel between Johns Hopkins and Howard. One was rich and white, the other poor and black. Johns Hopkins had nothing like the problems of Howard.

But at least the comparison allowed Just to be seen alongside a biologist with a budget approximately eight times the size of his own. Pearl operated on "about $50,000 a year," not including an equal amount for undergraduate instruction.[154] The fact was that at white universities, in general, zoology budgets were anywhere from four to ten times the amount allotted at Howard, even though their leading scientists were hardly four to ten times more distinguished than Just. The comparison made Embree see that it was appropriate for Just to be provided with a stipend of about $15,000 a year. This would be a significant increase in Just's current annual budget of about $6,000 or $7,000; it would make a real difference to him and the department.

The initial plan was for the Rosenwald Fund to share the cost with either the Rockefeller Foundation or the General Education Board, but this fell through; Rockefeller officials considered it unwise to deal "directly with an individual scientist."[155] Still, Embree worked hard throughout the summer and fall of 1928 preparing Just's case for the annual meeting of the Rosenwald Fund's trustees. Perhaps, he thought, the fund might assume full responsibility for the proposed grant. As president of the fund, he was anxious to move beyond the traditional functions and methods of foundation support. Most other American foundations were "too large and too formal" to attempt to help "individuals of exceptional promise," but there was reason to hope that the Rosenwald Fund would be "free and flexible" enough to provide the "unique service" of supporting "a few individuals during their creative periods."[156] Just was to be the test case. In Embree's mind, he had no claim to equal rank with giants like Pasteur, Koch, or Darwin, but he was certainly "a scientist of unusual accomplishment and promise, . . . an outstanding biologist regardless of color and unquestionably the leading Negro scientist of the world." Embree had been told this many times, mostly by Just himself. What he needed now to persuade the board was an objective appraisal of Just by a renowned scientist.

Ralph Lillie was asked. A physiologist and the brother of Frank Lillie, he had known Just at Woods Hole since the early 1910s and had recently expressed a special interest in Just's work on water loss in living cells. He was asked by Embree for a frank opinion of Just's work, that is, whether it was "really as good" as it was said to be and "first class . . . compared with any standard," or whether it was "simply unusual . . . for a Negro."[157] The old question of race still lingered, sure to color any letter of evaluation.

Ralph Lillie did not quite know how to respond. He knew Just to be "an exact, reliable and industrious observer in the field of experimental embryology, thorough and painstaking in his work, and unusually enthusiastic and persistent"; he considered Just's work on the physiology of early development and fertilization to be "a real contribution to . . . knowledge in this field"; and he appreciated Just as "a man of independence and calm judgment, of fine moral quality and idealism, of a good clear active intellect and essentially disinterested in his point of view."[158] The trouble was that in his own mind he could not reconcile this high opinion with Just's racial background. How could a black man merit such a recommendation? Ralph Lillie searched his mind to find the answer, to set the record straight. The result was somewhat clumsy, to say the least. According to him, Just did not have "the mental qualities and intellectual outlook of a negro, but rather of a white man," and was from all "appearance and other characteristics" a "mulatto . . . racially about three-fourths white." Though Ralph Lillie stated that the racial question was "immaterial to the question of continuing the support," it was one that he was unable to ignore. Finally, he suggested that Just should have a "position of security and standing in the institutions devoted to the education of colored people."

To Embree and the rest this was a glowing evaluation. Embree had the assurance he needed to lay the case before the fund's trustees at their annual meeting in November and push through a major resolution in favor of Just. The following motions were adopted:

> RESOLVED that the sum of Fifteen thousand dollars ($15,000) be and it is hereby appropriated for the Department of Biology of Howard University, under leadership of Dr. E. E. Just, for the year 1928–29.
>
> RESOLVED that the Julius Rosenwald Fund pledge itself to appropriate Fifteen thousand dollars ($15,000) during each of the four years of 1929 to 1933 for the Department of Biology of Howard University if Dr. Just continues to direct that department and to pursue scientific research and graduate instruction substantially in accordance with the program outlined on the statement above.
>
> RESOLVED that the sum of Five thousand dollars ($5,000) be and it is hereby appropriated, of which so much as may be needed shall be used in the purchase of scientific equipment, including books and journals, for Dr. Just's department during the five-year period, 1928–33.[159]

At last Just had managed to secure substantial backing, $80,000 over a five-year period—something no other black in science had ever done.

His persistence had paid off. For at least five years, perhaps even longer, he would have no financial worries. The grant was a major event, calling nationwide attention to him as a black scientist. The publicity pleased him, but nothing made him so happy as the knowledge that he could now make plans to fulfill his dream of traveling and working in Europe.

As he arrived in Woods Hole on that snowy evening in December 1928, Just felt overwhelmed by a quasi-religious feeling. He knew he was approaching a great divide. He had been caught in the middle all his life—caught between an ancestry that traced back to Germany and one that traced back to Africa, between his love for South Carolina and his ambition to move beyond it, between his duty to teach and his desire to do research, his loyalty to the race and his responsibility to himself. There at Woods Hole, as he wandered into the sacred laboratory of his mentor Lillie and nostalgically reflected on the past, he was again in between, on his way from Howard in Washington to the Stazione Zoologica in Naples. As he packed slides to take with him on the trip, and put other bits and pieces together, he wondered where his life would go now that he had the means to do what he wanted. For the present there was no way to sort out the choices and tensions. He would have to wait, and judge by experience. A host of possibilities flashed through his mind as he gazed out the window across the snow-covered dunes that lead to the sea.

Mary Mathews Just
ca. 1882

Charles Fraser Just
ca. 1882

Family Portrait
left to right: Inez, Ernest, Mary, and Hunter
ca. 1890

Editorial Board of the Kimball Union Academy newspaper
1903

E.E. Just at Dartmouth
ca. 1906

E. E. Just
ca. 1920

Ethel Highwarden
1912

E.E. Just
1912

Ethel and their three children
left to right: Highwarden, Maribel, and Margaret
ca. 1925

Margaret Just
ca. 1929

Franz Schrader and Sally Hughes-Schrader
ca. 1925

L.V. Heilbrunn
1928

Frank Rattray Lillie
ca. 1929

Just pitching horseshoes,
Donald Lancefield in background
ca. 1923

E.E. Just and Franz Schrader,
on board the *Cayadetta*
ca. 1927

Reinhard Dohrn
ca. 1928

Roger Arliner Young,
Just's student and assistant
ca. 1928

Main entrance to Harnack Haus
ca. 1931

Hedwig,
overlooking valley below
ca. 1934

Hedwig, picking quinces
ca. 1935

E.E. Just
ca. 1936

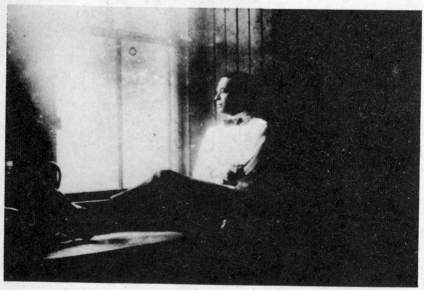

E.E. Just,
stretched out on window-seat in Marie Hinrich's room
ca. 1927

PART II

CHAPTER 5

Europe: First Encounters, 1929-31

Europe was to be a new world for Just. As he boarded the *Dresden* in New York Harbor one cold evening in early January 1929, he had a sense of fresh beginnings. The passage over would be the start of a new era for him, and his life—past, present, and future—would come into view from a different perspective. Once he had settled into his cabin, he began to feel more relaxed. He did not know exactly what awaited him abroad, but still he was looking forward to the experience. The trip took about a week, and he spent it mulling over what he had done and what he might do. He kept to himself on board, wrapped up in his thoughts. His sixteen-year-old daughter Margaret was with him, but he did not spend much time with her. Margaret occupied herself as best she could, knowing that her father wanted to be left alone when he was thinking.

So far Just's life had been busy, perhaps too busy. He was tangled in a web of activity, and the more he worked the more entangled he became. It was as if he were a pawn in some larger scheme. Though he had been moving all the time, he had been unable to do much about the direction in which he was going. His struggle for a career left him no leisure to reflect, to put the past and present together in his mind and use that to try and shape the future. He had been willing to accept, perhaps with too little thought, his strange experiences as a child in Maryville, his anxieties as a young man at Dartmouth, and his frustrations as a black scientist in a white world. Now at midlife, a little late perhaps, this was all changing: he was becoming more contemplative, turning inward to see if he could find any pattern in his life, to try and summon up all his spiritual resources in preparation for what might turn out to be a new life altogether. Questions, questions, and more questions flashed through his mind. But he knew he would have to wait some time for the answers.

When the *Dresden* arrived at Cherbourg, Just and Margaret traveled immediately to Paris. There they registered at the Hotel Cambon, right in the center of things, a stone's throw from the Jardin des Tuileries, and spent the next week taking in the sights. The city was even more than Just had hoped. It was alive with crowds of people who seemed happy and carefree, unaffected by the damp fogs rolling in off the Seine. Just and his daughter were captivated by this new atmosphere, so different from the more businesslike way of life they were familiar with in Washington. Their days were spent soaking it all in, strolling through the city from the Right Bank of the Seine to the Left and back again. They could not enjoy the sidewalk cafés, which were closed for the winter, but they stopped often for coffee and for wine and cheese in restaurants along the Champs Elysées, always sitting near a window and gazing out at the people milling around in the street. Shopping was a temptation they withstood for the most part, but they were unable to resist buying an ounce of some exotic scent or other at Lubin's, an exclusive and expensive perfumery on the Rue Royale.[1]

Occasionally Just would leave Margaret at the hotel and go off by himself to visit churches, museums, and scientific institutions. Montmartre and Sacre Coeur impressed him, and he was truly overwhelmed by Saint-Sulpice and the grandest cathedral of them all, Notre Dame. Going through the Musée de Cluny took him back to the time in college when he had studied history, and he thought how much more fascinating it was to see the actual relics and documents than to read about them in books. He did not really enjoy the Louvre—there was simply too much to see and too little time to see it—but he spent several hours at the smaller Jeu de Paume, where a new display of pictures by Monet and Manet had just been put up. The scientific menageries, especially the Trocadéro and Jardin des Plantes, contained more specimens and seemed to be better arranged than any he had ever seen in America. He had visited the Smithsonian Institution in Washington many times, but it simply could not measure up to its French counterparts. The Smithsonian was not much older than Just himself, whereas the Jardin des Plantes had been around for nearly two centuries. Paris made him reflect on how young and inexperienced America was compared to Europe. He became more and more conscious of the venerable grandeur of the Old World as he ambled along Avenue Louis Pasteur, wandered through the halls of the Sorbonne, and gazed from the Panthéon to the Eiffel Tower.

What continued to strike him most, though, was the liveliness of the city and its people. Wherever he went, things were happening—in the restaurants, hotels, theaters, and concert halls. For the moment he was content to observe rather than take part, for he was not entirely comfortable. The place was new to him, and the language was unfamiliar (apparently Margaret did most of the talking). Furthermore, even though he no doubt had read Du Bois's many editorials in *The Crisis* on "the lack of race feeling" among the French,[2] he was nonetheless a bit wary about putting himself in situations that might cause the kind of disturbance and embarrassment they would definitely have provoked in America. One day he passed by L'Opéra and heard the chorus rehearsing the finale of the last act of Mozart's *Die Zauberflöte*. He thought about buying tickets for the evening performance, but in the end decided not to. Never in his life had he seen an opera. He would have liked to go many times—but there were few theaters in Washington where he was welcome, and the only way he could listen to singers like Lily Pons and Ezio Pinza was on the Sunday afternoon radio broadcasts. Now in Paris he had the chance to *see* a live opera, to feel the excitement of costumes and sets and dancers, but he was still too ill at ease to actually do so.[3]

The week passed by quickly, and on or about 18 January Just and Margaret headed for Naples. What route they took is unclear, but it is likely that they chose the popular southern one through Dijon, Lyon, Marseilles, and Nice, and on through Genoa into Rome. Though perhaps more picturesque, the alternate Alpine route was shorter and did not offer the same opportunity for sight-seeing stopovers between connections. Just and Margaret spent the night in at least one of the southern French cities, probably Marseilles or Nice, and again in Rome when they arrived on 22 January.[4] The next morning they boarded a local train, and at about noon they had their first look at Naples—a colorful sight, with the bay sparkling in the background and Mount Vesuvius looming in the distance.

Reinhard Dohrn, director of the Stazione Zoologica, was on the platform to meet them. It was an unusual gesture—even for Dohrn, who was known for his warmth and hospitality. But the fact is that Just was a special case: a black scientist, the first ever to work at the Stazione, with an international reputation and glowing testimonials from top American science administrators such as Frank Lillie and M. H. Jacobs.[5] Things had to go just right, so Dohrn, even though he was fighting the flu, took the afternoon off from his duties at the Stazione to make sure that Just and Margaret got settled properly. The plan was for them to take rooms with the Baronessa Mascitelli, an eccentric aristocrat who had turned her mansion into a kind of high-class inn with reasonable rates for visiting diplomats and scientists. Heilbrunn, who stayed there the summer before, had written the baronessa in mid-December suggesting the arrangement.[6] The racial question caused some apprehension on his part but it did not bother the baronessa, who had been unfazed the previous year when a dark-skinned Egyptian came to visit, and there was no trouble getting reservations for Just and Margaret at her *pension*. Dohrn drove them over personally, and took the opportunity to show them a little of the city and its environs.

Naples was everything Just had heard and imagined. There were the narrow streets, occasionally as steep as staircases, winding up and down hills; the moldering medieval fortresses and cathedrals such as the Castel Nuova and the Chiesa di San Fernando; the quaint little outdoor shops and cafés, open on mild days even in winter; and, most of all, the people, with their animated discussions and bustling activities. But what was also striking was the poverty, something the American tourist is never quite prepared for. The evidence of filth and disease was everywhere, and the people, especially the hawkers who eked out a living by selling their wares on the sidewalks, appeared emaciated, overworked, and malnourished. Just reacted strongly to this curious mixture of the picturesque and the squalid:

> The babel of sounds, supposed to be music, issuing from the two sets of musicians of the rivalling cafés . . . thrust through my consciousness.
>   . . . The stench of water, the unkempt boats, the odors from the homes of basket-weavers, the barefoot female vendors of vegetables, the naked children sporting in the glare of the noon-day sun, the pool, the Castel Nuova beyond—all were a background of thoughts and feelings not easily gauged, never completely assayed. . . . The streets had a charm, but they were also like an endless maze of latrines.[7]

Flashbacks from his childhood came to him—the damp, dirty alleys off Inspection Street in Charleston, and the impoverished farmers and laborers on James Island. His memories there were of black neighborhoods and black people: never before had he seen whites living under such conditions.

Naples was a new experience altogether, one that would take some getting used to. Not so the Stazione Zoologica. Just felt completely at home there from the very beginning. The Stazione had legendary meaning for him. Its founder, Anton Dohrn, Reinhard's father, was one of his heroes in biology, along with Charles Darwin, T. H. Huxley, and Claude Bernard. The elder Dohrn was a nineteenth-century German zoologist in the Darwinian mold, traveling constantly and doing fundamental work on the origin of vertebrates. One of his most ambitious projects was to show that man had been derived ultimately from the *Annelida*, the group of wormlike organisms, including *Nereis* and *Platynereis*, that Just was later to take an interest in. On a trip to Messina, Sicily, in 1868, Dohrn got some of his best results by using portable aquariums to breed and observe the animals, and this gave him the idea of establishing a permanent laboratory. From a climatic point of view, southern Italy seemed to be the perfect site. Dohrn settled in Naples around 1870, and started planning his new venture. He confronted many problems, not the least of which was persuading city authorities to allow him to build the marine station in the Villa Reale, the most beautiful park in Naples. Once this difficulty had been overcome, there was the financial problem. But with the moral backing of scientists such as Charles Darwin, Dohrn managed finally to attract donations from various sources—including his father, Carl August Dohrn, one of the wealthiest and most socially prominent men in Germany. The Stazione Zoologica opened in February 1874.

Anton Dohrn was a flamboyant and colorful figure, always in some scrape or other, fighting personal and political battles and at one point nearly engaging in an old-fashioned duel with a city architect.[8] But despite the odds, and because of his energy, the Stazione Zoologica grew in size and prestige for over thirty years. By the turn of the century scientists were going there from all parts of the globe, to occupy "research tables" sponsored through endowment or rental by governments, universities, and foundations. When Dohrn died in 1909 the loss was deeply felt in the scientific community worldwide. Reinhard took over the directorship immediately and the Stazione continued to prosper, at least for a while. Then, in May 1915, at the height of the First World War, the Italian government took over the laboratory's administration; the Dohrns had always been perceived as German expatriates whose first loyalty was to the Vaterland. The Stazione remained open but its attendance declined, as much because of the incompetence of the new administration as because of the war. Even in peacetime things did not improve. The Italian government tried without success to locate a good director. Dohrn was out of the question, because the Italian academic establishment remained fiercely anti-German for some time after the war. He was finally reinstated in 1924, when the situation at the Stazione hit its lowest level. The Dohrn tradition of administrative excellence won out in the end despite, as one scientist put it, the "odds of stupidity" created by the Italians.[9] By 1925 foreign scientists were again going to the Stazione. Just was part of this revival, and so proud to be there that he offered to make a personal financial contribution as a way of showing his "high regard" for what the Dohrns had done for science on an international scale.[10]

Things got off to a good start. Dohrn, always concerned and helpful, had arranged to place Margaret in a special school for foreign students, and on the first day he even escorted her to class. At the Stazione itself, arrangements for laboratory space and equipment and introductions to the staff were ably handled by Dohrn's secretary, Margret Boveri. Almost immediately Just was working away on *Nereis dumerilii*—sending out orders for specimens, and sometimes going on the boat himself as he had done at Woods Hole; bending over his microscopes and slides for hours on end; writing up reams of notes on ideas and experiments. Everything seemed to go just right—"beautifully!!" in fact—from the moment he set foot in the Stazione.[11]

Outside his laboratory room, in the lobbies and lounges, Just met scientists from everywhere in Europe, and was struck by how much interest they showed not only in his work but in him. Compared to the MBL, where he felt he had often been treated more like a walking encyclopedia of scientific knowledge than a living, feeling individual, it was indeed an unusual experience. European scientists seemed to him to be "more human" than American scientists, to live "a more natural sort of life . . . gentle without being soft."[12] They were just as productive, but they produced with "less sound and fury" over their own research projects and more concern for each other as people. Also, their interests extended far beyond science—into cultural activities such as music and painting. During coffee breaks from their experiments, they would talk about the new Richard Strauss symphony or the latest canvas by Picasso. The tone was never glib, always earnest and sensitive.

On Sunday afternoons, Dohrn, whose family had been intimate with musicians such as Felix Mendelssohn and Jenny Lind, hosted concerts by professional groups in the main lobby. These were always well attended and warmly appreciated by the scientists-in-residence, and afterwards there was sure to be talk and more talk about art, life, and the human condition. The first concert Just heard at the Stazione was by the Busch Quartet, one of the top touring chamber-music groups in the world. The program included works by Mozart, Beethoven, and Brahms.[13] Just was charmed. The closest he had come to "good" music at the MBL, after all, was the crude choral get-togethers on the wharf and around the campfire. Understandably, the refinement and sophistication of the scientists at the Stazione, coupled with their genuine human concern, seemed truly wonderful.

Another aspect of life at the Stazione may have seemed a little out of the ordinary to Just at first, namely, the fact that it was run somewhat differently from the MBL—in a feudal rather than democratic fashion. A blue blood of sorts, Dohrn played the role of gracious host to the visiting scientists and of benevolent lord and master to the employees. There was no questioning his wisdom or judgment: what he said, went. True, Lillie had his own sort of aristocratic connections, and some believed he held his position at the MBL because of these, but he nonetheless encouraged a democratic kind of participation unheard of at the Stazione. In effect Dohrn ruled like a king, Lillie governed like a president. The contrast was marked, and no doubt struck Just as a reflection of the continuing tension between the Old World and the New.

The other differences Just knew about already; indeed, they were part of the reason he had come to Naples in the first place. Chief among them was the chance for relief from the crowd of students and young investigators that had plagued him at the MBL, forever asking elementary questions and generally making nuisances of themselves; and from the need to wait through the long, cold winters for the marine fauna to reestablish itself in the coastal waters off Woods Hole. At the Stazione it was all research and no teaching, and scientists could work there year-round.

Just was happy. The pressures and anxieties of life in Washington and Woods Hole seemed, momentarily at least, a thing of the distant past. In Naples everyone was kind and his research went like clockwork. Just felt so much at home that he began to strike up some close friendships, for example with John Runnström and Curt Stern, visiting researchers (from Stockholm and Berlin respectively) who spent a great deal of time confiding their personal and professional problems to him.[14] Also, the women scientists were captivated by his charm and good looks and flocked around their "black Apollo," never willing to let him out of their sight.[15] Prominent painters even urged him to come to their studios for portrait sittings.[16] Indeed, Just found the Europeans to be open and intimate and outgoing with him in a way that people at Woods Hole, even the Schraders and Lancefields, had never really been. The point was brought home with full, perhaps exaggerated force one day when an American couple deliberately snubbed him in the dining room at the Stazione.[17]

A relationship of special intimacy developed between Just and Margret Boveri. He was twenty years older than she, but that did not seem to matter. Both were new to Naples, both were in some sense escaping difficult situations elsewhere. Boveri had been studying history in Munich and doing quite well, but in the summer of 1928 she fell into a depression that made her feel dissatisfied with herself and fed up with life in general. A change, she decided, was very much in order. The Boveris were well-to-do financially—partners in the Brown-Boveri Company, a large industrial corporation with branches in Germany and Switzerland—so she would have had little trouble going wherever she pleased for as long as she liked. But her branch of the family was intellectual rather than entrepreneurial, and though she was tired of her history studies she still wanted to do something constructive rather than fritter her time away in some resort town. To her it seemed that the Stazione Zoologica would offer a perfect change of scene, new friends, and stimulating activity. She had known the Dohrns for years, ever since her father, Theodor Boveri, the noted embryologist, used to take the family to Naples for months at a time. With no difficulty she persuaded Reinhard to give her a position as his secretary. Even then things did not improve, however. In Naples she became suicidal, and at one point seriously considered taking the final plunge into the ocean from the rocky cliffs on the island of Capri.[18] Her only hope was to find a friend she could trust. Almost like a miracle, Just appeared on the scene.

At first it was nothing more than a business relationship—Just dropping into Boveri's office to ask a question about some administrative detail or other, Boveri giving the help any efficient secretary would. But soon they got to talking about this and that, and found they had a great deal in common. There was their interest in music and the theater, their enthusiasm for intellectual pursuits, the love of travel, and, last but not least, the unhappiness associated with the past. Not surprisingly, the relationship blossomed into something deep and intense—with more than a hint of the romantic about it.

More often than not, after a hard day's work at the lab, Just and Boveri would attend a concert or play and spend the rest of the evening talking and strolling around the city. There would also be weekend excursions—exploring Pompeii and the Herculaneum or the early Greek settlements around Cama and Baia, relaxing in the peace and quiet of the islands off the coast. Easter at the Dohrns' country home in Porte d'Ischia was particularly memorable, because it was then that they first really confided in each other. Just told how the poverty in southern Italy reminded him of his childhood in the South Carolina lowlands, how the atmosphere at the Stazione contrasted so favorably with that at the MBL, how the people in Europe had warmed his heart. Boveri listened attentively, and spoke with feeling about her own memories and her own problems. The rapport between them was deepening; in fact they were becoming spellbound by each other.

Just was careful not to allow distractions, romantic and otherwise, to sway him from his main purpose: getting some serious work done and establishing some solid contacts in the European scientific community. By mid-April he had done enough in the laboratory—more than he was able to do in several summers at Woods Hole—and it was time to travel a little and meet scientists at institutions other than the Stazione. He had a long-standing invitation to visit Friedl Weber, professor of botany at the University of Graz in southeastern Austria. Though he was not in the same field as Just, Weber was the managing editor of *Protoplasma*, a general biology journal to which Just contributed regularly. Because Just planned to spend two weeks in Graz, he had to forgo several other invitations, including one from the zoology department at the University of Cambridge and one from the Oceanographic Institute in Monaco,[19] so that he could get back to Naples and wrap things up there before leaving for America by the end of June.

Travel arrangements were settled quickly. Just had fallen in love with the Mediterranean, and he decided to go to Graz the long way round, by ship to Bremen, instead of taking the direct Alpine route through northern Italy. On 23 April he and his daughter Margaret boarded the *Stuttgart* for the two-day voyage around Gibraltar and into the North Sea. Like their Atlantic crossing it was an uneventful trip. Just again spent the time sorting things through in his mind. Stretched out on deck in the balmy spring air, he composed an ode to the beauty of the world around him:

Beyond the slowly moving purple sea
Diamond dusted in the sun
Hills now vaguely dimmed as mantled in the mist
They rise and softly etch the sky.

Standing off the coast of Spain
All my sails are spread.
I find the hidden lane
And every fear is sped.
The purple sea is home for me—
Harbors are my bane.

Laughing to the rising sun
Crooning moon at night
I dare my course to run
Each wave I love—a sprite.
O, purple sea I love but thee
Anchors are a blight.

I would run on for aye
Over this purple sea—
Crooning night, laughing day
Joy in the heart of me.
O, wine-dark sea, I hold to thee—
Take me utterly.[20]

The sea helped him think, and there was so much to think about. Why
did he feel so at home in this foreign land? How had he made so many
"beautifully fine friends" and done so much "*wonderfully . . .* good
work" in such a short time?[21] Would he, when the time came, have
second thoughts about going back to America? Only time (and more
thinking) would tell.

The train ride from Bremen to Graz was breathtaking. Just and
Margaret drank it all in—from the industrial smokestacks of Hanover
and Brunswick to the spectacular forests of Bavaria to the charming
hamlets and chalets along the Danube River. In Graz they spent a week
with the Webers, enjoying long walks through the woodlands and
meadows at the foot of the Alps. Just tried out his German with little
success; if it had not been for Heilbrunn's sister (who was also visiting
the Webers), the trip might not have turned out as well as it did.
Margaret knew French and some Italian, but Just had to rely on
Heilbrunn's sister to interpret for him in the Austrian shops and
restaurants. Like most educated Europeans (except the French), the
Webers spoke English with ease. Conversation at the dinner table was
always in English, as were the scientific discussions Just held with
Professor Weber and his colleagues at the university. Quite naturally,
however, Just was beginning to wish he could speak the language of

these people among whom he felt so much at home. Before leaving Graz he vowed to the Webers that the next time they saw him he would speak to them in French, German, or Italian—but definitely not in English.[22]

Back in Bremen, Just felt a touch of sadness. Margaret was scheduled to take the next boat home, and Just was about to lose his traveling companion. They had not really seen much of each other, especially in Naples, where Margaret was in school and Just in the laboratory all day. But because they had always been close as father and daughter, there was an unspoken understanding between them that they would help each other out if anything untoward happened so far from home. Margaret was sorry to leave, and Just was sorry to see her go. There was also some concern on his part about her traveling alone, but he felt better when the ship's first officer promised to look out for her, and Heilbrunn wrote that he would meet her in New York and put her on the train for Washington. On 9 May father and daughter said their fond farewells.

In a way, Just felt more relieved than sad. Something had happened to him—spiritually and romantically—and he was nervous about Margaret noticing the change, especially since he could not quite put his finger on it himself. With Margaret gone he was now free to explore and reveal himself as he saw fit. There was no more need for paternal decorum; he could say and do what he wanted without constraint. In his room that night, at the Hotel Europäischer Hof, he scribbled down his thoughts about love and the soul:

> Each heart in all this world of human things a plaything is: the gods above . . . make sport at will—discords stark arise and earth, erstwhile paradise, becomes a noisome place. But somewhere in this world of human things each heart its mate doth own. No mad desire, no fierce onslaught . . . can mar these twain if once they beat at one. Then . . . sterile earth a paradise becomes.[23]

This was rather vague and tentative, but it signaled the start of a more open and honest relationship with Margret Boveri. Now they could be more than close friends, they could be lovers.

Underneath, Just had felt a little cramped by Margaret's presence in a more general way. She seemed to represent a link to a world he wanted to forget, at least for the time being. He was captivated by Europe, the place and its people, and the more he thought about it the less he wanted to be reminded of America—a home that did not really feel like home. So, unconsciously or not, with Margaret out of the way he suddenly felt like abandoning all inhibitions and opening up his heart and mind to the experiences he had had for the last four months.

Just held deep affection for Dohrn, a lovable soul, cultivated in the best Old World sense and good almost to a fault. He dashed off an impassioned letter to Dohrn that he knew would probably not reach Naples before he did:

> I have an impulse to write you which I cannot subdue. . . . For myself, it is difficult easily to express what is in my heart. But this difficulty is lessened by the feeling that somehow you know how I feel toward you. I very much fear that from the first moment that I saw you I fell desperately in love with you—a feeling that has given me great satisfaction and happiness. More, perhaps, from a selfish point of view, has been the courage which you have given me and a sort of *elation* that persists and will persist. I know now after having known you what the French mean by *élan vital.*[24]

A close bond had grown between Just and Dohrn. It had begun with a kind of physical intimacy, from the first moment they saw each other at the train station. Though Dohrn was not as striking as Just in appearance, he had an attractive warmth and vitality about him, and like many Europeans he was demonstrative in his affections, always kissing and embracing. Before long the relationship had deepened, extending to the metaphysical. Thanks to Dohrn, Just began to think of his work in an intensely philosophical way, to sense that his "little inconsequential finger pokings at these problems of nature" were becoming transformed into a "peculiar . . . love for things human (and *earthy*) and a transcendentalist outlook, . . . a beautiful mixture of reality and idealism."[25] He came to revere Dohrn as a symbol of Europe—the wonderful atmosphere, the glorious experience that had enabled him to find "a more soul-satisfying attitude toward life" than he had ever dreamt of in America. He was opening up to the world and to himself, probing deep feelings that he had never been conscious of before.

Time was growing short. Just had slightly more than a month left at Naples to finish up the work he had come over to do. Actually, all that was left was for him to get his notes in order and make one or two final observations. The major experiments were completed, and the data he had gathered appeared to prove his hypothesis that the European sea worm *Nereis dumerilii* and the American sea worm *Platynereis megalops* were not one and the same species. He had held this hypothesis in opposition to one or two Woods Hole scientists ever since a study he had done in the early 1910s showed that the breeding habits of *Platynereis megalops* differed markedly from those described by Friedrich Hempelmann (1911) for *Nereis dumerilii*. After finding further differences at the Stazione he was eager to quickly put out a paper showing that these

European and American species were indeed distinguishable in more than name alone.[26]

Two other groups of experiments had also gone well. First there was the fertilization work, in which Just repeated his Woods Hole experiments on *Echinarachnius* and *Arbacia* with two European sea urchins, *Paracentrotus lividus* and *Echinus microtuberculatus*.[27] The purpose was to bolster Lillie's fertilizin theory, which had been widely criticized as lacking general significance because only American species were being tested. Just was pleased to find, after much eye-straining observation of membrane separation and the action of fluids upon the egg and sperm, that fertilization capacity and the level of the glutinous substance Lillie called fertilizin were directly proportional to each other in certain European species as well. Second, there was the morphological work—dissecting, embedding, diagramming to reveal form and structure—on the worm-like chordate *Amphioxus*, found in abundance underneath the wharves in front of the Stazione. Just enjoyed doing this in his spare time, as a relief from the taxing microscopic work on fertilization. No doubt he also liked showing off his techniques, the fame of which had traveled from Woods Hole. The *Amphioxus* work received such constant and favorable attention, in fact, that Dohrn joked about setting up and conferring on him the Stazione's special award, the Order of the Golden Starfish.[28]

The thought of going back to America did not appeal to Just at all. There were still many things for him to do and many places for him to see in Europe, and he did not feel as if he had even scratched the surface. Could he be bold enough, he asked himself, to do what he wanted and stay for the entire summer? Or should he feel obligated to sail for America? His family was expecting him at the end of June, the Woods Hole crowd was hoping to see him in July, and the Howard administration was planning for him to do some preliminary work for the newly founded Committee on Graduate Instruction in August. He would have been willing, if not happy, to follow his conscience and go home, except that the pressures and hostilities he knew he would face there made him feel anxious and upset. Not knowing what to do, he put the temptation to stay in Europe as far back in his mind as possible and made plans to return to America. The only change was that he would forgo Woods Hole and spend the summer in Washington.

But it turned out that it was not that easy to leave Europe behind, and the attempt to do so took a toll on Just's health. His old sinus trouble flared up again, in much the same way it had when he pushed himself in the fall of 1920 to get his paper ready for the conference at which he attacked Loeb's work on fertilization. The pain got worse and worse. Boveri talked him into going for a checkup with an ear-nose-and-throat specialist, but he shook so violently at the doctor's office that it was impossible to take an X-ray to see what was wrong.[29]

Frightened, he booked passage on the next available ship—the *Augustus*, due to depart 22 June—in the hope of getting the same Washington surgeon who had looked after him nine years earlier to perform any operation that might be necessary.

By the time he boarded the ship, Just was almost delirious. Boveri had had to go to Germany to visit her mother, but she had promised to get back to Naples a day or two before his departure. He had not wanted her to go, but now the important question was why she had not returned. Was it her way of telling him the relationship was over? His head throbbed harder. He sat on the bunk in his cabin—waiting for the ship to pull out, feeling as though he was being "choked and blinded by a blackness of noxious gases in a stinking dark room."[30] For a moment he felt "rigor mortis of the dead soul" setting in. The door opened and he looked up. Boveri was standing there, "a vision of carefree, detached, capable and cool loveliness." Hurriedly they embraced, and Just knew there was no way he could go back to America now. As they walked off the boat together, they saw nothing but each other, not even the rain falling in blinding torrents around them. An hour later they were on board the island ferry en route to Capri, for their loveliest weekend yet.

The following week they were off to Maloja, a quaint village in the foothills of the Swiss Alps just over the Italian border. Boveri had resigned her job at the Stazione, and now there was nothing to stop them from enjoying a month or two of complete freedom in each other's company. In Maloja, they stayed at the Post Hotel—sleeping till noon, wandering around till supper, reading *Lady Chatterley's Lover* and *Ulysses* to each other till midnight. It was the perfect hideaway. By an unfortunate coincidence, however, they ran into Boveri's Uncle Robert, one of the directors of the Brown-Boveri Company, who was in town on vacation.[31] Uncle Robert had them over to tea at his hotel, and showed his strong disapproval of his niece's new romance by refusing to exchange a single word with Just. But this did not upset the lovers, who were too wrapped up in each other to take much notice of Uncle Robert.

By mid-July Just was back "at par" and feeling like "a *real* person" once again.[32] His recovery would have been slower if he had had to worry about what people in America might think about the delay in his return, but he was able to rest easy, because Dohrn had covered up for him. An affidavit had been prepared to the effect that the physician on board the *Augustus* had refused to allow him to make the transatlantic crossing.[33] Also, in response to an inquiry from Just's assistant Roger Arliner Young, now the acting head of the zoology department at Howard, a telegram had been sent off stating that Just was staying behind to finish up some work at the Stazione Zoologica.[34]

In early August, when he could not postpone the departure any

longer, Just felt saddened but at the same time fulfilled. A final lunch
with Boveri under the trees outside the railway station in Maloja, and
then he would be gone. They sat in silence, afraid their feelings might
get the better of them. They had had a wonderful summer together, and
there was no point in spoiling it now with tearful farewells. After one
final kiss they went their separate ways—Boveri back to her studies in
Berlin, and Just back to Washington to spend a few days with Ethel
and the children before the start of the fall semester at Howard.

The fall of 1929 at Howard was routine, though Just's mind was as
much on the experiences he had had in Europe as on his teaching and
committee work. Whether in the lecture hall or conference room, he
could not help thinking about the people he had met and the places he
had visited three thousand miles across the ocean. He was far away
from Europe now, but he felt closer in spirit to Dohrn and Boveri than
to any of the students and colleagues he came in daily contact with on
the Howard campus. It was a strange feeling—living and working in a
place that had been home for nearly twenty-five years, but that now
seemed less like home than the place where he had just spent a mere
eight months.

So strong was his sense of being a part of Europe that he set himself
up as a kind of ambassador to America on behalf of the European
scientific community. While in Naples he had held long discussions
with Dohrn about Europe's desperate economic situation and the
terrible effect this had had on institutions such as the Stazione Zoologica.
One of his parting promises to Dohrn was that he would not rest until
he had tapped every conceivable source of funding in America. True to
his word, he spent much of September, October, and November writing
letters to various influential people on behalf of the Stazione.

Appeals went first to Lillie, McClung, and Conklin—three of the
best-connected biologists in America.[35] Just outlined his case clearly,
stressing the importance of the Stazione as a top international scientific
institution. As he saw it, scientists everywhere had an obligation to
help the Stazione survive the financial and political problems resulting
from the war, but on the whole Americans were not doing enough.
True, the International Education Board (a subsidiary of the Rockefeller
Foundation) was channeling two or three thousand dollars per year
into the Stazione, but this contribution, Just thought, was little more
than a drop in the bucket. By contrast, the Kaiser-Wilhelm-Gesellschaft
and the Italian Ministry of Education, despite serious financial troubles
of their own, were giving several times that amount. Why then could

not American foundations be persuaded to give their fair share to this important international enterprise?

Determined to leave no stone unturned, Just also wrote to Oswald Garrison Villard, the prominent NAACP official, who was also executive editor of *The Nation*, a liberal journal with a wide circulation throughout the United States.[36] He had not been in touch with Villard since the days of the Spingarn medal, but he hoped Villard would remember him and give him a sympathetic hearing. The cause of the Stazione was international and therefore sure to appeal to Villard, whose favorite slogan in *The Nation* was "universal brotherhood." Just went into the history of the Stazione in great detail, everything from Anton Dohrn's struggles to the administrative incompetence of the Italians and Reinhard Dohrn's fight to put the Stazione back on its feet. Perhaps, after reading all this, Villard would be moved to exercise his influence with one of the large New York foundations—the Commonwealth Fund, for instance.

Just carried the banner of European science with pride. Naples and Graz had been an almost mystical experience for him, and the letters he was writing now were, in a way, starry-eyed tributes to the scientists he had met and the scientific institutions he had seen in Europe. But there was also an ulterior motive. Just realized full well that if his appeals proved successful he would be able to deepen further his relationship with Dohrn and perhaps even secure favored status at the Stazione. Also, he had always felt a little ashamed of asking for money for himself, but now that he was doing it on behalf of someone else, and for an international cause at that, his own begging might seem more dignified.

Much to Just's disappointment, the responses he got were either negative or supportive only in a mild way. Lillie was a little brusque, pointing out that American scientists had enough on their shoulders already and should not be expected to "labor overtime" to improve conditions at the Stazione.[37] Conklin promised to use his influence to secure aid for the Stazione, but expressed uncertainty as to what he would do and when he would do it.[38] McClung did not respond at all. All this reinforced Just's suspicions that American biologists were provincial, indifferent to science as an international enterprise, and unable or unwilling to stretch their vision much beyond Woods Hole. He would have felt completely discouraged, except that Villard, the only nonscientist he had written to, showed interest in the cause and made an effort to see if anything could be done for the Stazione through the Commonwealth Fund.[39] Nothing came of it, but Villard then invited Just to publish a letter of appeal in *The Nation* on behalf of the Stazione.[40] Just was afraid that such an appeal might offend the International Education Board, which might in turn cut off the funds it was already providing.[41] A better plan, he thought, would be to arrange some general publicity, perhaps by publishing a review of the article on the Stazione that Boveri was about to bring out in one of the popular German journals.

Since his colleagues who had the contacts were not going to help scout out sources of funds, Just decided to see what he could do himself with the foundations. His one solid contact was the Rosenwald Fund. He had wanted to avoid going this route, for fear he might push himself too far with the people who were already supporting his own work at the rate of $15,000 per year. But the risk was worth taking for European science; and if his efforts produced results, further possibilities might open up for him at European laboratories. Just made his plea to Embree, describing the Stazione as a "very poor" institution and "an international venture which has served not only European biology, given an impetus to American institutions of a similar kind, inspired so many leading biologists who have worked there, but also in a large human way influenced thinking the world over."[42] He wanted the fund to set up an endowment at the Stazione of $100,000, but failing that, he hoped Rosenwald would find it possible to make a private gift of around $10,000.

Embree responded with an emphatic no: support for the Stazione fell entirely outside the scope of the Rosenwald Fund, and it would be unreasonable to expect Rosenwald to make a private contribution in any amount.[43] This had been Just's last resort, and he did not know where else to turn. He let Dohrn know how the situation stood, doing it in such a way as to leave some room for hope.[44] American scientists seemed interested in the Stazione, he told Dohrn, but they were bent on waiting for the political climate in Italy to settle before taking any decisive steps. The stock market crash had made it difficult to get funds from the foundations; if the financial climate had been better in America, Just claimed he would have had no trouble finding $25,000 for Dohrn somehow and somewhere. Perhaps a little stretching of the truth would keep Dohrn's confidence and friendship.

The cold fact was that American scientists were hardly concerned about the needs of fellow scientists overseas. Just was chagrined. What would American scientists expect the Europeans to do, he wondered, if the MBL at Woods Hole was on the brink of extinction because of financial problems? It was a good question, but too impertinent for Just to ask. Quietly, and no doubt prudently, he observed but kept to himself the stark contrast between what he perceived as the selfishness of American scientists and the humane, cooperative, giving attitude of the Europeans. He wished he were back in Naples.

But it would be a bit much to go back now, so soon after his last trip and just as he was getting the graduate program under way at Howard. Maybe it would be best to stay in Washington for a while. Under the terms of the Rosenwald Fund grant he had a duty to develop his department on home ground, as well as to further his research work wherever it might take him. Also, his apparent infatuation with Europe was causing some concern among the Woods Hole people. Staying in America for the academic year should set everyone's mind at rest.

The decision, however, was not that easy. An invitation had come from Max Hartmann, the foremost German embryologist, for Just to go to the Kaiser-Wilhelm-Institut as guest professor from January to June 1930. Just knew that he should go, but because he was already committed to Howard for the year he thought he had better seek out Embree's advice.[45] At issue was whether his duty to develop the graduate program should take priority over his research. There was no question in his own mind that research came first, at least for the moment, and that he should take this chance to finish up the studies begun at Naples. He needed to work in a productive and cooperative atmosphere and to cement further the contacts he had already made. Besides, taking up an invitation of this sort could do as much for the stature of Howard and other black educational institutions as struggling to develop a graduate program that would not win wide recognition for many years to come.

Just began making plans to sail, after hearing from Embree that he should decide on his own after consulting the Howard administration. The final go-ahead came from E. P. Davis, dean of the College of Liberal Arts, some time in late November or early December.[46] There was much to do before leaving—reservations to be arranged, departmental details to be wrapped up, papers to be proofread, notes and instruments to be packed, travel advances to be ordered, and so forth. In the rush he had to cancel an appearance at the annual meeting of the American Society of Zoologists. His excitement grew as Christmas drew near. Europe was an experience he wanted to explore in depth, and nothing could sway him from that purpose. At last he seemed to be choosing his own direction, finding his own niche.

On 2 January Just boarded the *Berlin* in New York Harbor, this time without Margaret or anyone else in his family. The next six months would be his and his alone; he would do whatever he wanted whenever he liked. No ties, no duties—except to himself and his research. As he strolled on deck, the brisk sea air made him feel alive and free. He asked himself, though, if what he remembered of Europe could be true, if it might not simply have been a pleasant dream. After a day or two on the *Berlin*, his doubts were dispelled. On 4 January he went to a concert in the ballroom, and listened to the orchestra play works by Grieg, Bizet, and Offenbach.[47] The music was light for the most part, but it took him back to the Busch Quartet concert and the coffee breaks he had spent discussing one symphony after another with scientists at the Stazione. Later that evening he had dinner with two charming German-

Americans, a Mr. Kuhler and a Miss Ulrich, artists on their way to Leipzig to do a commission, etchings for one of the city monuments.[48] They talked about art into the wee hours of the morning. To Just these artists seemed more German than American—warm, intelligent, and cultured, much like Dohrn and Boveri.

The dark, cold train ride from Boulogne to Berlin was less than pleasant, but Just did not mind. He stared blankly at the rain streaming down the train car windows and blurring any view he may have had of the dismal, wintry landscape—lowland France, Belgium, Germany— that slipped by outside. His mind was elsewhere. He was waiting for the moment when he would be back with Margret Boveri, when he would meet his new colleagues at the Kaiser-Wilhelm-Institut, when he would find warmth and peace once again.

Boveri was at the station, her short, stocky frame clad in a beautiful and costly fur. Things had been going well for her, despite the economic morass that the newspapers said was swallowing up Germany. Just was eager to know what had happened since they last saw each other. Indeed, there was much for her to tell and much for him to hear. Her letters to him had been few and far between, in order to avoid raising suspicions about his new female correspondent. Now, free to talk as she pleased, she chatted away in the taxi going home. She told how for a time she had lived with her mother in a hotel on the Friedrichstrasse, near a lovely park full of tall oaks; how she had begun to take courses at the Hochschule für Politik (a kind of polytechnic institute specializing in political science) and met many interesting people there; how she had been lucky enough to sublet a small house—"Friedenau," on the Rotdornstrasse—from two painters, a married couple, who were in Rome studying for the year; how she had managed to maintain a high standard of living by using the money her mother, who was now teaching zoology in the United States, had been receiving monthly as a pension from the University of Würzburg ever since her father died. All this chatter captivated and amused Just; he was not used to seeing Boveri quite so breathless with excitement.[49]

At "Friedenau" that evening, the tone was subdued and romantic. Together they prepared a grand Italian meal of pasta, veal, and room-temperature *vino rosso* to remind themselves of the time they had spent in southern climes the year before. The kitchen was small and cozy, as were the other rooms in the house. Boveri had set up a living-dining area in the painters' studio, circular and bow-windowed, a charming place to relax and eat by candlelight. After dinner they listened to Bach and Vivaldi on a gramophone Boveri had received as a gift from her rich Aunt Victoire the previous Christmas; then they danced away the evening to the music of Bessie Smith and Jelly-Roll Morton and his Red Hot Peppers.

There were two compact bedrooms at "Friedenau," with a bed, wash-stand, and bathtub in each. The first night, and every night thereafter, Just and Boveri rumpled the bedclothes and filled the washbasin in the spare bedroom. This ritual was for the purpose of fooling the house-keeper, Frau Becker, who had been told that Just was a paying lodger. Frau Becker was an inquisitive woman who used to pry into drawers and read personal mail, so it was not easy to keep the truth from her, but she never guessed what was going on.

Just and Boveri had other things to do besides play hide-and-seek with Frau Becker. As this was his first real visit to Berlin (he had merely passed through once before on the way from Graz to Bremen), they spent a good deal of time exploring the city. At night Berlin came alive with bright lights, burlesque shows, and beer gardens. Its people, half a million of them unemployed, were trying to drown their sorrows in an orgy of unrestrained gaiety. A job could not be had anywhere, but what did that matter as long as they were having a good time? The happiness was all on the surface, though—a kind of desperate bravado in reaction to the economic problems that had plagued Germany since the end of the war. By the late 1920s scientists and other professionals had managed to resume their former standard of living, but the ordinary man in the street did not have a share in this. Just may or may not have detected the frustration underlying the gaiety in the bars and night-clubs; probably he did, because he was an avid reader of German newspapers,[50] which were full of stories about social disorder, runaway inflation, and political murder in Berlin. But foreigners can be short-sighted, and it is possible that Just simply let himself be caught up in the hustle-bustle with no thought of what it all meant. Besides, much of his attention was focused on aspects of life in the city that did not necessarily have much to do with the mood of the ordinary working man. He and Boveri walked up and down the Unter den Linden, admiring the concert halls, the opera houses, and the statues of Ger-many's most famous musicians, artists, and philosophers, all lit up in the most magnificent manner. To the outsider especially, Berlin was still a bright city with a grand cultural tradition. Little did Just, or the Berliners for that matter, know that in three years Hitler and his storm-troopers would come marching through the historic Brandenburg Gate at the western end of the Unter den Linden to throw a shadow over Germany and change the course of world history.

Nothing could have been further from Just's mind than German politics and society. His mind was on his research and his career; he was trying to get started in a new and different environment. For up to twelve hours of every day he was at the Kaiser-Wilhelm-Institut. For the first week or so, he wandered from laboratory to laboratory, mar-veling at the elaborate facilities and meeting scientists—such as Nobel laureates Otto Meyerhof and Otto Warburg—whom he had known by reputation only. His host, Max Hartmann, introduced him around. He was a celebrity in his own right and was treated as one. A tea was given in his honor and was attended by the institute's top scientists. He was as impressed by them as they by him. What most caught his attention was that, despite their very real accomplishments, these scientists maintained an air of congeniality, cooperation, and intimacy at the Institut. The welcome they gave him made him "glow on the inside."[51] It did not matter that he spoke German only falteringly or that some of them, including Hartmann, did not know a word of English.[52] Communication, somehow, was not a problem. He seemed to fit easily into this close-knit scientific family, this most famed of famous research organizations.

Just was filled in by Max Hartmann on the details of the history and development of the Kaiser-Wilhelm-Gesellschaft. It had all begun in 1910, when, to celebrate the centenary of the University of Berlin, Kaiser Wilhelm II announced the formation of a society for the advancement of scientific research.[53] Wealthy industrialists had pledged upwards of twenty million marks to the project, and the Kaiser himself had donated some of his property in Dahlem, a fashionable suburb of Berlin. In 1911 the society, called the Kaiser Wilhelm Society in honor of its prime mover, met for the first time to plan the establishment of a research institute. Among its members were famous scientists such as Emil Fischer and August von Wassermann, and its president was Adolf von Harnack, the eminent theologian, who combined far-reaching interests with superb leadership qualities. With Harnack at the helm, the society built not one, but five institutes in Berlin-Dahlem—chemistry, physical chemistry, experimental therapy, biology, and physics— over the course of the next twenty years. Institutes were also established in other cities, including Cologne and Munich.

Most of the institutes were set up without trouble, but the one Just was attached to—the Kaiser-Wilhelm-Institut für Biologie—had a shaky beginning. By 1913 the building was completed and a director, Theodor Boveri, had been appointed. Everything seemed to be going smoothly. But when Boveri ran into bureaucratic difficulties with the Ministry of Education and at the same time fell ill with influenza (leading to his death two years later at the age of fifty-two), he declined the post and returned to Würzburg. Then World War I began. Undaunted, Harnack pressed forward with his plans and finally persuaded Carl Correns, the leading German plant geneticist and a rediscoverer of Mendel's law of heredity, to assume the directorship. In 1915 the Kaiser-Wilhelm-Institut für Biologie opened its doors to qualified, carefully selected researchers and technologists, each of whom was assigned to one of four *Abteilungen* (departments). Each *Abteilung* was run by one of four top biologists—Carl Correns, Richard Goldschmidt, Max Hartmann, and

Otto Mangold—who focused the work of his particular department around his own research interests and programs. Over the years the structure and personnel of the Institut remained fairly uniform.

With Hartmann's help, Just came to understand the Institut and what it was all about. Soon he was feeling at home. Early in the morning he would take the subway from "Friedenau" and walk the mile or so to the Institut across the Grünewald, a large public park covered by meadows and woodland. Right next to the Institut was Otto Warburg's house, and Just would tip his hat and exchange a few words with the old Nobel laureate, who made a habit of riding horseback every morning clad in the boots and spurs he wore as an officer in the German cavalry during World War I. Also next to the Institut was a garden, about an acre in size, which Carl Correns had set aside for his experiments in plant genetics. Even in the winter months Correns could be seen puttering around in the dirt with his spade. At the entrance to the Institut Just was invariably greeted by Franz Marquadt, a tiny but strong man whose job was to insure that no unauthorized person entered the laboratories and that certain needs of the scientists— carrying heavy equipment, arranging repairs and so forth—were met. Herr Marquadt, the story goes, once carried a lame two-hundred-pound scientist up three flights of stairs with no apparent effort.

Life at the Institut was informal, comfortable, homely. On the way to his own table, located in Hartmann's *Abteilung* on the third floor, Just would stop for coffee and a chat with scientists in the *Abteilungen* run by Mangold and Goldschmidt. As in Naples, the conversation touched on a wide range of common interests, scientific and otherwise. Just especially liked talking to Mangold's assistant, Johannes Holt-freter, who had fascinating ideas about painting and sculpture as well as fertilization and development.[54] Around one o'clock or so he would join Mangold, Holtfreter, Hartmann, and Hartmann's assistants— Karl Belar, J. Hämmerling, J. Gross, and Helgo Culemann—for the midday meal at the Harnack Haus, the social center for the combined institutes. Then it was back to work for the rest of the afternoon, with more coffee and conversation—perhaps also an early evening party or lecture at the Harnack Haus—before the return trip by subway to "Friedenau" at night.

Just was feeling creative, energetic, full of the spirit of adventure, and it showed in the way he buckled down to his work. Since his experiments at Naples he had become convinced that cortical (ecto-plasmic) behavior might contain the answer to all problems in biology. So far he had worked only with marine eggs, but now he was eager to assemble further evidence for his hypothesis by examining another kind of species, *Amoeba proteus*, a unicellular freshwater organism.[55] Studies on *Amoeba* might, he believed, confirm or disprove his line of thinking, equally important in either case. Hartmann had himself been working on *Amoeba,* so he was able to offer Just advice on difficult preliminary techniques such as isolating the species and determining its optimum physiological condition. The minds of these two scientists worked very much alike. Aside from the *Amoeba* interest, Hartmann was as convinced as Just about the value of correct experimental methods: in 1910, with the publication of his *Practikum der Bacteriologie und Protozoologie ("Handbook of Bacteriology and Protozoology"),* he had done for bacteriology and protozoology what Just had done for experimental embryology in the *Collecting Net* articles of 1928.[56]

Some of Just's time was spent poring over a large group of slides he had brought with him from America. These slides were the result of the ultraviolet work on *Nereis limbata* that he and Roger Young had begun at Woods Hole in the mid-1920s. He had not done much with them recently, but now he wanted to see if by close cytological study he could show something about the mechanism by which cancerous cells are created through exposure of eggs to ultraviolet light.[57] Several scientists at the Institut, especially Hartmann, Goldschmidt, Victor Jollos, and Mangold, were so impressed by this line of work that they loaned Just the service of their artists and technicians for several hours a day. Up to ten people might be studying the slides at any one time, drawing painstaking diagrams and making careful measurements. The medical importance of all this was obvious, but Hartmann and Goldschmidt urged Just to also use his results to corroborate his general theory on cortical behavior in much the same way he was doing with *Amoeba proteus.* Just's study of the ultraviolet slides suggested that there were important but quick and hardly discernible changes in the cortex during development that had not yet been dealt with seriously by embryologists.

German scientific and educational institutions had been isolated from the rest of the world as a result of the First World War, and the effects of that isolation were still noticeable at the Kaiser-Wilhelm-Institut. There was hardly a foreigner to be seen anywhere. Correns and his colleagues had grown used to working by themselves to build and maintain their own brand of science. They did not strive for international diversity. It would be wrong, they maintained, to reduce the working space of productive German scientists by cluttering up the laboratories with others for the sake of saying that this one came from Italy and that one from the United States. The only foreign scientists invited to the Institut were those who were thought to have something special to share with and gain from their German counterparts. The first American to have received such an invitation was Just.[58] Except during the war years, his articles had been cited regularly in German journals and he had received frequent requests from German scientists for information on his work.[59] He was admired in Germany. Other

American scientists had come to the Institut, but they had not been given the same privileges and they were obliged to pay large fees.

Just was proud of the position of honor he had been accorded by the Germans. Never before had he felt so respected by his peers, so confident in his work. He was rather amused that the other Americans at the Institut seemed jealous of his position. There was something strange in their voices—a note of disbelief that he was being lionized by the likes of Hartmann. One day Just bumped into Gregory Pincus, a visiting Harvard scientist, and was told that his work was out of date, that his scientific arguments were "encumbered with echoes of ancient controversy," that "new tides" were "rising in his old familiar sea, tides beyond his or anyone's control."[60] But this statement did not bother him, especially since high-ranking German scientists were urging him to continue along the lines he had always taken. Nor did he allow it to affect his relationship with Pincus, who, along with his wife, spent many hours "swapping stories" with Just in Berlin restaurants.[61] Backed by the Germans, Just could afford to ignore any negative remarks, including those that had begun to circulate in America about how "absurd," speculative, and philosophical his work was becoming.[62]

The Institut seemed to be a real community, a supportive environment in which everyone worked together and helped each other. Most striking, perhaps, was the fact that there was no apparent ethnic or religious prejudice.[63] Jewish scientists—Meyerhof, Warburg, Goldschmidt—worked side by side and in complete harmony with Christian colleagues—Correns, Hartmann, and Mangold. This was not typical of Germany, and Just knew it. In general, the German academic establishment was anti-Semitic. A Jew had great difficulty securing a university appointment of any kind, and if he was lucky enough to find one he was often rebuffed by colleagues and students alike, in the laboratory and the lecture hall. The Institut was different. Harnack had made sure of this in the planning stages—much to the consternation, no doubt, of the anti-Semitic Kaiser Wilhelm. To bring together the best minds regardless of background and race was in Harnack's view the surest route to scientific progress. From all indications, his experiment was a great success. Only later, when Hitler introduced the infamous Aryan laws, did it fall apart—and even then most of the Institut's "racially acceptable" scientists, adhering to Harnack's principles, refused to go along with the purge of their Jewish colleagues. Mangold was perhaps the only high-ranking scientist at the Institut to espouse the Hitler cause.[64]

Though there was none of the obvious kind of prejudice that Just was familiar with from his experiences at Woods Hole and elsewhere in America, one incident—one that he did not understand the full meaning of—suggests that the Institut scientists were not quite as open and liberal as he may have imagined. In late January, after Just had been in

Berlin a few weeks, he was invited to be a live-in guest at the Harnack Haus for the remainder of his stay. The grand old man himself, Adolf von Harnack, presented the invitation in person. Just was flattered, and there was no question but that he would accept.[65] He was elated by the possibilities—of meeting more scientists, getting more work done, using the time spent coming and going by subway every day to better effect. He would still, of course, be able to spend some evenings and most of each weekend with Margret Boveri.

What he did not know was that there was a reason behind the invitation. Scientists at the Institut had been talking about him and Boveri, and were upset by the living arrangement at "Friedenau." They did not want to see the Boveri name tainted. The memory of Theodor Boveri, one of Germany's top biologists and the original choice as director of the Institut, was sacred, and it was simply too scandalous for his daughter to live openly with a married man, a black one at that, a member of *"dieser falschen Rasse"* ("that debased race").[66] How could a separation be finessed? Hartmann spoke to Harnack and came up with the idea of inviting Just to move out of Berlin and into the Dahlem neighborhood. This diplomatic solution did not break up the lovers' relationship altogether, but it at least helped make the affair a little less visible.

The Harnack Haus provided all the opportunities Just had hoped for, and one that was a complete surprise. He and Harnack became close friends. The old theologian had been one of Just's idols during his Kimball Union days, when Harnack's *What Is Christianity?* had touched a deep chord, allowing Just to reconcile his religious fundamentalism with his growing interest in the world of man and nature. Since then he had lost his faith in God, what with his single-minded devotion to his experiments, teaching, profession, and so forth. But now, under Harnack's benign and powerful influence, he felt it all coming back to him. It was a kind of reconversion experience, a "great event" to be cherished always.[67] The spiritual intensity he had shared with his mother was surging through him once again.

Just spent many evenings with Harnack, taking tea and talking. Though physically frail, the old man was alert and intellectually penetrating. The usual topic of conversation, one that Harnack had spent a lifetime pondering, was the relation between science and religion.[68] Harnack showed Just that to find an answer to the important philosophical questions in life—man's origin, his direction, his goals —it would be necessary to bring science and religion into greater harmony. These two fundamental areas of knowledge had much to offer each other and were far from being as disparate as the fanatics on both sides would have people believe. The scientific method could help eliminate dogma from religion, and spiritual belief could add a dimension of mystery to science. Man was meant to both scrutinize and stargaze, to combine reason and feeling, logic and faith. A balanced outlook would make him happier and more productive.

Just benefited a great deal from these discussions, which formed the basis of a new philosophical outlook that would eventually carry over into his scientific work. He did not, however, cloister himself away in the rarefied air Harnack breathed. There was too much happening around him in the flesh-and-blood world—and, in particular, there was Margret Boveri.

The relationship did not cool, despite Just's move to the Harnack Haus.[69] Boveri came over every weekend and some weekday afternoons to play tennis, eat lunch, and relax in the Boveri Room, a lounge dedicated to the memory of her father. She did not like Just's scientist friends too much, mainly because they always flocked around with questions about her father and his work. The only ones she could tolerate were Herr and Frau Gross, two Austrian scientists who seemed to be less painfully conscious of "the Boveri tradition" than were their German colleagues. Occasionally she and Just would linger in the early evening to listen to Herr Gross play chamber music with an amateur group comprised of two or three other Austrians, nonscientists, living in the neighborhood. Most of the time, though, they went back to "Friedenau" in time for dinner.

"Friedenau" was not always the peaceful, romantic retreat Just might have liked it to be. Candlelit evenings had given way to politics, politics, and more politics. The house was strewn with Boveri's notes from her studies at the Hochschule für Politik and pamphlets on the Social Democratic Party. She talked incessantly about the latest lectures by her mentors at the Hochschule, Onken and Hoetzsch; about her progress with the thesis she was writing on British foreign policy during World War I; about what she thought the present government in Germany should be doing to establish friendlier ties with other nations. At times, when she saw that Just was dozing off, she would try to arouse his interest by comparing her own concept of international relations with, say, the experiment Dohrn had been trying at Naples on a much smaller scale. If this did not work, as often it did not, she would change the subject and put on a Bach or Vivaldi record.

Boveri's friends were political too, and Just took to them even less than she to his at the Institut. They were left-wingers for the most part, students and lecturers at the Hochschule or members of the Social Democratic Party. Arnold Wolfers and Fritz Berber, two of her teachers, held forth constantly about the international situation—and it was clear from what they said that she had gotten many of her ideas from them. Then there was the Communist Party group: Gisela von Dehn; Charlotte Gräfin Stenbock; her husband, Alexander Graf Stenbock-Fermor; Tania Kurella and her two brothers, Heini and Alfred. Doctrinaire communists every one, they talked about the need to radicalize the party; about the book Stenbock-Fermor was working on, *Deutschland von Unten* ("Germany from Below"), a Marxist analysis of German society; and about a monthly pamphlet, *Rote Hilfe* ("Red Support"), that they were planning to publish to help revolutionize the working class. Their regular meeting place on Saturday evening was "Friedenau." Just did not care for them or their views, but out of deference to Boveri he never missed a meeting, although always making certain that he was well tucked away in a corner of the living room. He was a master of the art of retreat.

Boveri knew a few right-wingers as well, people who were not attached to any political party but who were dissatisfied with the republican government and fed up with the rantings of the socialists. Like many Germans, they had a hard life. Elsbeth Krell struggled in a part-time job to support her unemployed husband and school-age children, and Bob Beckmann, though he was trained as an engineer, had to buy bread for his family out of the measly wages he earned as a handyman. Then there was Carl Gerhardt, who, though he held a steady job as organist at one of the city cathedrals, was disgusted that he could only afford to live in one of those cramped city apartments the state was putting up to cope with the housing shortage. They ranted as much as the left-wingers. Their hope was for some kind of a purge, a new and strong leader to rise up and take over the reins of government. At the moment Hitler seemed little better than a hooligan to them, but over the course of the next year, as their frustrations grew, they thought seriously about joining the Nazi Party.

Everybody seemed to be caught up in a wave of political extremism. Even Frau Becker, the old cleaning lady, walked around the house muttering "*Wir sind alle Kommunisten*" ("We are all communists"). More and more, when Just was in Berlin, he wished he was back in Dahlem talking philosophy with Harnack, music with Gross, or art with Holtfreter. Only one of Boveri's friends provided any relief. Doris Heider, daughter of a zoologist at the University of Berlin, used to come over to "Friedenau" and suggest that the three of them take sight-seeing tours around the city. Her special interests were architecture and automobiles, and as a result they often ended up admiring the houses on the Argentinische Allee or visiting the showrooms and fiddling with the gears on the latest Mercedes model. It was fun, but even then there was no escaping politics. To walk out on the street meant being accosted by nasty pamphleteers, whether leftists, rightists, or centrists.

Just needed a break from all this. Around the middle of March he left Berlin and headed for Naples. He was anxious to see Dohrn again, finish up some observations he had begun the previous year on *Nereis dumerilii*, and find a little peace and quiet in the "pure air" of the Italian countryside.[70] Naples was cold and rainy, uncomfortable for collecting or taking long walks, so he spent most of his time inside talking to Dohrn. It was pleasant, restful.

Early in April, toward the end of his stay, he had an experience that made him look forward to returning to Berlin. He was in the center of town one day, taking in the sights, jostling for a place on the crowded sidewalk, when on the other side of the street he saw someone whose face looked familiar. It was the former crown prince of Germany, the son of Kaiser Wilhelm II. Pushing through the crowd, Just crossed the street and shook hands with him. They took to each other right away and talked for a while about Germany and the Kaiser-Wilhelm-Institut. Before they went their separate ways, Just received a warm invitation to visit the royal household in Potsdam, a few miles from the center of Berlin.[71]

Back in Berlin he was teased by the Institut scientists and scolded by Boveri's friends for rubbing elbows with a member of the fallen House of Hohenzollern. He took the gibes in stride, retorting that since the republicans, socialists, fascists, and so on, seemed unable to cope with the nation's problems, perhaps it might be better for all concerned to reinstate the monarchy. How serious he was about this is unclear. If, however, he did subscribe to the small-scale monarchist movement centering on the crown prince, more than likely it was a romantic impulse rather than a reasoned response to the political situation. He had read about the Hohenzollerns from the time he was at Dartmouth, and had always thought of them as representing a kind of triumph of the spirit in the way they had risen from obscurity to make something of Prussia and build a prosperous German empire. He knew about the military campaigns, the princes and kings and emperors. The Hohenzollerns seemed to him to symbolize German culture at its best. How grand were their achievements compared to the ineffectual bickerings of modern-day politicians, plebeians with little soul and even less sense.

Just went to visit the prince in Potsdam some time in late April. The "court" was bustling with friends, toadies, hangers-on, genuine monarchists, and three or four newspaper reporters. After dinner everyone gathered for political discussion in the grand ballroom. The mood was light, the tone nostalgic. No one talked about monarchist coups or anything so dull and practical; rather, the focus was on bygone days, past glories, faded dreams. Just was comfortable in this group of merry eccentrics, and instead of simply sitting and listening to the proceedings, he chimed in with a few comments of his own about how much he had always admired the House of Hohenzollern.

The prince did not care too much for politics, though he and some of his followers were members of the Stahlhelm ("Steel Helmet") organization, a monarchist-tinged right-wing veterans group viewed by the republican government more as an anachronism than a threat.[72] His first love was wine and women. One of his acquaintances was Grethe Alter, a woman with a bubbling personality and a unique blend of beauty and brilliance, and at the end of the political meeting he introduced her to Just and left them to talk. They had a common interest in anthropology, it turned out. Alter had been in Africa doing research on musical forms among certain remote tribes, and she asked Just to help her get funds to bring out a book on the subject and to form a German choir devoted to the art of African music. As he had with Dohrn, he promised to see what he could do when he returned to America.[73]

That time came sooner than he expected. May passed quickly and by early June he was on the ship sailing home. He spent two or three days in Washington before going on to Woods Hole. He went to Woods Hole for one purpose: to attend a celebration honoring Lillie on the occasion of his sixtieth birthday.[74]

Just arrived early in June, when most workers had already settled in for the summer. The place seemed different to him, and he could not figure out just how. He had decided to stay and work for the summer, though he was expecting numerous interruptions, just as before. After finding a room, he joined Heilbrunn to tie up the details of the celebration.

On the evening of 27 June the members of the laboratory gathered in the main lecture hall for a seminar and reception in Lillie's honor. E. B. Wilson presided, setting an atmosphere in which "one could not help but sense the sincere appreciation of Dr. Lillie's work on the part of his associates, his co-workers, and his students both past and present."[75] After Wilson's introductory remarks, four of Lillie's students stepped forward one by one to deliver brief research reports. Heilbrunn was too excited about the occasion to spend much time on his chosen topic, the effect of ultraviolet rays on *Arbacia* egg protoplasm. Instead, he entertained the audience with anecdotes about Lillie. W. C. Young and B. H. Willier stuck more closely to the program, but they too could not help showing affection for Lillie. Young spoke on the post-testicular history of spermatozoa and reproduction in the male guinea pig, Willier on the developmental relations of the heart and liver in chorio-allantoic grafts. Then it was Just's turn. All attention focused on him, Lillie's most prized student and perhaps his best friend. He talked for ten minutes on problems of fertilization, but his mind was not on the topic. Still in a state of euphoria over his European experience, he stepped off the podium and said to the audience: "I have received more in the way of fraternity and assistance in my one year at the Kaiser-Wilhelm-Institut than in all my other years at Woods Hole put together." A peculiar silence fell over the room.[76]

Just left Woods Hole without saying good-bye to anyone, not even Lillie. He never returned. After that summer, he did his research in Europe. He had found a new world, perhaps a new life abroad.

❧

The last thing Just wanted to do was suffer through a summer of boredom in hot, sticky Washington, with only the memory of Europe to fall back on. For some time he had been thinking about attending the Eleventh International Congress on Zoology to be held in Padua, Italy, from 4 to 11 September, and he was thrilled in early August when an official invitation arrived for him to be a keynote speaker at one of the main sessions.[77] The Howard administration refused to sanction the trip, but that did not change his mind.[78] Quite apart from his love of Europe, he needed to gain more intellectual exposure, to further his reputation and thus increase his leverage for securing financial support in the future. He had missed the congress in Budapest three years earlier due to lack of funds, and he was unwilling to let anything like that happen again. If necessary he would pay for the trip out of his own pocket rather than from the Rosenwald grant. The time he took for the Padua conference would be personal summer vacation leave, for which he would be accountable to no one.

Just arrived in Padua around noon on 3 September. After settling into his room at the Hotel Storione, he joined the other visiting scientists at an informal get-together in the hotel ballroom, the Sala del Laurenzi.[79] Several of his old friends from Naples and Berlin were there—Dohrn, Hartmann, Otto Mangold. Together they sipped sherry and caught up on all the news. From a distance Just glimpsed some Woods Hole people, but he was determined to stay as far as possible out of their way. Perhaps the only American he spoke to at the party was Oscar Riddle, a scientist who, interestingly enough, had never set foot in Woods Hole.[80]

After the party Just wandered around the neighborhood, marveling at the high level of organization with which the Italians had arranged the congress. The main square, brightly lit, was set up with four long "language tables" run by students from the Fascist University Organization. All a congress member had to do was walk up to the appropriate table—in Just's case the "English table"—and he would have a young man or woman fluent in the language at his disposal for as long as might be required. The students were there to perform almost any service, from secretary to tour guide. Just was pounced on by an enthusiastic young man who took him on a fascinating tour of an exhibit of original sixteenth- and seventeenth-century Italian zoological illustrations, specially put together for the congress members at the university library.

The Italians were out to impress their visitors with the great strides the country was making under Mussolini. This became very clear at the opening ceremony of the congress, which was presided over by government officials and party sympathizers. Professor Ferrari, the government-appointed vice-chancellor of the university, spoke about the scientific Renaissance—Vesalius, Fabricius, Galileo, Morgagni—and how Italians were coming to see it as a symbol of nationalistic pride and integrity. The minister of education, His Excellency Signor Rocco, stressed the official government stand on zoology as a field with strong social implications and practical possibilities. Neither Ferrari nor Rocco mentioned Mussolini by name; that privilege was left to Paolo Enriques, president of the congress and the highest-ranking Italian scientist to be a member of the Fascist Party. Enriques gave a real political harangue, linking the pre-Fascist period to a *"senso di inquietudine penosa"* ("feeling of unbearable pain") and the Fascist period to the notion of *"volontà e . . . parola"* ("will and wisdom").[81] He admitted frankly that one of the main reasons Italy had offered to stage the congress was so that *il Duce,* the nation's great and glorious leader, could show the world what fabulous social improvements had been made under his leadership. Fascism, he also told the audience, held the promise of good will for people of all nations.

Much of this probably passed over the heads of the visiting scientists, who were anxious to get down to the business at hand—reading and discussing their papers. Just was no exception. Like everyone else, his main concern was to present his scientific material as clearly as possible and see what sort of reception it drew.

There were a few minor mishaps leading up to his lecture. Originally he had been scheduled to share the session on 6 September with Hartmann, but Enriques asked him to wait until 8 September since Hartmann ran overtime.[82] Then on 8 September the session began with a lecture by a Frenchman, Charles Gravier, who was so boring that everybody had either fallen asleep or left by the time it was over. But Just woke up the audience. His talk was on "the role of cortical cytoplasm in vital phenomena." Although he was a little nervous— this was the first time he was airing publicly his new theory about the importance of the ectoplasm—he gave what was generally considered to be the best lecture at the congress. Afterwards everyone flocked around the dais to ask questions and comment on the fascinating possibilities Just's theory held. Next day the Padua newspaper carried a summary of the lecture and a drawing of Just captioned, much to his amusement, "scientist from India."[83] This kind of attention, as always, flattered him no end.

Just dropped in on several of the other sessions, especially those in experimental embryology and invertebrate zoology. Many of his friends from Naples and Berlin were scheduled to present papers, and he wanted to be there to give them moral support (much as they had given him) and to hear what they had to say about their latest research on membranes and echinoderms. One session he attended, comparative anatomy, was outside his field altogether. He went because his old teacher William Patten was scheduled to speak, curiously enough on the same problem—mosaic patterns in the head of arachnids—they had worked on together at Dartmouth twenty-five years earlier. Sitting in the auditorium, half listening and half dreaming, he let his mind drift back to the old days when Patten had been an inspiration to him, rather like "a cool shadow" in "the heat of life's struggles."[84] And clearly Patten had not changed. Up there on the dais he exuded the same aura of intelligence, generosity, and composure. Strange how different he seemed to be from other Woods Hole scientists.

The congress stretched over eight days, three days longer than usual, because a variety of social events had been planned. Again, this was part of the government's effort to please and impress. The Committee of Fascist Leaders hosted a tea on the grounds of one of Padua's medieval castles, and the *podestà* (a mayor appointed by Mussolini) put on an elaborate banquet at his plush mansion in the suburbs of the city. There were also afternoon excursions to Fusina, Venice, Torcello, Brusegana, and Rovigo. Each of these towns had something of nationalistic importance—splendid medieval monuments or progressive farming cooperatives—that the tour guide eagerly pointed out. What struck the scientists most, though, was not the country's zealous politics but its breathtaking scenery. On one occasion they stood speechless for half an hour on a bridge over the River Po near Rovigo, gazing out over the twilit landscape.[85]

In all this, Just stayed close to Dohrn. Only rarely were they not seen together. Even during spare moments outside the round of conferences, parties, and tours, they passed the time talking about life and what it all meant. With Dohrn, Just felt like a young man again, "free and fresh," ready to grasp life and live it fully.[86] But he was leading such a strange double existence, and everything was starting to appear dreamlike. Europe seemed a pleasant fantasy, America a terrifying nightmare. At times neither place seemed real. What could he do? Dohrn, not knowing how to respond to this, said in a half-joking way that the best thing for Just to do would be to prolong the fantasy and stay away from the nightmare—at least until reality came into clearer focus.

Just took the suggestion seriously. He had to get back to America some time, but there was no reason why he could not spend an extra two weeks enjoying Europe. Right away he wired Boveri and arranged to meet her at her family's country home in Höfen, a small town in Germany near the Belgian border. Now as the congress drew to a close he did not feel the depression he might have felt with nothing but the return voyage to look forward to. At the farewell ball he was in high spirits, and perhaps even joined in the rousing chorus of "Giovinezza," the Fascist hymn that Enriques considered an appropriate way to end the congress proceedings.[87]

Höfen was beautiful, a quaint little town surrounded by quiet countryside. Everyone knew everyone else, so Boveri had taken the precaution of inviting her respectable middle-aged friend Hildegard von Weber to make up a threesome at the house.[88] That way, she thought, there might be less gossip about her dark-skinned foreign gentleman visitor. In the daytime all three of them would walk for miles through the fields and meadows outside the town, and then at night, after dinner, Hildegard would turn in early and leave the lovers sprawled out in front of the grand old fireplace, sipping wine and talking.

After two days it became clear that Boveri's plan to use Hildegard as a kind of screen had not worked. Uncle Robert called with the news that he was on his way to Höfen from his home in Bamberg. The year before, when he had turned up in Maloja, was an unfortunate coincidence—but this time someone had sent him word about what was going on at the Boveri chateau, and he wanted to come and see for himself. There was nothing to do but put up with Uncle Robert. He was a bore as well as a busybody. In the evenings he would sit in a corner of the living room directly under a portrait of one of the Boveri matriarchs (no doubt as a hint to his niece about family honor) and hold forth in a dull monotone on the most banal subjects. Though fluent in English, he always spoke in German and addressed himself only to Margret and Hildegard.

Margret was upset. Why, she asked herself, could Uncle Robert not be as tactful as her other relatives—all of whom left her alone to carry on the relationship as she pleased? Without further ado, she suggested to Just that they get on the train and spend some time with Doris Heider in Austria. Just agreed, not so much because he was upset by Uncle Robert but because he wanted to spend what was left of the summer in an atmosphere free of tension. They packed up and left poor Hildegard to listen to Uncle Robert fume about his niece's "*mésalliance*."[89]

The mountain air of Austria made them feel free again. Now they would not have to spend their last days together contending with hostile relatives. The Heider chateau, Schloss Thinnfeld, was a perfect retreat, a charming and peaceful old house in the tiny Alpine village of Deutsch-Feistriz, not far from Graz. Doris welcomed the lovers with open arms, and then left them to their own devices. When some other visitors arrived Just and Boveri had to take separate bedrooms in the interest of propriety, but this was a minor snag in what turned out to be their most beautiful time together.

After four days at Schloss Thinnfeld they left for Bamberg. Boveri wanted to get back to Germany to see what was happening with the national elections. News had reached her that the Nazi Party was making incredible gains and might end up with as much as twenty percent of the popular vote. She could not quite believe this, but one of the best places to observe political trends was Bamberg, in the heart of Bavaria just thirty miles north of Nuremberg, a center of Nazi fervor. Germany was going through an intriguing period, and Boveri did not want to miss any of it.

Just complied with the move, though he was not all that interested in German domestic politics and would rather have spent what was left of his vacation in the peace and quiet of the Austrian Alps. Still, he found Bamberg interesting. While Boveri was at the newspaper offices or the polling stations, he walked around admiring the old bishops' palaces and the turreted cathedral where Pope Clement II was buried. He thought about Harnack, about the many hours they had spent discussing ecclesiastical history, about how much of a loss it was to himself and the world that the old man had passed away earlier in the summer. From Harnack, his spiritual benefactor, his mind wandered back to America and the man who had given him the material wherewithal to come to Europe and have all these beautiful experiences. At the hotel he wrote Rosenwald a long letter of thanks.[90]

On 26 September Just and Boveri waited for separate trains at the Bamberg station. They were both northbound, but he was headed for Bremen to catch a ship and she for Berlin to resume her studies. This farewell was not as pleasant as the one in Maloja. He wanted to sit quietly and reminisce about what a lovely month he had had; she wanted to talk about the Nazis and what their election gains might mean for Germany in the future. It was not a meeting of minds at all. They kissed good-bye tenderly, but the idea ran through both their minds that in the end it might turn out they were not made for each other.

There would be no chance for Just to return to Europe again until the spring of 1931. Under pressure from all sides, he wanted to get the zoology department organized on a better footing. He would have to keep his research in the background, catching spare moments here and there for studying slides and making notes. But in the long run, he thought, such a sacrifice would be worthwhile. He would work hard at

Howard, and when the time came for his next research trip to Europe he would not be made to feel guilty about leaving.

Life at Howard was not pleasant. Just did his best with the limited resources at his disposal, but the administration seemed bent on squeezing every last drop of energy out of him. They expected too much for too little. Finally he gave up trying to satisfy them, and spent more time daydreaming about Europe and planning for his next trip. Each morning he sat in his office waiting for the mailman to bring in letters from friends across the ocean. Boveri was writing him more frequently now, and so were his colleagues at the Kaiser-Wilhelm-Institut. Their letters were warm and witty, promising more happy times ahead if only he could break away and come back. Holtfreter, in particular, tantalized Just with talk of food and women and artists and vacations in Italy.[91]

The Woods Hole people sensed Just's uneasiness. They knew that if something was not done soon he would drift out of their orbit permanently. A concerted effort was made to bring him back into the fold. W. C. Curtis had him serve on the National Research Council's new Committee on Radiation Research; R. A. Budington invited him to lecture at Oberlin and arranged for him to be a consultant to the college administration on tenure decisions; the Society of Zoologists put him at the head table for banquets at the Cleveland meetings right after Christmas, asked him to chair important sessions, placed him on the podium to make keynote addresses, and elected him to the Committee on Nominations. On the personal side, Heilbrunn tried to renew their intimacy, gossiping to him about science and scientists, and at one point even promising to get him an invitation to a party given by Theodore Dreiser, the novelist, on the front lawn of a posh Philadelphia club. And most touching of all was Lillie's hope that he and Just, despite all, would remain friends "until the end of the chapter."[92]

Nevertheless, Just's mind turned to Europe once again in early March 1931. His annual report on the zoology department was due. He did not feel like putting it together, but he knew that once it was finished he could leave for the summer. It was a strange report.[93] Clearly his full powers of concentration were not focused on the department. He began by listing activities and accomplishments and problems, but toward the end he wistfully quoted Goethe: *"Es gibt nur zwei Möglichkeiten; entweder ich habe Zeit, oder die Zeit hat mich"* ("Man has two choices; either he grabs time or time will grab him"). Europe beckoned. Over there were rich experiences still to be tapped, while in America life seemed to be no more than a constant waiting and hoping for the impossible to happen.

The Howard administration refused to permit him to leave in late March as he had planned.[94] Though annoyed, he resigned himself to another month in Washington. A few weeks here and there did not really matter, as long as he could get over to Europe some time soon. Besides, one of his European friends—Helgo Culemann, an assistant of Hartmann's—was working in the Howard laboratory on fertilization experiments of interest to them both. Culemann had come over the previous fall, on Just's invitation, to lead a graduate seminar in physiology. With Culemann there Howard at least had a little of the flavor of Europe, and things were not quite so tedious and depressing for Just. They worked and talked and ate together, much as they used to do at the Institut. In mid-May they sailed for Europe.

Back in Berlin, Just settled down to work. The year before he had had to leave without finishing up his *Amoeba* experiments, in order to get back to Woods Hole for the Lillie birthday celebration. This time, however, he was going to make certain nothing got in his way. For all practical purposes he had wasted a year at Howard on what seemed to him to be petty administrative procedures. Now it was time for some serious research, away from the distractions of students and university officials.

Still, his social life at the Institut continued to be active. There were the early morning discussions in the laboratory, the lunch breaks, the evening coffee hours and cocktail gatherings. He was becoming so friendly with his German colleagues that his contact with them began to extend beyond the walls of the Institut. Ever popular and always the perfect guest, he was invited to their homes for drinks and dinner. One evening he arrived at the Mangolds' house with a mysterious package that he put in the closet along with his coat and umbrella.[95] After dinner, when the Mangolds' ten-year-old son was saying good night to each of the guests in turn, Just excused himself, went into the hall, and brought back the package—a gift for the child. Excited, the little boy undid the wrapping and pulled out an American Indian costume. He put it on over his pajamas and insisted on wearing it to bed, beaming all the while. This gift of Americana was never forgotten by the Mangolds.

More than ever, Just felt good in Europe. No longer did he avoid other visiting American scientists. He was becoming too self-confident for that. Ironically, he began taking a few of the younger ones under his wing, as he had done at Woods Hole. Partly it was his new sense of "belonging," partly the knowledge that Americans back home were showing a little more respect for his professional position. He gave special attention to Heilbrunn's former student Lester G. Barth. One evening he took Barth to the most elegant restaurant in Berlin. The young postdoctoral fellow gazed in wonder as Just deliberated on wines and courses and was fussed over by the maitre d', waiters, and stewards. Such deference to a black man would never have been dreamed of in America. The two men joked about the difference, but there was a serious and disturbing dimension to the joke that did not escape their notice.

Just continued to see Margret Boveri, but not as much as before. Again he was staying at the Harnack Haus, in part because he enjoyed the company of the scientists there but also because his feelings for her had cooled. She, on the other hand, was as ardent as ever. Before his arrival she had taken great pains to find a place large enough for them both to live for the summer.[96] She had had to give up "Friedenau" when the painters returned. With the acute housing shortage in Berlin and stiff rules enforced by the government on the amount of space an individual renter could occupy, her only chance to get a two-room apartment was to be engaged or married. She worked out some such "official" arrangement with the brother of her right-wing friend Elsbeth Krell and moved into a nice apartment on Neuchateller Strasse, near the corner of Hindenburgdamm and Händelplatz. All this trouble was on Just's account, and she felt a little piqued when he decided not to stay with her.

Just felt himself drifting away from Margret. They had few friends or interests in common, and so the relationship had nowhere to go. They had first been attracted to each other in Naples, when they were two lonely strangers trying to find a place for themselves in a foreign land. It was a relationship built on fear of the unknown, without a basis for anything long-standing. Now the differences between them were becoming painfully obvious, at least to him. He was a natural scientist, devoted to his experiments and only political in a theatrical, romantic, passive way; she was a social scientist, dedicated to finding practical measures for righting the wrongs of the world. Their only common interest was music, but an evening concert here and there could not sustain a love relationship.

For the most part Just stayed with his scientist friends around the Institut, but he also found the opportunity to renew his acquaintance with the crown prince and his circle.[97] He was again invited out to Potsdam for dinner and political discussion, and this time he was also urged to come prepared to give a lecture on the world situation. He declined the invitation to lecture, pleading that he had no expertise in political affairs, but gladly agreed to go for dinner and discussion as he had done the previous year. It would be pleasant, he thought, to talk some more with eccentric monarchists about the glories of the Hohenzollern dynasty. Much to his regret, the atmosphere around the "court" had changed, and there was a new tone of ugly frustration similar to the one that characterized Boveri's right-wing extremist friends. The monarchists were now advocating Nazism, in the hope Hitler might consider reestablishing the Hohenzollern throne if he came to power. Some of them wore brown uniforms and swastikas. The crown prince seemed his same old wine-and-women-loving self, but there were hints that he too was swaying with the tide. His brother, Prince Augustus William, had been quoted as saying, "Where a Hitler leads, a Hohenzollern can follow," and in less than a year the crown prince was in fact to allow himself to be used by the Nazis in propaganda displays of various kinds.[98]

Just always tried to keep away from extreme politics. He sought out Grethe Alter and took her into a corner to talk about anthropology. He was still fascinated by her "wonderful work on African songs," by the courage and seriousness with which she had set out to be "the first white person . . . ever . . . to record certain tribal music, especially that used in various ceremonials."[99] They discussed the possible relationships between African music and American jazz, and agreed that one of the most essential needs of modern anthropology was for "a really expert and scientific analysis of African backgrounds of American Negro culture in all phases, not only in music and other arts, but also in religious and tribal customs." Perhaps Just could help Alter begin such an analysis, as he had helped Herskovits a few years before. It was an exciting prospect, and the end product would be ample reward for the time and effort involved in canvassing foundations.[100]

The *Amoeba* experiments at the Institut were going well, but Just needed a rest. In mid-June he went south for a little vacation with the Dohrns. They entertained him at their home and their country house in Porte d'Ischia. He enjoyed the sun, the quiet—so different from the gloom, political and otherwise, of Germany. Just strolled around the marketplaces, relaxed on the beach. He did not register for a table at the Stazione, though he did go out with the collecting crew a few times to look for curious specimens. To him zoology could be a restful pastime as well as a serious enterprise. He was invigorated by the smell of the sea air, by the sound of the nets splashing in and out of the clear blue water. He never knew what the nets might bring up, but each time there was sure to be something of interest—a multicolored fish, an eel with large fins, a larva with more or less than the usual number of arms or legs.

Just returned to Berlin in late June, but he would rather have stayed in Naples. He was growing tired of the German political situation. Though it did not bother him when he was at work or among his scientist friends at the Institut, it made him nervous to go out after dark. Violence was increasing. He had been warned by a friend months earlier, before leaving America, of the way life had deteriorated in Berlin: ". . . the Communists and Fascists are both making energetic whoopee. Perhaps they'll make history . . ."[101] At the time he thought this an alarmist statement, but seeing was believing; clearly his friend had not been far wrong. All sense of moderation seemed to have gone out of the Germans. The government, weak and precarious, was caught between the extremist factions, powerless to do anything. Walking out on the street meant having to be alert for the sound of gunfire and shattering glass. The next step, apparently, would be revolution or anarchy.

Just planned to cloister himself in his laboratory for the rest of the summer. He wanted to finish up the *Amoeba* work before leaving for America in time for the start of the fall semester at Howard in late September. But Boveri would not leave him alone. At the end of June she gave a party in his honor.[102] Gregarious and unconventional, she was anxious to show off her black lover to her friends and colleagues. And sensing that Just was growing more distant toward her, she hoped a party of this sort would bring him back to his old self.

Just did not want to go. He knew what kind of friends Boveri had; they were loud, doctrinaire people with little understanding of the things that interested him. When he arrived, deliberately late, the apartment was full of smoke and noise. To be polite he drifted from group to group, but his mind was blank. More than once he was offered a pfennig for his thoughts. Smiling to himself, he realized how insulted these people would be if he told them the truth: that he simply was not interested in anything they had to say.

Then, over in the far corner, he saw a young woman who was also bored by the noisy political chatter going on all around. She was standing by a bookcase, smoking one cigarette after another.[103] Tall and elegant, she was dressed in white, like "an angel." There was an air of superiority about her, as if she were above the world and all its sordid realities. At the same time her eyes, which pulled him in almost like a whirlpool, seemed soft, sensitive, generous, intelligent.[104] Here was someone worth meeting and talking to, someone who might have sound ideas, deep interests.

He went over and she introduced herself as Maid Hedwig Schnetzler.[105] A philosophy student, she had become friendly with Boveri while attending one of the seminars at the University of Berlin, and her father was managing director of the Mannheim branch of Brown-Boveri. Although she was about to finish up her dissertation, she had not yet decided what she would do afterwards. Perhaps she would get a teaching job in one of the high schools, perhaps she would simply go home. For a woman in academia the future was bleak. Hedwig had the knowledge and intelligence to shine alongside the best minds of the day, but she had not yet found out how and where she might be able to make use of her intellectual energy.

Just listened and watched intently. He spoke about some of his own hopes and aims, but his attention was fixed on the woman standing in front of him. She seemed brilliant as well as beautiful, a perfect combination of the best human qualities. There was no cant or pretense about her; she was a deep and honest thinker with a warm, giving personality. Before he knew what was happening, he was, as the Germans say, *verschossen*—madly in love.

Hedwig felt much the same way. She appreciated Just's sensitivity, his intellectual depth, and last but not least, his exotic appearance. She had always been attracted to foreigners with swarthy complexions. A year or two earlier she had had an affair with an Indian who was studying at the university. Now here was a man as attractive as her other friend, and far more intelligent. Maybe, just maybe, she had found what she had been looking for—a lover who could stimulate her in every way.

Alone together, they talked away the evening. Boveri did not notice. Two days later she was pleased to hear that they were planning to go to the opera, and she even offered to make the reservations for them. Little did she know that that night at the opera would signal the end of her own affair.

The tickets were for *Die Zauberflöte*, the same opera Just had decided against seeing in Paris the first week of his first trip to Europe. He was too captivated by his companion to remember anything about that, however. After the performance they strolled down to the end of the Unter den Linden and sat under the trees near the Brandenburg Gate. There, in the moonlight and the balmy air, they embraced and kissed. It was the start of something new for them both.

Just and Hedwig were together for the rest of the summer, going to concerts, strolling in the woods, lying in the grass near a lake, discussing views and ideas of mutual interest, peering through the microscope at his slides. Something important was afoot, and every minute counted. Time rushed by all too quickly. In early September Just had to say good-bye and take the train for Bremen. On 10 September he boarded his ship for America.

For Just the next decade was full of trauma and excitement. Europe brought to the surface many of his anxieties, his negative feelings about life in America, but at the same time opened up new possibilities for him and sparked his creative energies. Both his science and his personal life underwent radical changes. Many of his childhood dreams resurfaced: his underlying awareness of his German ancestry, his religious intensity, his penchant for fantasy. Though approaching middle age, he was not too old to change. Europe made him a different person.

Hedwig Schnetzler became his sole passion, while Ethel and the children faded out of his life. After the summer of 1931 he made seven trips to Europe. He wanted to make more, even to remain there permanently. That was impossible, but at least he had the chance to go back and forth between Europe and America, to live a kind of nomadic existence he had known before as a child in Charleston and on James Island. Thoughts and feelings about his life and career in America that had been with him since his first days at Woods Hole, but until Europe had been simmering inside, only boiling over on rare occasions, now began to emerge. With Hedwig's prodding he began to think more clearly, to unravel his thoughts and come to decisions as to who were his friends and who were not. His European experiences helped him develop the self-confidence to take a new look at people like Schrader and Heilbrunn, to question their motives, to keep as friends only those who were genuine or who could be helpful to him. Questioning the relationship with Lillie was most painful to Just, since the two men had been close for so long, first as mentor and protégé, then as collaborators and friends. But Hedwig was evocative, drawing insights from Just that he had never before been conscious of, much less expressed. Perhaps Lillie had not been as open or as helpful as he might have been; perhaps he had used Just's loyalty and scientific skill for his own ends.

After Europe Just began to stand up for his rights as a scientist, an oppressed black man, a human being. At Howard he became openly antagonistic to the administration, which now under Mordecai Johnson seemed even more like a trap than it had under Durkee, holding Just in a viselike grip that grew tighter the more he struggled to escape. The pressures were harrowing. Teaching and committee work kept him on campus up to fourteen hours a day. His research was suffering. To continue it effectively, he had to leave Howard. But no white institution in America would hire him.

As the Depression took hold, money grew tighter. After Rosenwald's death in 1931 the Rosenwald Fund decided to drop the grant in support of Just and his department at the end of the initial five-year term in 1933. Desperately he ran from source to source—the Carnegie Corporation, the Rockefeller Foundation—looking for money. He had become unpopular, perhaps even despised in America for his love of Europe, and as a result his funding efforts met with severe opposition. He grew more aggressive, developing a pathetically strident tone in his approach to the foundations. This did not help matters. As a last resort, he tried to establish contact with wealthy and influential people in all walks of life—Lady Nancy Astor, John Barrymore, Benito Mussolini, and Charles Lindbergh, to name only a few of the most prominent. He became embittered and exhausted by the struggle to continue his research, to move to a more favorable environment, to begin a new life. But his heart was hard, his spirit unbroken.

More than anyone else, the Woods Hole scientists did not like the new Just. Europe had opened up his science, had given him the confidence to take daring steps in his scientific work. He moved away from experimental details to larger philosophical views; he searched for a new world view through some grand alliance of the various fields of science, history, and philosophy. Perhaps these theoretical disputations were too speculative, with insufficient evidence to make them deserving of serious scientific attention in America; perhaps, too, these speculations reeked of arrogance.

His other goings-on in Europe, in particular his affairs with white women, became stock-in-trade gossip among American biologists. To their minds he had gone too far, plucked too many discordant notes, and he could not be forgiven. Angrily, he met their opposition; he refused to let them dictate either his professional direction or his personal life. His work became stinging in its criticism of them, as scientists and as people. But he could not afford to alienate their sympathies completely. He needed the help of the most influential among them, Lillie and Ross Harrison for example. He made public gestures of respect toward them, while secretly he came to loathe them and covet the opportunities and positions they had received on silver platters and he had had to struggle for. Though sometimes close to paranoia and even schizophrenia, he managed to remain for the most part clear, precise, and rational.

In 1938, almost penniless, he began a self-imposed exile in Europe. Washington had become an impossible place to live, what with his futile efforts to get Ethel to agree to a divorce and the deteriorating situation at Howard. Europe was a refuge. There he and Hedwig lived openly together for a year before marrying in the summer of 1939. In France they completed what he considered to be his crowning achievement, a book entitled *Biology of the Cell Surface*, based on the theories he had recently developed and the experimental results of his entire career. Because Europe had little in the way of professional opportunities, he fought to keep his professorship at Howard, his only source of income. When he refused to go back to Washington permanently, his salary was cut off. He stayed in Europe, preferring to struggle there in poverty than to live in America. In mid-1940, however, he was forced out of his laboratory room at Roscoff Station in Finistère, France. The Nazis had taken over, and the station closed its doors to all foreign scientists. Just tried without success to get out of France. He was detained and interned by the Nazis for a short time. After his release, there was only one place for him to go, and that was back to Washington. But as the novelist Thomas Wolfe, whom Just read and admired, so aptly put it, "You can't go home again."

CHAPTER 6

Howard University: Continuing Struggle, 1929-31

When Mordecai Johnson assumed the presidency of Howard in the fall of 1926 the university was, in a word, a mess. Its faculty morale was low, its financial status shaky, its academic standards abysmal. Howard, with its diffuse system of colleges and junior colleges, its unaccredited professional schools (the exceptions being the schools of medicine and dentistry), and its makeshift graduate programs, was a long way from being a major university. Its reputation had gotten worse under Durkee, and in many ways the educational standard was far below what Just first encountered in 1907 under Thirkield.

The upgrading of Howard was Mordecai Johnson's major goal. He moved quickly to eliminate the junior college program, start up permanent graduate programs, and acquire accreditation for the various professional schools. Toward the end of his first year as president he clinched a formal annual appropriation from Congress for Howard, totaling $368,000 in 1926-27.[1] The appropriation had been in the works for some time, but it is unlikely that Congress would have committed itself to a school and a cause with a leader of less vision and drive.

Johnson came to his position by way of the ministry, having served for several years as pastor of the First Baptist Church in Charleston, West Virginia.[2] He had been educated at a black preparatory school, Atlanta Baptist College (later Morehouse College), and like Just he had done college and graduate work at major white universities, earning his bachelor's at the University of Chicago and his master's in theology at Harvard. As a student he had distinguished himself as both an activist and a scholar, serving in one or another official capacity on the International Committee of the Young Men's Christian Association and delivering a key address, "The Faith of the American Negro," at his Harvard graduation exercises. But Howard's first black president, while he was well known in the black community, had few connections beyond it. Nevertheless Johnson's energy and dedication gave promise that he would be able to quickly make contacts and establish himself as a valuable and highly respected administrator.

Johnson had ambitious plans for making Howard "a power in the life of the Negro people and in the nation."[3] Among the programs on his drawing board was the expansion of graduate studies to include all departments—and especially zoology, which boasted a scholar of world-wide reputation. He did not present a proposal to Just for the zoology department right away, preferring instead to speak in general terms about "extending" encouragement for Just's scholarly aspirations.[4] A careful politician, he knew he should proceed cautiously because of Just's earlier reservations regarding the feasibility of graduate work at Howard. Within a year, however, he had sufficiently gained Just's confidence to raise the issue in a more straightforward manner.

Just was inspired by Johnson and his talk of improving black education. In many ways, he saw a reflection of his mother in the new president. Johnson had the same kind of strength and authority necessary to get a job done. He also possessed a similar missionary zeal, a fire-and-brimstone dedication to the cause of Negro education. Later Just came to feel that Johnson was suffering from a "Messianic complex," much like Martin Luther centuries before, but in the early years the president seemed to Just to have all the qualities of a good, perhaps even a great leader.[5] Just was awed. He agreed to listen to Johnson's plans, think over his views on a graduate program in zoology, and reconsider the larger question of graduate education at Howard.

Doctoral work had never been offered at Howard, but masters' degrees had been awarded at the school from its inception. In the late nineteenth century a graduate could earn the master of arts (M.A.) or master of science (M.S.) degree simply by undertaking three years of "professional, literary, or scientific studies" outside of the university and then submitting a sketch of his occupation and the line of study in which he was most interested.[6] Toward 1900 a thesis was added as part of the requirements, and year by year after that the requirements were stepped up to include comprehensive examinations, a year in residence, specific scholarly aims, and the like. There was no real program, however, and few degrees were awarded; only a highly motivated graduate would stay to work under a favorite teacher who might or might not be willing to spend the extra time, in addition to a heavy undergraduate course load, directing advanced thesis work. Most of the degrees awarded were M.A.'s in history or education. The Master of Science was never awarded, and in 1907, when Just first went to Howard, it was abolished because of lack of interest.

For ten more years graduate study at Howard remained in limbo, a part of university life but not integral to it. Then, in the 1918–19 academic year, the first Committee on Graduate Studies was established. Under the chairmanship of the new dean of the College of Liberal Arts, Carter G. Woodson, the committee was to try to develop an awareness of the importance of training more black scholars in the various subject areas. In addition to the university deans there were two faculty members on the committee, Just and Alain Leroy Locke, and a year later a third was added, St. Elmo Brady, chairman of the chemistry department. These men, eminent scholars with doctorates from prestigious institutions, were the pride of the Howard community. Their job was to create a viable graduate program at the master's level. It would be unwise, they reasoned, to go about the thing in a big way; with the university's limited financial and personnel resources, the primary focus had to remain on undergraduate education. Be that as it may, in the catalog for the 1920–21 academic year a specific graduate program was outlined for the first time, with the promise of teaching fellowships for candidates qualified to help out in their departments. By 1925 there were eight graduate students in residence. President Durkee proudly announced this development in his annual report that year. The number of M.A.'s awarded increased twofold. The first M.S. was given in chemistry in 1923, and in 1925 two more chemists took their degrees.

Chemistry and history were the first two departments to establish graduate work within the new guidelines. Even though Just was a permanent member of the Committee on Graduate Studies and was no doubt asked to plan a master's program in zoology, the zoology department remained unchanged. In his view the department had neither the personnel nor the equipment for a graduate program, at least not one up to his standards. With only one professor with a Ph.D. and almost as few microscopes for instructional purposes, the department had to work overtime merely to maintain its high-quality undergraduate program. Just was interested in the idea of training a few black biologists, but only on a limited scale. Instead of running a full-scale graduate program at Howard, he directed his efforts to finding graduate fellowships at other universities for his most promising undergraduates. He got Roger Arliner Young into the master's program at the University of Chicago, and Louis E. King and Greene C. Maxwell into programs at Columbia.[7] As far as Just was concerned, the zoology department at Howard had to keep its primary emphasis on premedical education for undergraduates. It was only practical to do so, since there were next to no career opportunities for black biologists, whereas black doctors were in great demand in the South and the urban centers of the North.

Just had mixed feelings about graduate studies in zoology at Howard. Working in the forefront of science, he knew that the budgetary demands for a graduate program were substantial and exceeded those Howard could meet, that it would be best to continue to focus on zoology as integral to a sound preprofessional education for under-graduates. On the other hand, he was thinking seriously about passing on to more students "some of the many problems" he could not possibly finish up alone.[8] He had tried to work with undergraduates, at one point even writing papers with them for presentation at a scientific congress, but this was a "haphazard" arrangement at best.[9] There was always Young, of course, but she had too many teaching duties and only had time to follow up one aspect of his work, the ultraviolet, during a month or two at Woods Hole each summer. From several points of view, then, it would be good to start a graduate program, not a high-powered one requiring major overhauls in the department but a limited M.S. course tailored to his own work. Just could set his students to work on problems he was interested in but did not have the time to work on himself, and meanwhile he would be contributing to John-son's plans to strengthen the university.[10] Most importantly, there was a chance that an expanded commitment to the cause of Howard and black education on Just's part might make Rosenwald willing to renew his support in a larger way. Just was not a person to let a financial opportunity slip through his fingers, especially one that he had courted from the start.

It was in the summer of 1928 that Just submitted his eight-page proposal for setting up a Rosenwald Institute of Zoology,[11] where an M.S. degree would be earned in one year by regular students, in two by students with research or teaching assistantships. The full proposal was considered too grandiose by the Julius Rosenwald Fund,[12] but it helped win Just an arrangement whereby $80,000 would be provided over a five-year period to support two aspects of his proposal: de-veloping his own research program and training advanced students in zoology.[13]

The Rosenwald grant meant money and money meant power, espe-cially in a community like Howard where funds were scarce. Many a payday teachers opened their envelopes to find IOU's from Howard, certificates hardly negotiable in the black community, let alone the white. A grant of $80,000 to a single individual or department at Howard was unheard of. So when Just and the zoology department received this money, all eyes flashed that way. "Who will spend the money?" everyone wondered. "Who will control the allocations?" For this was at a time when most academic insitutions, including Howard, had hardly begun to work out ways of handling the complicated procedures involved in grant proposals.

Johnson wanted complete control over every department, from rou-tine administrative matters to curriculum modifications to the broader goals and direction of the various programs. He had set about gaining this control by placing clear limits on the activities of the officers of his administration and the heads of academic departments. He appointed some new administrators who seemed likely to comply with his wishes, and dealt with those he had inherited from Durkee's administration and even further back one by one. He was not long in attacking the venerable Kelly Miller, the former dean, for instance, who would not toe the line.[14] Just was an "inheritance" too, but he was in the prime of his career and, as Howard's most famous faculty member, required more delicate, if not devious, treatment than some of the others received. Johnson was prepared to work this out over the course of time, but first he was going to make certain that the Rosenwald money came under his jurisdiction and not Just's.

Just had other ideas. Having had some experience with outside funds, and a great deal with Howard University, he knew that to insure correct, efficient administration of the grant he should set down on paper the process for allocating the money. Howard had a treasurer, Emmett Scott, but Just wanted absolute control of the expenditures, such that his signature—and his alone—would be all that was necessary for a requisition. Within a week of hearing news of the grant, Just drafted a letter to Embree, the Rosenwald Fund's president, outlining the procedures he planned to follow. He showed it to Johnson, and was confronted with a reaction he did not quite expect. The president began to plead that Just not be so rigid about the terms of the grant and that he not send the letter. Johnson's arguments are easy to imagine. As the new president, Howard's first black leader, he needed to have a role in the running of affairs. How could he ever gain respect from national foundations and funding agencies if he did not acquire experience handling money? And there were larger university concerns that had to be taken into consideration, ones that only he, as president, the focal figure, could identify and deal with properly. Besides, Just was a scientist—a brilliant one at that—and should devote his attention to furthering his career and not worry himself over petty administrative matters. They were black brothers together, Johnson most likely said, and the time had come for them to trust and help each other in their separate roles. Just listened without saying a word. Then, as he was about to leave Johnson's office, he held out the letter and slowly ripped it to shreds.[15]

Just got straight to work trying to revamp the undergraduate courses and organize the zoology graduate program. At the time there were only four undergraduate courses offered; that had to be increased to at least eight, since one of the university prerequisites for admission to graduate study was eight undergraduate courses in the particular field. The number was actually increased to ten, and there were five additional upper-level courses (for undergraduates and graduates) and nine courses for graduate students only. As a result the zoology department was offering six times as many courses in 1928–29 as in 1927–28, with no increase in the teaching staff. There was no cause for immediate concern, Just figured, since it was sure to take at least a year for students to show interest in the new program. The crunch would probably begin in 1929-30, and there was time in the interim, he hoped, to think about expanding the staff. In any case, what was uppermost in his mind was not the graduate program but the fact that these new funds meant increased opportunity for him to continue his own research, perhaps at one of the top European laboratories.

What was uppermost in Johnson's mind was control of financial affairs at the university. He liked to make that control apparent to all by showing the faculty his power over their salaries, even using salaries as a means of rewarding friends and punishing enemies. Just came under special scrutiny because of his outside funding sources. At the time the Rosenwald funds came through, he had been receiving a grant from the General Education Board of $2,500 per year. This grant was not part of his salary; it was a continuation of the old Rosenwald grant that had been a supplement to support his summers of research at Woods Hole. It was administered by the National Research Council and not Howard University, and the checks were mailed directly to Just's home at 412 T Street. Johnson thought the General Education Board grant should be regarded as part of Just's salary, leaving Howard that much less to pay. He was as disturbed about the amount of the grant as about the arrangement. With the money from the General Education Board, Just's income ran about $5,000, higher than that of most members of the Howard community, and with the Rosenwald grant it could increase still further unless close tabs were kept on budgets and expenditures. In Johnson's opinion, Just should earn more or less the same as other faculty members—and certainly not more than the president himself. It did not occur to Johnson that Just might deserve more; after all, he had achieved a higher level of recognition than anyone else at the university.

Johnson used the salary issue to try and deflate the image of power that Just was beginning to acquire. The conflict between the two men deepened during the winter of 1928-29, as Just prepared for his first transatlantic voyage. News coverage of Just's trip to Naples was extensive. An article in the *New York Times* was headlined, "A Negro Biologist Has Won Distinction in his Field."[16] All the attention focused on Just, only secondarily on Howard, and never on the university's president. Furthermore, while pretending to dislike publicity, Just secretly courted it. He told his life story to Mary Church Terrell, the black writer, who was so fascinated that she wrote a long article on him for the *Washington Post*.[17] Johnson no doubt felt slighted, perhaps a bit jealous, and wondered what approach he should take, what attitude he should assume toward the booming success of Howard's finest scientist. Would he support him or would he fight him? One trip to Europe was all right, he must have thought, a new thing for a Howard faculty member—but subsequent arrangements of this sort would have to be more strictly monitored. He would get a tighter rein on the allocation of money and keep a close watch on Just's future travel plans.

In January 1929 Just left to spend six months in Naples, and while he was there he did not have much time to think much about the Howard zoology department. Still, he wrote to Embree about hiring a part-time outside instructor for the graduate work the following fall, a stopgap measure until one of the two young instructors in the department, Roger Young and Louis Hansborough, earned a Ph.D. and took over full time.[18] The letter was also an attempt to gain some leverage in his struggle with Johnson. Just hated to complain, and if we can believe him, he did not want to approach Embree as a "wayward child appealing to his godfather." But the situation was bad and growing worse, and Just felt he had no alternative but to detail the disagreements that were brewing between him and Johnson about not only salary but broader educational issues, particularly the new graduate program in zoology.

Conflicts among blacks were often difficult to solve internally, within their own community. They had little control over their lives when it came to financial arrangements, but at the same time they hated to appeal to the white man as the arbiter. Just was no exception. Should he fight Johnson or should he go along? At first he thought about waiting to see how things would work themselves out, but he could not allow himself to continue that course indefinitely. He had the capacity to be decisive and pointed, something we have seen in his dealings with Flexner and others. So when it seemed clear to him that there was no other way out, he went to Embree. The move was unusual. Johnson had created a stinging fear in the black Howard faculty, had made them feel guilty about going to whites behind his back, about tattling on Howard's first black president. Scared or loyal, perhaps a little of both, they kept quiet. Just, however, was not afraid of Johnson, and his earlier feelings of loyalty and admiration were evaporating.

No more was said or done about the salary issue or the graduate program while Just was in Europe. His life there was too full of excitement, scientific and romantic, for him to be concerned with such mundane matters. Besides, he was hardly in financial straits; checks from the General Education Board were still being sent to him each month.

On his return to Howard in September Just threw his energies into getting the graduate program under way. One of the first orders of business was to hire another staff member to help lead the seminar work. Heilbrunn was a possibility: he had been fired from his post at the University of Michigan, and was in need of a job. His colleagues advised him against accepting the Howard post, however; Franz

Schrader said that the whole idea was "intolerable" and that Heilbrunn should wait until something better turned up.[19] In early October Heilbrunn declined Just's offer in order to take up an appointment at the University of Pennsylvania.[20]

Just next thought of John Runnström, a Swedish scientist he had met in Naples in 1929 who was also having job difficulties. Runnström reacted positively to the idea of going to Howard, but stressed that he had no real desire to leave Europe and would only do so as a last resort. His current appointment was scheduled to terminate in June 1930, so he would not know for sure what the situation was for at least another year. Because he was a first-rate scientist, Just promised to keep the appointment available and hire a part-time instructor for the 1929-30 academic year. Johnson did not appreciate this, and suggested that Just was "temporizing" in an attempt to hold positions open for his "friends."[21] It was not, Just quickly pointed out, so much a question of friendship as of scientific quality. Hiring men like Heilbrunn and Runnström would help establish the reputation of the Howard graduate program. Just did not really expect Runnström to come, but if there was even a slight chance he might do so then it made sense to keep the appointment open. In the end, Runnström did not come; he was promised substantial backing by the Rockefeller Foundation to build his own laboratory—backing that, ironically, Just helped him secure.[22]

Just knew that there would be trouble getting a scientist with any kind of reputation to come to a place like Howard, a black institution with poor facilities and badly prepared students. He did not hold his hopes too high or his breath too long. By mid-October he was looking around for a good junior man with a Ph.D. Johnson had suggested "borrowing" someone from another university, but Just replied that no university was in "a lending mood" and that Howard's best hope was to use the advantage of being in the nation's capital and get someone who was doing scientific work with the federal government to come over from time to time, in the late afternoon or early evening if need be.[23] Just contacted T. C. Byerly, a young zoologist whom he had met at Woods Hole in 1926 and who had come to Washington to work with the government after earning his doctorate. Byerly's work fitted in perfectly with the main thrust of the program Just wanted to develop at Howard, covering the fields of general physiology and experimental embryology.[24] If Byerly could arrange time off with his supervisor in the government, the Howard graduate program would be off to a good start.

Byerly agreed to teach two or three seminars beginning with the January term. Everything was in order. Just would devote his full attention to the graduate students in the fall, and while he was in Europe during the winter and spring terms Byerly would take over. Two additional instructors, Greene C. Maxwell and Lawrence A.

Whitfield, had been hired to pick up the slack with the undergraduate courses, since Just did not have time for undergraduate teaching now and Young and Hansborough were both due for study leave. The personnel situation was touch and go, but if everyone chipped in, Just told himself, the department should be able to carry on.

There were many applications for admission to the graduate program, though none of the applicants had the depth of preparation and seriousness of attitude Just had hoped for.[25] Five students were admitted, but only two of them, William Bright and Hyman Chase, had satisfied the minimum requirements laid down by the university for admission to the program. Just did not want to accept the other three at all, but was pressured to do so by Johnson, who thought that a program with only two students registered could not be considered bona fide. One of the other three, surnamed Reid, left to teach after a couple of weeks; another, Baxter Goodall, stayed for most of the year and then went into journalism. The third, Wallace Wormley, was committed to his work but had so much basic material to catch up on that he had no hope of earning the M.S. in one year.

Throughout the fall Just worked with his graduate students *"every single day."*[26] The seminars went well, and he even managed to launch his two star students, Bright and Chase, on thesis work that would earn them their degrees in the summer of 1930. Without Bright and Chase, however, things would have seemed dismal. Just grew impatient with the other students, though he was careful not to show his feelings to them. Meanwhile the undergraduate program was foundering because of lack of equipment, and Just threatened the administration that if something was not done about it he would cut out the seven new courses, thus making it impossible for Howard undergraduates to satisfy the admission requirements to the graduate program.[27] He was furious when the administration used the Rosenwald grant to purchase the necessary equipment for undergraduate course work. The terms of the grant did not call for it to be used in that way, and in any case departmental expenditures of that sort were supposed to be covered by the general budget of the university.[28]

To an extent Bright and Chase helped offset Just's anger, and made him feel better about the department and the whole business of graduate education. By December he was even ready to project for the zoology department at Howard "a program of research of significant proportions and of wide implications."[29] His enthusiasm peaked shortly before he left for Naples in January. He wrote a glowing recommendation for Bright for a post in biology at North Carolina State College, stressing his policy of accepting only the "best men possible" into the Howard graduate program and his hope that everyone who took the M.S. under him would "reflect credit on this department and enhance the possibilities of extending the scope of our influence in the scientific world."[30] He considered Chase's work better than that of the average Ph.D. at white universities, and also recommended that he be admitted to one of the summer courses at Woods Hole. It was the first time he had written such a recommendation.[31] Earlier, it will be remembered, he had refused to do this for S. Milton Nabrit, on the grounds that he had to be very careful about ascertaining the quality of a student's work prior to writing any kind of endorsement.

The fall term was particularly hectic because of Young's absence. Though she was his right hand in the department, he had persuaded her to go off to Chicago to further her studies. In the meantime he had accepted an invitation from Max Hartmann and was getting ready to spend two-thirds of the academic year at the Kaiser-Wilhelm-Institut. Neither he nor Young would be at Howard to teach courses and administer the department during the winter and spring terms. No matter how unwise, even selfish this may have seemed to the university community, he was convinced that the move was a judicious one because of the long-term benefits it would produce. International exposure would keep him in public view and perhaps put Howard on the map. By the same token, it was important for Young to expand her horizons. Most likely she would prosper in Chicago as he had done in 1915, and return to Howard, Ph.D. in hand and ready to assist in running the department and assume the professorship created by the Julius Rosenwald Fund. His plans for Young did not work out, however. Hers is a sad story of frustrated hopes and blighted talent.

After working under Lillie's supervision at Woods Hole during the summer of 1929, Young went straight to Chicago to prepare for her qualifying examinations and fulfill the university's one-year residency requirement. She seemed to be a good candidate for the doctorate. In 1926 she had secured the master's degree at Chicago and been elected to the science honors society; since then she had worked closely with Just on research problems and published noteworthy work of her own. She seemed motivated. At Woods Hole she had become part of a circle of ambitious young biologists who were working on their Ph.D.'s in the hope of securing faculty appointments. Young was different from her Woods Hole friends, though, in that she had assumed an assistant professorship, the second position in the Howard zoology department, right after obtaining her bachelor's degree. This quick professional advancement had started rumors about a possible relationship between her and Just, and the rumors only increased when the two were often seen together working late at night in the laboratory. One way to quiet speculation, Just believed, was for Young to earn her degree and thus appear more deserving of her position.

Chicago was not new to Young, and she had few problems settling down to work. Courses kept her busy the first semester. She performed well, though not up to the same level as four years earlier, apparently because her eyes were "bothering her somewhat."[32] The week before she was to take her comprehensive examination, Just departed for Europe. Whether she needed his presence in America as some kind of symbolic support for her efforts we shall never know, though it would not be unreasonable to expect so. She took the examination on 10 January, and the results were disastrous. She could not answer the simplest questions, many of which she had handled with no trouble in her course work, in her teaching at Howard, and in her research with Just. She was rejected by the doctoral program.[33]

Everyone was upset. Young left Chicago, and no one knew where she had gone. Anxious, Frank Lillie informed Johnson about her "nervous state" and the "severe mental strain" she had suffered.[34] Johnson could shed no light on the matter. All he knew was that she was gone from Howard for the year, and that, with Just in Berlin, the zoology department was left with few teaching resources.

Young had in fact left for a rest in the hope of pulling herself together to continue her studies. Confused and tired, she tried to explain to Lillie what had happened:

> This note may be as disconnected as my work has been but I hope I can make clear to you how very grateful I am to you. The examination was quite fair and easy—in the oral examination with you I could not think of but one paper out of the five in my report.
>
> The trouble is, that for two years I've tried to keep going under responsibilities that were not wholly mine but were not shared and the weight of it has simply worn me out. I have forced myself on so long that I automatically accepted arrangements for examination which I knew the first of last August I would fail unless there was some relief. Instead of relief the situation has become worse since I've come here. It is not exactly and [sic] outside thing—it does concern my work at Howard—if I took an examination a month from now under the same conditions I'd more than likely do worse. I could go on as long as there was any hope of satisfying Dr. Just by getting the degree before my return. I could keep on trying, but now that I've made that impossible I think I would gain more and create less embarrassment by giving up the whole thing for a while at least.
>
> Again may I thank you for your real kindness. I'm sorry I couldn't talk to you but I seem to have lost my grip all around.[35]

She was frustrated and angry. She felt that she had been forced to shoulder too much of the administrative responsibility for the Howard zoology department and that, as a result, she had not had adequate time to keep up her work in the field. In her mind, Just was to blame. He had gone off to Naples in 1929, requiring that she give up a fellowship she had received from the General Education Board and take on the burden of the department for eight months.[36] Also, if there was a romance—as gossip in the Howard community had it—Young must have felt abandoned on that count as well. The year before he had not written her once from Naples, and her pleas to Dohrn to tell her Just's whereabouts had proven futile. To add insult to injury, when he returned in August he had chosen to stay in Washington rather than join her at Woods Hole. Then in September she had gone off to Chicago, and soon after that he had begun planning another trip to Europe. They seemed to be growing apart.

Science had become Young's livelihood. She had devoted every spare moment to her work, spending her evenings in the laboratory and her summers at Woods Hole on top of her regular teaching duties. She had even done permanent damage to her eyes while performing experiments using ultraviolet rays—a sacrifice, in a way, for Just and for science.[37] Though she had received some money from the General Education Board and from Rosenwald through Just, she had been obliged to rely on her Howard salary to support herself and her invalid mother and to supplement her research and study expenses. She could not allow her job to be threatened by not having a Ph.D. Johnson had been pushing Just to hire Ph.D.'s and Just was pushing Young to get hers. In effect, Just was almost obsessed with the idea that Young had to acquire professional status and gain a more prominent image for herself in the scientific community—even though he played a part in keeping her at Howard and persuading her not to accept job offers at other institutions, notably Spelman College in Atlanta.[38]

Young's failure in the examination made her worry more about her job and her financial future. To make matters worse, she was penniless when she returned to Chicago in March. Desperate, she borrowed money from Libbie Hyman, an instructor at the University of Chicago whom she knew from Woods Hole.[39] She registered for two more courses in the spring term, but did not perform well in either. The best thing she could do, it seemed, was resume her teaching at Howard and postpone, for a while, her work toward the Ph.D.

Young and Just never got along after her failure at Chicago. Though he never mentioned the incident explicitly, he must have felt let down. After all, he had promoted her and her work, paraded her at Woods Hole, gone to bat for her at Howard and with the foundations. He had put his own reputation—one thing he was especially sensitive about— on the line. He had been teacher, mentor, and companion to her, and she simply had not measured up to his expectations. Still, there was no question that he needed her services around Howard. It would be best, then, to ride with the tide and hope that things would work themselves out.

Whether or not Just and Young had been romantically involved before, Young now began to act the jealous woman. In the fall of 1930 she noticed that Just was receiving more mail than usual, and more from Europe—from Margret Boveri, in particular.[40] Just had asked Boveri to use his Howard address rather than the one at T Street so that his wife Ethel would not be able to put two and two together. Usually he was careful to be in his office when the mail was scheduled for delivery, but with his hectic schedule this was not always possible. Young started going through his mail and intercepted one of Boveri's letters. With proof in hand she headed straight for Johnson's office. She knew that Johnson and Just were drifting apart and that a disclosure of these romantic carryings-on in Europe would only widen the gap. Only later, about a year or so, did Just find out what had happened.

While he was in Europe in 1930 Just had again found little time to think about the graduate program and the direction it should take. In any case, he felt comfortable about the way things were progressing. Perhaps—and he could never give up this idea—there would even be a chance some time in the future of organizing a larger-scale program, something like the Rosenwald Institute of Zoology he had suggested two years earlier. For the present, however, it would be best to go slowly, considering the limitations in staff, equipment, and student body, and to make sure not to ignore the research end of the Rosenwald stipulations. Just was determined to maintain his scientific reputation. If he stopped producing, there would be no way for him to attract either funds or students.

The question, though, was how a graduate program could get along, let alone flourish, in its crucial first years with the head of the program away in Europe. Was Just living in a fantasy world to think that the inexperienced staff he had left behind could handle the necessary administrative business, the delicate negotiations so important for a program just getting under way? Hansborough was then too inexperienced, really no more than a teaching assistant, with no clout and none of the courage needed to face up to Johnson. Byerly was white and working part-time, both of which prevented him from having much influence, should he have wanted it—and it is not certain that he did. Young might have managed to do something—when Just was in Naples she had taken over administrative duties for the department and even ventured to make contact with funding agencies such as the

Rockefeller Foundation—but she was in Chicago for the entire 1929–30 academic year, and was in no condition to help when she came back to Washington early in the summer. Besides, she was disgruntled about Just and was in no mood to cooperate with him in any of his enterprises.

There were other rumblings, both at Howard and the Rosenwald Fund. Neither Johnson nor Embree was satisfied with the progress of the department, and told Just so when he returned to Washington in July 1930 after six months in Berlin.[41] As far as they were concerned, he had shirked responsibility by taking off for two-thirds of the academic year and leaving seminars and thesis direction to a young part-time instructor. Byerly's seminars had had to be scheduled for nighttime hours, after his regular job, and this schedule took its toll on the students. At the end of a hard day's work in the laboratory the students were hardly happy to spend four or more hours in the seminar room at night. If Just had been around, they reasoned, there would have been no need to schedule so many nighttime classes.[42] Embree and Johnson may also have thought that more conscientiousness on Just's part might have allowed all four of his graduate students, instead of only two, to take their degrees at the June commencement.

Embree needed outside leverage to bolster his criticisms, so he tried to muster the support of Abraham Flexner and Frank Lillie. Flexner had just been appointed a member of the Howard board of trustees and had some power to influence decisions concerning Just, and he was known to have fixed ideas about Just's work and career goals. Perhaps contrary to expectations, he suggested that it was a good thing for Just to have had a chance to spend some time doing research in Europe. He also made it clear, however, that now that he had had that chance Just should "return to his post and do . . . his job, namely, that of a professor in a graduate school, who, while carrying on his own work, trains a few persons, among whom there will be someone who can succeed him, and others who can introduce modern biology into places like Fisk and Morehouse."[43] Lillie was more sympathetic toward Just, even though Just had been abrupt at Lillie's Woods Hole birthday celebration. Lillie knew the conditions Just worked under. As far as he was concerned, it was pointless to compare Just to professors in other graduate schools, since Howard had problems of space, quality of students, and equipment that professors elsewhere would have been horrified by.[44] Besides, the two had spent hours at the beginning of the summer and before the birthday celebration talking about the problems at Howard, and Lillie had found Just to be "fully cognizant" of his responsibility to build a first-rate graduate-level department. Just had pledged to do his utmost in the 1930–31 academic year; he had no plans for further research trips until the summer of 1931. Lillie warned Embree that much of the responsibility lay with Howard, that Just's accomplishments depended in large measure on the administration's active cooperation in raising physical conditions in the zoology department to a minimally functional level.

Though burning with anger inside, Just substituted reason for venom, trying as he was to keep the peace not so much between himself and Johnson as between himself and Embree. Through the good graces of the Rosenwald Fund he had managed to go to Europe twice in less than two years, and a third trip, to the International Congress in Padua, was in the making. He did not want to jeopardize his chances for continued support needlessly, so he made an effort to explain to Embree, calmly and clearly, the difficulties he faced at Howard.[45]

One major problem was the physical condition of the department. The Rosenwald grant had been conditional on additional space being provided, but Howard had provided none. Graduate students had to be placed in the classrooms used for undergraduate work, and sometimes even in Just's office. This overcrowding caused problems all around— for the typist, the students, and Just. High-power microscopy was impossible, and research output had ground to a halt. For years there had been plans to erect a new chemistry building with government funds, but each year they were put off. So zoology and chemistry remained squashed into Thirkield Hall, competing fiercely with each other for every inch of space, including closets and hallways. In any case, Johnson had already promised all vacant space that might turn up to departments other than zoology.

Johnson favored the chemistry over the zoology department. Chemistry was without a strong leading figure like Just, but it appears to have been Johnson's plan to develop the young black scientist Percy Julian into one. If some other black scientist were to attain stature equal to or greater than Just's, Johnson reasoned, Just would not be able to demand and get so much attention. Also, such a development would represent quite a coup for the president himself. With the administration's help, the chemistry department had a good small library well stocked with research journals, even beyond its needs. The department used the basement of Thirkield Hall for a laboratory, and, for routine teaching, had half the second and all the third floor. In addition, contracts supposedly were in the works for a new half-million-dollar chemistry building.

The physics department, on the other hand, could not begin to compete with zoology. It was very weak. The head of the department, Frank Coleman, was just beginning to try for his Ph.D. at the University of Pennsylvania. The department had one elementary laboratory which doubled as a lecture room; there was also an advanced laboratory, but it was rarely used. For some reason the physics library had a good collection, consisting of five hundred volumes, including research journals which the department could hardly have used. The botany department was better than physics, but not much. Its instructors were fairly good but not very qualified. C. S. Parker, the head, was not an outstanding research man, and the department's quarters were in general "cramped and unkempt."[46] But the department maintained a fairly good laboratory, in the kitchens of an abandoned dormitory.

As is often the case in science departments, the problem of space, space, space plagued the departments at Howard. Over the years Just had learned how to improvise, but as the leading scientist at the university he no doubt felt the crunch more severely than most. He recalled how in 1910 the then newly constructed Thirkield Hall had raised physical conditions for the science departments to a merely adequate level. Now, twenty years later, in spite of the rapid expansion of the science programs no new facilities had been arranged. It was a sorry situation indeed.

There was also the problem of students. In general the Howard zoology department could attract only poorly qualified applicants, since the better students went into medicine. Just understood the reasons for this, and he was honest with his own undergraduates about the gains and drawbacks of being a scholar versus a physician: while scholarship promised intellectual enrichment, medicine promised status and financial security. It had become a matter of course for Howard students to choose the latter, and Just knew it. To develop a feeling for the importance of scholarship would take time, and the place to start was in the undergraduate program. Though zoology was probably Howard's strongest undergraduate department, it was still weak from a broad academic point of view. Just wanted to focus on the undergraduate program as the first step in building a quality department overall, including, eventually, a graduate program to train scholars for zoological research. Difficulties had arisen, however, because Johnson was forcing Just to rush into graduate teaching without having established a basis of quality education among Howard undergraduates.

Then there was the problem of staff development. Just felt more hopeful about solving this one, since it was mostly a question of "laying hands" on good people.[47] The process would not be without snags, however, for Johnson was opposed in general to hiring whites.[48] He insisted on the importance of encouraging black applicants for faculty positions. That was all well and good in theory, Just used to retort, but where were qualified black zoologists to be found? Johnson pointed out that the chemistry department, with its several black instructors working towards advanced degrees, seemed to be having no problems in this direction. But there were problems, and Just knew it. One of the top professors in chemistry at Howard, Jacob Shohan, was white, and the department could not do without him, at least for the present. Yet to prove his point Johnson began a campaign to remove Shohan from his position. Some time in the early 1930s Shohan was abruptly dismissed. Sickened by "the ugliness, the bestiality, the savagery of the thing" and by the fact that it was Johnson who had arranged it, Shohan raised such a furor with the trustees that he was reinstated promptly.[49] This only added fuel to the fire. Johnson enlisted the help of the department head, Percy Julian, to humiliate Shohan. One day Julian ordered Shohan to clean up the laboratory; then the following day he told him to take over the clerk's duties for an hour while the clerk took shorthand elsewhere.[50] Shohan decided that under these circumstances he would not stay. His one consolation was the support Just gave him throughout the ordeal. He thanked Just profusely, suggesting that if some of the "yellow-bellies on the faculty" would stand up to Johnson the way Just did, things at Howard would improve.[51] Just later went out of his way to find Shohan another job, writing strong letters of recommendation to Emanuel Celler, the congressman from New York, and Mrs. Annie Nathan Meyer, a cousin of Supreme Court Justice Cardozo.[52]

Howard, a poor black school that could offer its faculty little in the way of benefits, tenure possibilities, or pensions, was an unattractive career option for the average white Ph.D., and many whites would not have wanted anything to do with a black institution in any case. Those who went to Howard, more often than not, were people who could find jobs nowhere else, men and women with third-rate minds or personality problems or both. In a sense, Johnson's attitude was justified. But Just felt that Johnson went too far, that he had closed his mind to whites who had something to offer as well as to those who were mere hangers-on. Just knew whites, high-quality scientists, who would be glad to come to Howard to give of their expertise. What was the sense of turning them down just because of their skin color? They could contribute a great deal until such time as more qualified black scientists became available.

The fact was that the Depression was on, and many able white scientists were unemployed and would have given anything for a job anywhere. When Heilbrunn lost his position at Michigan, he seriously considered going to Howard because the job situation nationwide was grave. And then there were the Europeans. Many of them were losing their jobs either as a result of economic cutbacks in the laboratories or, in the case of Jewish scientists, because of the rising tide of anti-Semitism, especially in Germany. Just wanted to hire two or three of his colleagues at the Kaiser-Wilhelm-Institut on a temporary basis in the 1930–31 academic year. But it was the same old problem. Johnson balked because these scientists were white, and also because they were friends of Just. What they had to offer seemed not to count. Of course, both Just and Johnson were aware of one difficulty, namely, that the cultural differences between these scientists and black Americans might prove to be an insuperable barrier, or, at the very least, might require an extended period of adjustment, especially to accents and customs. Further, while European scientists could be relied on to maintain a high level of excellence in research, most were not likely to have had much experience with teaching. Their institutions were oriented toward research, not education. But Just had faith that the pros outweighed the cons, and he managed to get the administration to hire one of the applicants he had recommended, Helgo Culemann. Culemann taught a couple of undergraduate courses and took over some of the seminar work Byerly had led the year before. It worked out very well indeed. Unfortunately, when Just later got Culemann a job at Amherst questions were raised about Culemann's ability as a teacher and honesty as a research scientist. How true the accusations were is unclear; they may have had something to do with anti-German feeling on campus and a personal disagreement between Culemann and another Amherst professor, Harold H. Plough.[53]

The problems of space, students, and staff at Howard could be worked out, but it would take time, cooperation, and patience on the part of everyone involved. Not much could be accomplished overnight.

The cooperation Just was hoping for did not materialize in the 1930-31 academic year. Instead of helping Just, the administration seemed bent on needling him. Every piece of equipment, every animal specimen Just requisitioned for his research or teaching work had to be justified with lengthy explanatory memoranda. Not that the administration had any way of independently evaluating the requests, however. For example, Just once sent in an order for 500 *Amoebae* and had his request denied simply because the dean, Emmett J. Scott, worried that that *Amoebae* would be "swarming like a herd of elephants" around the Howard campus.[54] Just only discovered the misunderstanding by chance, in conversation. Amused, he informed Scott that an *Amoeba* was only 300 microns in diameter and that 500 could be placed in a thimble. The episode gave him the whimsical idea of publishing an article entitled, "Putting *Amoeba* out to Pasture," but he never did.

In 1930-31 the students presented more problems than they had the previous year. Along with Wormley, there were two new M.S. candidates, Chauncey Parker and Alphonso Warrington.[55] Warrington left at the end of the year without taking a degree. Parker and Wormley had aptitude but were not dedicated to zoology in the way Bright and Chase had been. Both took their degrees at the end of the year, but neither stayed in zoology. Wormley went on to study education to qualify himself for a high school teaching position, and Parker accepted an administrative post with the Department of Playgrounds and Recreation of the District of Columbia.

Again, one of the things that kept Just going was his admiration for Chase's work. After earning his master's in 1930, Chase had decided to stay at Howard to do his doctorate under Just, with a thesis on respiration and cell division. There was no guarantee that he would get a degree at all, since the Howard trustees had not yet passed legislation allowing for the awarding of a Ph.D. That he wanted to stay at Howard rather than go on a fellowship to one of the large white universities was flattering to Just. A little over a year later he was forced to go to Stanford to finish his degree, when the Howard administration decided it was too early to institute a doctoral program.[56] For both men disappointment was mitigated by the knowledge that they had worked well together for three years.

Curiously, Just was beginning to think that a doctoral program might be easier to put into effect than a master's.[57] A successful master's program required a large number of courses and a level of faculty supervision that a place like Howard could not afford. A doctoral program, on the other hand, did not need to be as highly structured or as well staffed. The doctoral student would be finished with course work and require a minimum of direction, and in most cases would work on a specialized project tailored to the professor's main research interest. Of all the teaching levels, in fact, the doctoral was the least time-consuming and the least expensive—a point articulated by Just but not readily understood even within the academic community itself, certainly not at Howard. Besides, apart from the basic methods articles he had published in the *Collecting Net,* Just's experiments and observations were too subtle to be passed on readily to any but the most serious, motivated, and experienced students. The science he was doing required workers at the doctoral level or beyond.

Just had been swamped with applications to the graduate program from students as far away as Mobile, Alabama.[58] The main problem, as usual, was the lack of preparation of the applicants, and Just spent hours writing letters of rejection that somehow had to not be discouraging. He was also beginning to receive applications from students, black and white alike, at major white universities, Indiana and Ohio State, for example, in response to rumors that a vital new doctoral program was getting under way at Howard.[59] Even H. J. Muller, then at the University of Texas and later a Nobel laureate, talked about sending his students to work under Just's "authority" for a while.[60] In all these cases, Just wrote back explaining that the program was limited to the master's degree, and that there were few if any funds available for fellowships.

About this time, too, the Howard administration started pushing Just to forget his research plans and dedicate himself to making "carbon copies" of himself among the master's-level students who came under his direction.[61] It was the last straw. Harassment was one thing, control of his life's work and ambition quite another. Just did not see how he could continue to work at an institution where the importance of his research was not understood, where he was viewed as some kind of teaching machine. The hope of finding a post on a research staff elsewhere came to him again. He was realistic, though, and knew that it was most improbable that he would be offered a job at a white institution. His Woods Hole colleagues had always been aware of the problem. Between 1927 and 1929, with W. C. Allee's support, he had been considered for a post at Brown University, but the Brown administration and perhaps also the faculty balked, although Just would have been "quite ideal except for his race."[62] Franz Schrader made an attempt to see if anything could be done for Just at a New Jersey college, but realized that the effort was doomed to failure from the beginning.[63] Heilbrunn, on the other hand, seriously thought something could be done at the University of Pennsylvania, if Rosenwald were to put up the money, and in mid-1931 he came up with a strategy that included Lillie, Curtis, McClung, and the Rosenwald people.[64] But this scheme also fell flat. The chance of finding another position was not merely slim; it was nonexistent. Just knew that he would have to stay at Howard. But at the same time he was determined not to succumb to the pressure there. At the end of a hard year of teaching he would go off to Europe as usual, to continue his research. His responsibilities, as he saw them, extended beyond the confines of Howard, into the larger world of science.

Yet the more Just wanted to expand his horizons, the more he became tied to Howard, bound and constrained by the financial arrangements Johnson had made with the Rosenwald money. When Just wanted to check grant expenditures, he could never get financial statements from the bursar; he did not know who was spending what. Was the money being channeled into other accounts? Was it, perhaps, being used to embellish the salaries of Johnson and other administrators? Just was adamant: he wanted the Rosenwald money to be spent only for the Department of Zoology. Though well aware that administrators often used funds intended for one purpose to finance another, robbed Peter to pay Paul, Just felt that he and the zoology department had already been robbed too much over the past twenty years and that this state of affairs should not be allowed to continue. Nonetheless, Johnson never permitted the department to use its full annual allocation from the Rosenwald Fund. At the end of the five-year period of the grant, Howard found itself underexpended by $12,000. For Just, this was glaring evidence of Johnson's stupidity.

ℒ

If Just was confronted with problems at work, he had to face as many at home. His marriage had become little more than a domestic arrangement. Since the early twenties he and Ethel had not shared the same

bedroom, but it was around 1930 or so that the strain in the relationship really made itself felt. The spark that had ignited them twenty
years earlier was faint now, almost gone. Just's long absences from
home, half of each year, his late hours in the laboratory, and his total
absorption in his work were hard on Ethel. He was remote, different
from other husbands and fathers in the Howard community. His one
close friend at Howard, Alain Locke, unmarried and an avowed homo-
sexual, was hardly the epitome of a family man. In fact, most of Just's
friends and colleagues who could be described as family men were
white, and were part of the group which had rebuffed Just's own
family at Woods Hole. As a result, Ethel made sure to stay as far as
possible out of their way. Whenever they came over she would disappear
into her room. The atmosphere became so uncomfortable that Just
stopped inviting them inside altogether. Instead, he sat outside talking
until midnight, often in the car with the motor running if the air was
damp or cold.[65] His professional life demanded that he keep up a
certain camaraderie with other scientists; he could not change that no
matter what Ethel thought or felt, nor could he summon up the
willpower to bring his career into closer harmony with the personal
side of his life.

In 1928, while planning his first trip to Europe, Just no doubt had
thought of taking along Ethel, perhaps even the whole family. The
cost was too high, however, and he had had to settle for one traveling
companion. Margaret was chosen, as Ethel could not leave the other
children and her aging mother behind. Besides causing jealousy among
the other children, Margaret's role as Just's companion perhaps created
some resentment in Ethel, who had long since lost that role herself.
The trip to Naples must have reminded her of the honeymoon she had
given up back in the summer of 1912 so that her husband could get on
with his career.

Always concerned with family appearances, Ethel never admitted to
anyone but herself that the marriage was a shambles. She knew that
Just was spending long hours, into the night, with his young female
assistant in the laboratory, but she ignored it. She would not discuss it
with anyone. She had few friends and none in whom she could confide.
For the most part she considered the people around her, including Just,
her social inferiors. Divorce was out of the question. Quite apart from
her desire to avoid embarrassment, Ethel realized that Just was a good
provider who allowed her and the children to live in exquisite comfort
—something very rare for black families during the Depression. She
was determined, then, to rise above the indignity of her situation, to
hold her head high through thick and thin.

Just cared less for appearances than for his work. His mind focused
on his scientific prospects, his professional image. To make something
of the marriage, to share more of himself with Ethel and to encourage

her to share more of herself with him, would have taken too much energy. To leave altogether, though, would have taken even more. Such things were not done at that time without jeopardizing one's professional status. The easiest thing to do was let the situation go on as it had been. There was no clear advantage to changing anything.

Though he gave little to the relationship with his wife, he put in a fair amount of time with his children. Margaret, the oldest, was considered his favorite. More inclined to the humanities than to the sciences, she shared with Just an interest in many areas—music, languages, literature, the cultural anthropology of the black race. They grew even closer intellectually in Naples in 1929. Just was pleased with his daughter's brilliance, her poise, her maturity. That fall he began to make a special effort to insure that she went to an excellent college. He asked the Schraders about Bryn Mawr, Barnard, and Wellesley. They told him that there would be problems at any of these places, especially at Bryn Mawr, where the color line was strictly enforced, but also at New England colleges such as Wellesley, where sororities did not consider blacks for membership.[66] None of this was new to Just, although it was clear that the social situation for blacks at white schools had if anything deteriorated since he was at Kimball Union and Dartmouth. In any event, he was certain that Margaret had the self-confidence to handle whatever problems she encountered. His main objective was to get her out of the black community for a while, especially away from the atmosphere of confusion and nonproductivity enveloping Howard. But Ethel, with the Woods Hole experience fresh in her mind, preferred not to see her daughter forced into a compromising position at a white school. Undeterred, Just kept up his inquiries, and even persuaded Ethel to go with Margaret to see one of his Woods Hole friends, J. W. Wilson, at Brown University in the fall of 1931. Much to Just's disappointment, and confirming Ethel's feelings about whites, Wilson informed them that blacks were not allowed to live in the dormitories at Pembroke College, the women's affiliate of Brown.[67] In the end Margaret went to Emerson College in Boston.

Highwarden, Just's only son, was much like his father in temperament and personality, though he did not share a taste or talent for science. The two enjoyed each other's company, and spent time together swimming and playing tennis. They joked about all kinds of things; for example, Just used to tease Highwarden about parading around in his ROTC uniform at Dunbar High School.[68] School was a serious issue, of course, and Just was careful never to push science as a "must" subject, always to let his son develop his own academic interests. Highwarden, with his little sister Maribel tagging along, made one trip a year (on New Year's Day) to the laboratory at Howard—a custom that had been started in order to bring them closer to Just and his work, not to pressure them into following in his footsteps.[69] It was more important, Just felt, to set a general standard of excellence for Highwarden than to chart out a specific career for him. Just was himself viewed everywhere as an embodiment of excellence—a brilliant student, a first-rate teacher, a leading figure in Omega Psi Phi, the fraternity that upheld the highest ideals of black manhood. Highwarden looked up to him on that account.

As the baby of the family, Maribel was special to Just. In 1930 she was not as refined as her older sister, but she had begun to show intellectual promise. Her mind was quick and full of curiosity. Just wanted her to go away to school. Though Dunbar was excellent, and sent its graduates off to white New England colleges such as Dartmouth, Amherst, and Harvard, Just remembered Kimball, his strong academic preparation there, and the benefits that travel and a new experience had given him. He was anxious for his youngest daughter to have what he had had. He made inquiries,[70] but for some reason nothing came of them, and Maribel went to Dunbar instead. Just continued to encourage her to develop her mind and particularly her talent for music. She was given a piano and sent for lessons.

The question of race plagued not only Just's marriage but also his way of bringing up his children. Like many of the growing black bourgeoisie, he wanted to shield his children as best he could from the scars he had himself received in coming up against the white world. The black community helped provide this shield, and was apt to be especially protective of the children of a prominent figure like Just. As long as the children stayed around Le Droit Park and the Howard community, they were safe. But when they ventured downtown or tried to use public facilities or wanted to eat lunch in a public café, the problem confronted them head on. By and large the Justs and other black families engineered their lives around this problem, yet even so it was impossible to avoid confrontations with racism. For example, Just had to take Highwarden to Philadelphia for eye treatment with a Dr. Spaeth, as all the "competent" occulists in Washington kept separate waiting rooms for blacks; the black occulists, in Just's view, were not capable.[71] Though long and tiring, the train ride spared Highwarden the humiliation of sitting in a segregated waiting room. Of course, there would have been no preventing some poor Appalachian white (or Tidewater aristocrat, for that matter) from rolling down his window and shouting "Nigger" at Highwarden as the train pulled out of Union Station.

Just's attempt to shield his children was not a sign of weakness on the issue of race. He struggled valiantly with the problem throughout his professional life—at Woods Hole, in particular. Yet while he knew his children should be made aware of the obstacles that awaited them in American society, something made him want to protect their tender young minds from that society's psychological poison. And he was divided in another sense too. He wanted his children to have ambition and strive for success in a white man's world, but he insisted that they maintain respect for the black world where their ultimate roots (and protection) lay. The line was a fine one to walk.

It was less fine for Ethel. She did not care to mix, and never did. Her dislike for whites went far back. The incident at Woods Hole made matters worse, but it was not what had made them bad in the first place. When she was a little girl, she and her mother had been rejected by her father's people for being too dark. The Highwardens were light-skinned, many of them white. The mixture confused young Ethel. Though her hair was straight, she was dark-skinned. She took pride in her white relations, reciting her genealogy back to a signer of the Declaration of Independence, but at the same time she defended her black relations and sided with her mother Belle, who represented the best of the black subculture that had emerged from years of oppression by whites. In Just she saw someone who was also a mixture, who shared with her a common racial and cultural heritage. She must have thought, when they first met, that with him she could experience something of the white world while living and working in the black. Little did she know that Just's life would draw him and his family more and more away from the black world and into the white—a world that was to drive a wedge into their relationship.

At Howard, it was not long before the hostility between Just and Johnson reached a point where dialogue, indeed communication of any kind, was difficult if not impossible. Confusion reigned, not only in day-to-day administrative matters involving applications for leave and requests for equipment, but also in regard to broader issues of funding. In June 1931, when the General Education Board grant came to a close, there was a question as to whether the board would renew the grant, whether the Rosenwald Fund would step in with $2,500 more per year, whether Howard University would foot the bill, or whether Just would lose the income altogether. Everyone was working at cross-purposes, or not working at all. The National Research Council tried to get the General Education Board to further its commitment; the board stayed aloof, saying only that it would consider the question later; Howard remained vague, and no one was sounded out as to what should be done. Finally the Rosenwald Fund heard about the problem thirdhand from Lillie and tried to explain to the council and to Howard that there was no need for the board to be called on for further assistance, that the additional money could be taken from the fund's $15,000 annual appropriation.[72] The reason the process was so circuitous was that Just and Johnson had created a maze of charges and recriminations against each other. Just had persuaded his friends at the council that Johnson was "a slippery customer," while Johnson had suggested to officials at the fund that Just was less than responsible in his approach to the department.[73] The council and the fund each failed to deal with the matter openly and directly, and the General Education Board became a spectator in an unpleasant game of politics.

Just was not the only faculty member at odds with Johnson. There was a growing feeling at Howard that Johnson had overstepped his bounds, had become "dictatorial and inflexible," had developed "a presumption . . . nothing short of psychopathic." There were even suspicions that Johnson had used the annual appropriation from Congress to benefit himself and his friends.[74] Many faculty members, even though they did not like what Johnson was doing, kept quiet out of fear that they would lose their jobs, or out of loyalty to him as Howard's first black president. Those who complained were either fired or forced to resign. In 1931 Johnson and the faculty of the dental school engaged in a "violent conflict" which resulted in the departure of several teachers, and at about the same time the dean of the law school tendered his resignation because of what he considered to be Johnson's "autocratic and ruthless dictatorship." Morale on campus was plummeting. The atmosphere was one of uncertainty and disunity, even worse than during the notorious Durkee regime. As a result, Howard received a good deal of unflattering publicity in the *Washington Post* and other prominent newspapers.[75]

Johnson was not without his share of supporters, especially among powerful administrators outside the university. Embree, for instance, told Rosenwald that Johnson was blemish-free, that the responsibility for the confusion at Howard lay with "an envious group" of dissenting faculty members.[76] Though he did not say so, Embree probably thought of Just as a part of this group of troublemakers. Another Johnson supporter was Abraham Flexner, recently appointed the chairman of the board of trustees at Howard. He had his hands full trying to still the grumbling against Johnson, even among some of the trustees. He certainly saw Just as a troublemaker, though he never made a formal accusation. He was deliberately cool when Just tried to discuss with him the "desperateness" of the Howard situation. Flexner said he was too busy to talk, and when Just pressed him further, suggested that it was improper for him, as a trustee, to discuss internal Howard matters with any individual, even a faculty member.[77]

Though he could not do much in the way of research amid the confusion, Just found time to get on with his teaching. There was nothing to be gained, he decided, by sitting around and indulging in "bitter expressions and bitter thoughts."[78] For the graduate work in

1931-32 he enlisted the able assistance of two young, white government scientists, Herbert Friedmann and Benjamin Schwartz.[79] Friedmann led a morning seminar in experimental embryology four times a week and Schwartz led one in physiology in the afternoons. In the interim, Just spent up to seven hours each day in the laboratory with the graduate students, helping them with conceptual problems and instructing them in techniques of embedding, sectioning, and analyzing material. Aside from Chase, who had yet to hear that he would not be able to take his Ph.D. at Howard, and Warrington, who was finishing up his thesis, there were four graduate-level students: Leona Gray, Caroline Silence, Dorothy Young, and Helen Smith. Just was inspired by their diligence and enthusiasm; to him their attitude was "both a justification and an ample reward" for the effort he was putting into the graduate program.[80]

The students were doing so well, in fact—working from eight o'clock in the morning to six o'clock in the evening and turning out material that would probably be "worthy of publication in any one of the best zoological journals either here or in Germany"—that Just considered the possibility of expanding the graduate program to include as many as twenty degree candidates.[81] The problem, though, was still lack of space and facilities. Not only had the Howard administrators failed to come up with additional laboratories and offices, but they were also taking an unreasonably long time to meet the zoology department's most basic needs, including more electrical outlets, new microscopes, and beaverboard partitions to allow each worker a certain amount of privacy. The department's "shabby" appearance made Just "heartily ashamed," so much so that he thought twice about inviting outsiders to come in and see the work he and his students were doing.[82] Those he did invite, he made sure, knew him and his scrupulous approach to science well enough to realize that he could not possibly be responsible for the poor physical condition of the department. Even then he found the situation "very embarrassing," but he tried to submerge the feeling so that his students could benefit from the visits of important people connected with funding agencies and scientific institutions.

One of the people he was instrumental in bringing to Howard to meet his graduate students was George Arthur, an administrator (his title was "Associate for Negro Welfare") of the Rosenwald Fund, whom he had met through the Bentleys in Chicago in 1916. Arthur wanted to interview Just for a biographical piece to be published in *The Crisis*. Even though Arthur had snubbed Just one day on the way out of Johnson's office—because Johnson had prejudiced him, Just felt—Just was still anxious to have him come. He wanted to use the occasion to see if he could arrange more fellowships for his students.[83] Also, Just felt the urge to clarify to a fund official some of the problems caused by the Howard administration's failure to live up to its part of the agreement on the Rosenwald grant; in other words, Just wanted to tell his side of the story.

The meeting went well. Arthur promised to send grant application forms not only for the graduate students currently at work in the department but also for some of Just's former undergraduate students, including Louis E. King and Lillian Burwell, who needed money to further their research careers.[84] He also said that he would speak to Embree about using some of the surplus funds in the Rosenwald grant to help the students directly, rather than allowing Howard to let the money sit idly or be used for purposes other than those for which it was intended.

Just went out of his way to help Chase, Gray, Silence, and Dorothy Young. He was still uncertain about the prudence of encouraging blacks to try to establish themselves as scientists, but, after seeing the enthusiasm of his new group of students he had decided that, whatever the problems, each individual should be encouraged to go into whatever line of work gave him or her the greatest "intellectual satisfaction."[85] If that turned out to be science, then so be it. There was little chance that his students would find jobs anywhere other than black high schools, but he was determined to use his contacts at black colleges to see what the possibilities were there. He persuaded R. S. Wilkinson, president of South Carolina State College, to do his utmost to hire Gray, Silence, and Dorothy Young.[86] The Depression, of course, had reduced the already small number of job possibilities for blacks with a background in science, so the fellowship route was still the best one to follow. Just tried to arrange scholarships for summer research at Cold Spring Harbor and fellowships for graduate work at the University of Chicago and Northwestern University.[87] He was not always successful in these efforts, but it was not for want of trying. Once a week he took the students out to lunch at his own expense, often inviting along scientists like Austin Clark of the Smithsonian. He wanted not only to expose his students to the best minds in biology but also to give them a chance to meet people who might know of job possibilities, or at least be willing to spread word about "the outstanding work" being done in zoology at Howard.[88]

Though Just had never been so inspired by his teaching work, he was anxious to get back to his research. The catalyst may have been the death of Julius Rosenwald on 6 January 1932. Just recalled the pleasure Rosenwald had taken in helping to demonstrate to the world that a black could do brilliant research, and the understanding he had always shown for Just's aims and ambitions. Now, with Rosenwald gone, there was a danger that forces opposed to or uncertain about the research end of Just's career might take the upper hand. The Howard administration had been stepping up its pressure on him to abandon his experiments and theories and devote himself to full-time teaching. Worse, Embree was siding with them on this point, having developed an unsympathetic attitude toward Just and his research.[89] The possibility of having to give up the part of his life that gave him the most satisfaction, of having to be stuck yearround in an environment that he was growing to hate, was disturbing to say the least.

Just stepped up his efforts to try and make Embree understand the difficulties of operating in a place like Howard. When he sensed a certain coolness on Embree's part (Embree could not seem to find a spare moment for an appointment),[90] he decided to take a new approach and stress the positive side of the Rosenwald grant. Perhaps Embree was tired of hearing complaints, perhaps he would do more to correct the problems and renew the grant for another five-year term if he had some good news for a change, Just must have thought. He sent Embree a copy of the scientific article he was most proud of—the one that had resulted from years of thinking about the cortex, that had elicited such a favorable response at Padua, that some German biologists were saying constituted "the most important generalization in biology appearing in the last thirty years," that had been painstakingly translated into German by Boveri for publication in *Naturwissenschaften*.[91] The article was testimony, Just felt, to nearly five years of productive research under the Rosenwald grant.

Embree did not seem too impressed. He complained about not being able to make any "headway" with the German version of the article, and proceeded to lecture Just on the value of writing about scientific matters as H. G. Wells, T. H. Huxley, and Julian Huxley did— "accurately and yet within the interest and comprehension of the average man."[92] A little piqued, Just responded that his theory was too new to warrant a popularized treatment, but that that would come in the future.[93] While agreeing with Embree in principle on the importance of simplicity in science writing, he urged Embree to reconsider his admiration for Julian Huxley, who, in his opinion, had undoubted "literary ability" but was at the same time a poorly trained scientist, prone to frequent inaccuracy. In any case, this was a side issue as far as Just was concerned. The particulars of Embree's criticism were not as disturbing as the fact that his letter was another in a long line of negative statements on Just's work under the Rosenwald grant. The need for a face-to-face interview was greater than ever.

A meeting between them finally took place in the zoology department at Howard toward the end of January 1932. Just recapped his accomplishments and difficulties under the Rosenwald grant, though he did not belabor the point of Howard's failure to provide adequate space and equipment. He put the issue on a sensitive and sophisticated plane—that of educational philosophy. He discussed the need for a "principle of orientation," a firm grounding in basics prior to implementing advanced work; he analyzed the fallacy of "reaching for the end without being willing to pass through the intervening stages in their proper sequence."[94] In his view, the pressure that had been put on him to produce creative biologists out of thin air was misplaced; there was no way to fulfill that expectation without the proper material—in particular, talented, well-grounded students. He had done the best job possible, but the program under the Rosenwald grant had not been a raging success from the educational point of view. The greatest potential success lay with his research, but even there he had not been as productive as he might have been in a more favorable environment than Howard. Since 1929 he had published fifteen papers and he was in the process of working up several other manuscripts for publication. But there was still a lot to do. He wanted to concentrate more intensively on the research side of his career, not out of a selfish wish to renege on his "duty as a Negro to the cause of the Negro" but out of his "great desire to make a substantial contribution to biology." In the long run, this was a higher and more important aim. If it meant leaving Howard, he would not hesitate to go.

Not only was Just happy to have had the opportunity to explain the situation, but he was ecstatic about Embree's favorable response. He sensed a renewal of mutual confidence, a return to "the old, original basis of friendship" established way back at the Negro Problems Conference in New Haven in 1927.[95] He could not blame Embree for having lost sympathy earlier, considering "the power of the forces . . . playing upon him." The main thing was that Embree had come to listen to the other side of the problem, that he was "fair-minded and just enough" to weigh both sides and no doubt make the right decisions in future. Just had confidence that the decisions would now go in his favor, perhaps to the point of relocating him—grant and all—at some other institution.

The frustrations, the "heart-breaking handicaps" of life at Howard did not go away.[96] Even a glowing letter that came from Wilbur Thirkield, Howard's old president, describing the great "joy" he took in watching Just build up "a genuine department of science," make "valuable discoveries," and achieve "an honored place among the leading biologists of the world," did not change Just's feelings.[97] Actually, it may even have depressed him, by bringing home the fact that the administration he had first worked under at Howard was the most sympathetic he had known there—and the most sympathetic he was ever likely to know. The Thirkield letter, while inspiring in some sense, no doubt also made him feel that his Howard career was one of increasing obstacles and declining productivity. The only answer seemed to be to find another position "as *soon as possible*."[98] It was a good thing, in a way, that the press was dragging Howard over the coals, so that Embree and Lillie and Schrader and everyone else could see clearly what Just had been trying to show them for some time—that "a change of location" was essential for him.[99]

In September 1932 the Julius Rosenwald Fund was undergoing its worst financial crisis of the Depression and the outlook was bleak. Embree indicated that while he fully appreciated the problems at Howard, he was convinced that anyone, Just included, who had any kind of a job—whatever its "drawbacks and shortcomings"—was indeed fortunate.[100] Moreover, there was little likelihood that the fund would renew the grant. There were larger, more important commitments that had to be met. The time had come for Just to depend less on the fund for research support and for Howard to take over financial responsibility for the zoology department.

It was a double blow for Just. Not only was his chance of escaping Howard slim, but his financial support was about to dry up. Runnström wrote glowingly about the "little nice laboratory" the Rockefeller Foundation had set up for him in Stockholm.[101] Just's colleagues seemed to be well set. Why then, he wondered, could he not have a nice little laboratory somewhere too, instead of having to struggle to find a moment of peace and to worry about where his next paycheck was coming from? The real question was who to turn to now that Embree and the Rosenwald Fund had abandoned him.

CHAPTER 7

The Search for a New Life, 1931-38

Not long after he met Hedwig Schnetzler in the summer of 1931, Just decided that he would spend the rest of his life with her. For him this meant Europe; it meant giving up America and finding some livelihood abroad. Life at Woods Hole had proven torturous; work at Howard was slowly killing his creative spirit; his domestic arrangements were bad and growing worse. To simply hang on was unthinkable, especially after seeing a glimmer, indeed a ray, of hope of a possibility elsewhere—in Europe, a place that seemed to Just so free and loving. But changing life, changing surroundings and mates, is easier thought about than done. Forty-eight years old, Just was at an age when people often undergo mental and physical transformations but find it difficult to adjust their lives accordingly. He needed a new atmosphere, new experiences, in which to grow as a person and as a scientist. It turned into a slow and traumatic struggle that lasted more than six years.

Just's relationship with Hedwig is best portrayed as a nexus or crux, a crossing where many roads come together. From that point of focus Just gained a unified perspective on the diverse activities and feelings and experiences of his life—his science, his disdain for Woods Hole, his troubles at Howard, his unsatisfying family life, his growing love for Europe. Twenty-five years younger than Just, Hedwig was only twenty-three when they met in 1931. Energetic and curious, she took an immediate interest in his work. She wanted to know how this black man had attained the level of success he had, especially since the reports she had heard of racism in America told her differently. Finally, she was determined to have some impact on it all, as his lover or his wife.

ॠ

Even before he met Hedwig, Just had begun thinking about the larger implications of his work, prompted mainly by his colleagues at the Kaiser-Wilhelm-Institut. Until 1928 his published work centered on elementary problems of cell biology, mostly experimental embryology dealing with the problems of parthenogenesis and cell division. His studies were particularistic, noteworthy for their careful execution, and considered valuable contributions to biology. During the first part of his career Just was more interested in establishing a substantial body of data capable of vertification than in conducting theoretical disquisitions; he thought that it would be wrong to attempt generalization on the basis of a preconceived theory, that it was best to wait for a sharp point of view to be gained by patient labor before attempting to establish a theory. But in 1928, about the time of his final disillusionment with Woods Hole, Just began to evaluate his work in terms of "a theory which fitted the facts."[1] While in Europe, in Naples but mostly in Berlin, he tried to develop a theory on the importance of what he called the "ectoplasm" in development. The Padua conference gave him an opportunity to air his new ideas publicly for the first time. After this successful presentation, he published three articles on the subject: one entitled "Die Rolle des kortikalen Cytoplasmas bei vitalen Erscheinungen" ("The role of cortical cytoplasm in vital phenomena") in *Naturwissenschaften*, which Boveri translated into German for him, and two in the *American Naturalist*, "On the Origin of Mutations" and "Cortical Cytoplasm and Evolution."[2]

When Just first met Hedwig in the summer of 1931 his basic ideas on the theory had already formed, though in many ways they were still embryonic. Central to his theory was the notion that the cortical cytoplasm (the "ectoplasm," in Just's terminology), the outer part of the cell, is more vital to life processes than had been generally recognized, that it was a crucial link between environment and animal, constantly reacting, shaping, conditioning. He had thought about this for a long time: even at Chicago he had argued in class for the interpretation of the nineteenth-century German biologist, J. A. Hammar, who had insisted on the importance of ectoplasm, and later he had supported Lillie's work on fertilizin, a substance situated in the ectoplasm. Hedwig listened carefully to what he had to say. The theory, in his opinion, "aimed at nothing short of revolution" and would provide "a new basis for the understanding of vital phenomena."[3] Hedwig was bright, though she knew no biology as such, certainly none at the research level; a humanities student interested in science, its philosophy and history, she regarded her conversations with Just as an opportunity to learn a new field and to infuse it with her own ideas from philosophy and politics.

In the winter of '31, after Just returned from Europe and his first encounter with Hedwig, they kept up their close relationship through letters, often very long and always intense and involved.[4] Science, profession, romance became intertwined, and defined their affair. Just's time away from Hedwig was spent daydreaming—fantasizing as he had done as a child on James Island. He longed to be with her. Through her, he thought, it was possible for him to feel deep within himself, and, by extension, to work more efficiently on his science. After a struggle with Johnson, Just managed to persuade Howard's board of trustees to give him time off to make another trip to Europe, his fifth in three years. On 20 March 1932 he sailed for Europe, to spend a few months in Berlin working at the Kaiser-Wilhelm-Institut.

Throughout the spring and summer he and Hedwig worked together. She studied hard to learn all she could about embryology and its history. His laboratory became hers, and he became her teacher as well as her lover. They set as their primary aim the completion of a book on his new theory of vital phenomena. Every day they worked toward that goal, even when they were not in the laboratory. They took frequent trips to Heidelberg, Königsberg, Spreewald, Sanssouci, and Bremen. These moments alone together, away from the hustle and bustle of the laboratory and the chatter of colleagues at the Kaiser-Wilhelm-Institut, gave them even more time to delve into science, as well as to share their pasts and to make plans for the future. They refined their thoughts on the book and charted out a strategy for completing it.

As the summer drew to an end they decided to go to Naples and work there for a while. Just still had fond memories of Dohrn, and Hedwig was a distant relative of his—her great-grandfather, Gustav Wendt, was married to Anna Dohrn, Reinhard's aunt.[5] The trip to Naples had to be strictly professional, at least in the eyes of other scientists at the Stazione. Just and Hedwig went there not as lovers but as coworkers—he the principal investigator, she his assistant. Just occupied a table for a short time during August and September, but he and Hedwig also spent time visiting small towns in the countryside, such as San Marco, Riccione, Verona, and Bozen. The fact that Hedwig spoke Italian made the trip even more pleasurable.

Just's relationship with Margret Boveri had been less open than this new one with Hedwig. For one thing, there was no one like Uncle Robert in Hedwig's family. When Just went to meet them at their home in Heidelberg some time in the summer of 1932, he was greeted warmly by everyone—her parents, Karl and Elisabeth; her two brothers, Karl and Otto (Hedwig's twin), and her sister, Ursula. By and large the family accepted the relationship at the beginning, though they recognized the problems that Hedwig would have to face as the lover of a black man, and a married one at that.

But Just and Hedwig were not ready to reveal the situation to everybody. In Naples, for the sake of propriety, they took separate living quarters. Dohrn knew the truth, however, and it was not long before other scientists, the Americans included, somehow caught on. Perhaps Just was less circumspect than he had planned; perhaps he felt so good about the relationship that he did not care who knew, or, better yet, was proud for them to know. Just seems to have been gathering the courage to tell the world about his new assistant, his new lover, his new life. By the end of that summer he had decided once and for all to leave America and find a new life in Europe. He knew that transplanting himself would require money, and he resolved to find some. He was confident he could, despite the fact that the most drastic economic depression in history was then at its peak.

Hedwig was helpful. Just was clever at persuading donors, but she helped him put together a more strategic, comprehensive plan—one that would bring not a stipend for a two- or three- or even five-year term but a steady income that would set them up together for life in Europe. They sat down to devise a plan of action. The Rosenwald money had only a year to run, and though it seemed unlikely that any new money would come from that source, there would be no harm in trying again, and again if necessary. But there was no time to be wasted. Just needed big money. He would try the Rockefeller Foundation, the Carnegie Corporation, and a host of others. He would also seek out rich individuals, preferably millionaires many times over, and get into their good graces; he might be able to find a patron. All of his transatlantic voyages would continue to be in first-class quarters so that he could meet the world's wealthy in the dining room, at the roulette table, or on the shuffleboard deck. And he would make an effort to rekindle his friendship with those he had already met.

Hedwig and her family loved and courted the company of the rich. Though not aristocrats, both sides of the family were firmly implanted in the *haute bourgeoisie*. Karl Schnetzler was the son of the mayor of Karlsruhe-Baden; Elisabeth's grandfather, Gustav Wendt, was a distinguished philologist and educator, and a close friend of the composer Johannes Brahms. One connection, however, caused the family some concern: Elisabeth's father, Karl Eller, was of Jewish descent. But he too was a prominent professional, a distinguished lawyer who had served for a time as president of the District Court of Konstanz.[6]

Karl Schnetzler, Sr. had studied engineering and electrotechnics at the Technische Universität in Karlsruhe, the oldest and most prestigious technical institute in Germany. After working for two years in Vienna he joined the industrial firm of Brown-Boveri in Baden, Switzerland.

He was active in mechanical design and at one point received a patent for work on the repulsion motor and the three-phase alternating-current locomotive. Resourceful and hardworking, he was appointed manager of the plant factories and offices in Baden responsible for the production of heavy electrical machinery. Aside from being a brilliant engineer, he was a man with a taste for painting, oratory, and acting. He married into a cultivated family, the Ellers, and settled into a comfortable and prosperous life as a prominent citizen of Baden. In 1922 he was transferred to Mannheim to head the German branch of Brown-Boveri. There he persuaded the company's Swiss directors to buy out smaller companies in the area in order to expand Brown-Boveri's operations to include the development of refrigeration machinery, electrical furnaces, high voltage transmission lines, and plastic molding materials and laminates. His sons followed him into the business and became brilliant engineers in their own right; Otto was sent to the United States at one point to represent his firm's manufacturing interests there, and Karl eventually joined the General Electric Company in London. Hedwig followed in the footsteps of her mother's side of the family, pursuing questions of history and philosophy at the university level. Ursula went into the theater, a bent inherited from her father. At the time acting was still not an entirely respectable profession, especially for a woman, so Ursula was a kind of "black sheep" in the family. Some consolation was taken in the fact that she was a serious dramatic actress, not a burlesque queen.

The Schnetzlers had many of the social contacts and much of the social adroitness that Just needed in his campaign to find a new niche in Europe. He already knew how to deal with people, but the Schnetzlers, Hedwig in particular, gave him the courage to be bolder and more persistent. His hopes for a positive conclusion to the campaign rose.

In 1932, after Just's old friend Julius Rosenwald died, the future looked bleak. Sears stock crashed and money dwindled. The Rosenwald Fund was still under Embree, but many of the old people had gone. Just had few contacts there, and Embree, his original contact, was no longer helpful. Johnson had courted Embree's friendship, even inviting him to give the graduation speech at Howard one year, and had won him over to his side. Johnson and Embree had been against Just's European travels, and their mutual annoyance served to bring them closer. When the grant ended in 1933, the Rosenwald Fund refused to go any further.[7] Embree and Johnson even banded together to refuse to allow the unexpended balance (unexpended mainly because of Johnson and other Howard administrators) to be applied to Just's research.[8] His subsequent private appeals to the Rosenwald family, especially Rosenwald's son Lessing, proved equally futile.[9] But Just kept trying elsewhere, on his own initiative and with the prompting of Hedwig.

The Rockefeller Foundation seemed to be a good fallback source. The Depression had hit nearly all foundations, even the Rockefeller, but money was still to be had. The Rockefeller had just appointed a new administrator, Warren Weaver, to head up the Division of Natural Science, and an effort was firmly under way to promote research programs throughout the country and abroad.[10] Weaver, a mathematician with a background in the natural sciences, was interested in furthering scientific research; later, in the 1930s and '40s, he played a major role in the development of molecular biology. Just was well aware of the moves that the Rockefeller was making in the direction of the natural sciences, and he knew of Weaver's special interest in biological research. He was correct in predicting the overall influence Weaver would have on science, biology in particular, but dead wrong in assessing his own chances for funding.

Just contacted the foundation in October 1932.[11] He knew no one there personally; since he had last been in touch with the Rockefeller office, the administration had changed. Abraham Flexner was no longer with the General Education Board, having left in 1930 to devote his energies to developing the Institute for Advanced Study in Princeton. Just did not assume that Weaver and his colleagues knew about his case. In fact, considering the conflicts that had arisen over the administration of previous grants, it was to his advantage that they did not.

Weaver proceeded to review the history of funding to Just, looking at the efforts of the General Education Board, the National Research Council, and the Rosenwald Fund. After meeting with Just in New York on 28 October he laid the case out in front of him.[12] Apparently he understood, even sympathized with, Just's assessment of the difficulties with the Rosenwald Fund grant—the unwarranted pressure from the Howard administration to develop a graduate program at the expense of Just's time for research; Just's lack of control over expenditures, and the consequent misuse of the grant for purposes for which it was never intended; the gross underexpenditure; the continual conflicts, jealousies, and recriminations concerning Just's motives and duties. Just was persuasive. Weaver seemed ready to do something positive, though, as Embree had done in 1928, he stressed that he could make no promises.

Over the course of the winter, Just and Weaver corresponded regularly and met personally at least once. In early March 1933, right before Just was to leave again for Europe, Weaver began to have doubts. He could not make sense of the situation. Just was black, he was a scientist; he was a scientist, he was black. Weaver could not make those two facts fit together, any more than many other whites Just encountered over the years had managed to. He knew Just to be, without question, "the outstanding Negro biologist of the world."[13] But what did that mean? Could it have meant he was not *merely* the outstanding Negro biologist of the world? This was profoundly perplexing to Weaver, especially since Just himself never gave consistent clues to the answer. In Weaver's opinion, Just was "so queer a mixture of modesty and confidence" that it was difficult to tell how important his work really was. In conversation, Just had not been assertive; in fact, he had been "somewhat confused and almost shy." In correspondence, on the other hand, Just had pushed himself, had insisted almost to the point of being gauche that his work was important and deserved support. Weaver needed another opinion to make sense of what he perceived as a kind of schizophrenia—Just's arrogance under some circumstances, his diffidence under others.

As usual, Lillie was sought for counsel. Perhaps no one else, not even Hedwig, ever knew Just as intimately. His response was precise, pointed, and perceptive:

. . . . I would like to give you as objective a view as I can concerning Dr. Just. This is perhaps a little more difficult for me than it would be for some other zoologist because he was my graduate student and assistant for several years before he received his doctor's degree under my direction in 1916. After that date he assisted me in research work at Woods Hole and we published at least one article together.

I think that we have two questions in this particular case; first, the question of his scientific ability and status, and second, the very difficult psychological question. I shall deal with these separately.

As regards his scientific work, he has been easily one of the most productive investigators at Woods Hole for the last twelve to fifteen years. His studies have been characterized not only by their care and precision, but also by a very considerable degree of scientific imagination. I do not think that any qualified zoologist would maintain that his work was not strictly first class. He has also been a real leader among the younger group of biologists at Woods Hole, and has been extremely helpful not only to them but to many physiologists who came to Woods Hole better equipped with ideas than with knowledge of marine organisms. He has indeed occupied a rather unique position there. The details of investigations to which I refer have led him to certain generalizations ranging far into the cytological field and applicable as he thinks, to a great variety of fundamental biological problems. I believe that his thinking is quite sound, though I have sometimes felt that he might go a little out of his depth. Considered on the basis of his scientific worth he is entitled in my opinion to very special consideration.

The psychological problem arises from the fact that he is of a mixed negro and white descent. This creates a series of difficulties to which I feel he has on the whole reacted with great credit. On the one hand, his negro descent makes it impossible for him to secure a position in a large university such as he would be entitled to on the basis of his scientific standing. On the other hand, his special eminence as a negro scholar has led to perhaps an over pronounced appreciation of merit, especially abroad. He was the first recipient of the Spingarn medal. Early in his career Mr. Rosenwald became interested in him and through the National Research Council provided funds for his research. This gave him a very special position at Howard University. At that time Dr. Just felt that he had a special duty to perform towards his own race educationally, and he has tried to discharge that at Howard University with results that have been disappointing to him in many respects. I think that all this helps to explain both his modesty in personal conversation and also his rather exaggerated self-appreciation in his correspondence.

Biology is a subject in which schools of thought develop sharply defined controversies. Dr. Just is a rather ardent adherent when his convictions are fully formed. Thus I would say that he is against the extreme Mendelian school and also the extreme mechanistic school of thought in biology. I quite fully agree with his views on these subjects, so far as I understand them. He has, on occasion, I think, been a little needlessly antagonistic to representatives of opposite schools of thinking. His remarks about Jacques Loeb illustrate this point. To some extent he may be motivated by feelings of personal loyalty, which are strongly developed in him.

In view of the fact that Mr. Rosenwald and to some extent, I understand, the General Education Board through the National Research Council, have been quite largely responsible for his personal development, I feel that some satisfactory provision should be made for his future. I think it could be made on a more modest basis than he has apparently suggested in his correspondence with you. Frankly, I am a little afraid that some of his experiences, more particularly in Europe, have a little unsettled his balance on the matter of self-appreciation, but I do not believe that this should be allowed to weigh appreciably against his accomplishments; and I feel that those who are able should attempt to aid his future development in the most friendly possible way, not only for his own sake but also for the sake of the race that he represents.[14]

How Weaver took this is hard to tell. He continued to collect information on the merits of Just's case.

Meanwhile Just was off to Europe, to get away from the drain of Howard and the pressure of seeking funds. He needed a more relaxed and easygoing atmosphere to be productive. He boarded the *Berlin* in New York and arrived in Boulogne around 21 March. He went straight to Paris, where Hedwig met him and they took a hotel room. They renewed their friendship, shopped around the city, visited museums, and went to concerts. After about two weeks in Paris they left for the countryside, enjoying the picturesque volcanic scenery around Auvergne and taking in the fresh spring air. Their ultimate destination was Berlin, where Just planned to finish up some work in the laboratory at the Kaiser-Wilhelm-Institut and then spend most of the summer working on the book.

By the spring of 1933 Berlin was undergoing profound political changes, however. The Nazis had fast gained political power and were making the day-to-day lives of Berliners, Jewish and non-Jewish alike, unbearable. Although Just had always doubted the rumors he had heard of the possibility of a Nazi seizure of power, he could not ignore the sights around him that spring.[15] Inflammatory placards and graffiti went up everywhere, windows were smashed, people were maimed. It was a nightmare, and a deeply disillusioning experience. How could these cultivated people—the makers of fine Dresden china, the descendants of Bach, Goethe, and Schiller—be engaged in such filth, such crime? The white Southerners in the United States? Well, they were different. They had little culture and therefore little humanity, Just thought, except that which the black man provided: it was hardly a wonder that lynching would be a style of life with them. But how could such brutality exist among the Germans? This confused Just no end. But he was too impatient to search for the answer—his immediate goal was to leave Germany as quickly as possible. He was too upset to work productively at the laboratory, and even his romantic life was threatened. The Schnetzlers did their utmost to hide the fact that Hedwig's grandfather was of Jewish descent, but the Nazis seemed to have their ways of finding out. Just and Hedwig stayed in Berlin only two or three weeks. In mid-April they left for Naples via Switzerland. The Nazis did not issue the Nuremberg Laws enforcing the "racial purity of the state" until 1935, but the writing was already on the wall.

While Just and Hedwig were fleeing to Naples, Weaver was puzzling over what to do with regard to Just's proposal. The matter had bogged down in an administrative mire, and Weaver was coming closer and closer to simply acting on the old familiar policy: no help to individuals, only to institutions. To bear him out he needed only to point to the problems that had occurred with Just's grant from the Rosenwald Fund, of which he was well aware. So far he had only been in touch with Just (whom he could not figure out), but now he began insisting on wholehearted interest and support from Howard authorities for Just's plan. It was the old triangle, and it was sure to make things even more complicated.

Just informed Johnson of the approach he had made to the Rockefeller Foundation and asked for his support. The Howard president assented, but he made no promise that he would not modify the request, shift its emphasis, or restructure the grant. Johnson in fact asked Weaver for funding for the basic needs of the zoology department (salaries, scholarships, and so on); Just's research needs came second. But the grants the Rockefeller Foundation was offering were not for the kind of educational programs Johnson was interested in, as Weaver was quick to point out.[16] Whereas Just had originally made a request for grant money to be tailored to his own research, without backing from Johnson, he was now in an even more tenuous position: now he had Johnson's support but for a different proposal. Weaver, a clever man, detected the disharmony between Johnson and Just at once.[17] To say that he exploited it would perhaps be going too far. But this is what it must have looked like to Just in Europe.

What could he do? After counsel with Hedwig, he knew where he had to turn. But first he reviewed the situation carefully. He had decided not to shake himself permanently loose from Howard, but rather to attach himself to a foreign institution on a limited tenure. This was strategy, pure and simple. Just knew that the Rockefeller Foundation supported other American workers abroad, usually by providing traveling fellowships to those with a specific research goal. The foundation could have no excuse, he reasoned, for not supporting his work. It was not, after all, his fault that he could not obtain affiliation with a major white university where administrators, so he thought, would be more sympathetic to his work than Johnson had been. He would stay at Howard, but only three months out of the year. Ten years earlier he had accepted a more stringent arrangement without question: Howard half the year, Woods Hole the other half. But in those days he had had no options, no clear goals. Now he had a driving urge to get on with his work, he had Europe and its laboratories, and he had Hedwig. He wanted America and its institutions to pay—to pay for the frustration and pain they had caused, to make reparations for the abominable conditions under which he had been forced to work. They had an obligation to set him up in Europe and give him a new lease on life.

What was in Just's mind was not in Weaver's, however, as Just was beginning to suspect. This white man was not moved in the least by Just's plan to work in Europe. It did not seem to matter that there were precedents for the proposal, that other Americans were receiving Rockefeller funds in Europe. Besides, the parallel was not strictly drawn, anyway. Those workers had intentions of returning to America, but it was not clear to Weaver that Just did, though he had no hard evidence to back his suspicions. But the real stumbling block was the scientific community's unanimous agreement that Just's place was in America—Howard in the winter and Woods Hole in the summer. Weaver heard this echoed time and time again, from people like Heilbrunn, Schrader, and, perhaps most important, Frank Lillie.

Just understood what was going on. But he needed help and he had no one but Lillie to turn to. Moreover, he felt that his mentor had an obligation to help. Knowing Lillie's power and influence, he wrote from Europe and asked him to intercede with Weaver.[18] If ever Lillie could pull strings with the Rockefeller Foundation, or anybody else for that matter, now was the time. Just stressed the urgency of the situation, perhaps as he had never done before. But Lillie was not moved by Just's plea. Politely, but not too politely, he explained that he thought this was as good a time as any to set some things straight.[19] First, he was getting old, and whatever his inclinations might be, he had neither the energy nor the time for a struggle of this sort. No one but Just could fight Just's battles. Second, he did not have the power Just attributed to him. He was in no position after all these years to take an unyielding stand with foundation officials and make them go against their own policies. Just would simply have to take up arms and fight the war himself. Further, in order for Just to have a chance of winning, he would have to make an appearance on the battleground. Lillie thought it was a mistake, indeed a disaster, for Just to be away from America so frequently and for such long periods of time: "Out of sight, out of mind," Lillie reminded his old protégé. Just should return to America in time to spend part of August and September at Woods Hole, Lillie advised, to renew his associations and be on hand when the case of Howard University came up at the Rockefeller Foundation in the fall.

These were things Just hardly wanted to hear. He was spending midsummer in Tuscany (Naples was too hot) working ardently on his book. A wrist-slapping from his mentor was the last thing he needed. But he tried not to let it disturb him too much. He and Hedwig continued to work overtime every day, rising early and retiring late, and taking walks through the countryside only in their spare moments.

Just did not follow Lillie's advice. He returned to America only in mid-September, and did not set foot in Woods Hole. He was on the scene in time for the Rockefeller Foundation's decision, but by then the chances for a positive conclusion seemed slim indeed. Weaver had asked Just and Johnson to submit one common proposal. What he received instead was a series of separate letters from Just and Johnson— letters that repeated requests for money without showing any sign of agreement as to how the money should be spent. Weaver, feeling that he was constrained to do so by foundation policy, finally turned down what he called "an unsatisfactory and conflicting mixture of two proposals."[20] Whatever else he did for science in America, he gave Just not one penny.

There could not have been a worse time than the fall of 1933 for any American scientist, and especially a black one, to try and locate funds. The Depression was still on. It was still common to see men and women standing in long relief lines waiting to get soup for themselves and their families. Blacks were especially hard hit by the economic disaster. The fight of a single black scientist to keep his work afloat and to change his personal life went on against a backdrop of hungry people struggling for food and desperate for jobs. True, Just was earning $10,000 a year from the expiring Rosenwald grant and other sources—or at least he had that much at his disposal. Only in minor ways was he inconvenienced by the Depression, insofar as comforts for himself and his family were concerned. But he was suffering too. His life was his work, and to carry on with his career he needed funds from foundations and philanthropists who were not only tightening their belts but demanding more in the way of justification for research grants.

The situation was frightening, but Just did not give up. Next he turned to the Carnegie Corporation. Ten years earlier he had made an appeal to the embryology department of the Carnegie Institution of Washington and had been refused help. This time he went straight to the top.

The Carnegie Corporation, like the Carnegie Institution, was part of the philanthropic empire set up by Andrew Carnegie at the turn of the century.[21] Carnegie is perhaps best known as the man who put the modern public library system on its feet in the United States, but his philanthropy was more far-reaching than that. In 1901 he set up his first foundation, not in the United States, curiously enough, but in the land where he spent the first thirteen years of his life, Scotland. The Carnegie Trust for the Universities of Scotland sought to promote opportunities for scientific study and research among qualified students in that country. Carnegie was always interested in science, the result, perhaps, both of his upbringing in a household where the name of Joseph Priestley was revered and of his later role as the founder of the iron and steel industry in the United States. In 1902 he set up the Carnegie Institution of Washington as a trust devoted to scientific work, one that would "encourage, in the broadest and most liberal manner, investigation, research, and discovery, and the application of knowledge to the improvement of mankind."[22] Other foundations were established in quick succession: the Carnegie Hero Fund Commission, set up in 1904 to recognize heroic acts on the part of people whose duties do not necessarily require them to perform such acts; the Carnegie Foundation for the Advancement of Teaching, set up in 1905 to promote higher education (it was under the auspices of this foundation that Abraham Flexner did his report on medical schools); the Carnegie Endowment for International Peace, set up in 1910 to further understanding between nations; and finally, the Carnegie Corporation of New York, set up in 1911 as the culmination of Carnegie's philanthropic programs, with a basic endowment of over $100 million and a broad mandate to advance knowledge and improve universities. Carnegie's trusts were no less grand in concept and design than Rockefeller's.

Just went directly to the president of the Carnegie Corporation, Frederick P. Keppel. It was not their first meeting. Just had visited Keppel in late 1926 or early 1927, for what reason remains unclear but most likely to see if anything could be done to reverse the negative decision on Just's appeal for funds that had been reached by the Carnegie Institution around that time.[23] They kept up their acquaintance and at one point Just even recommended Keppel for the presidency of the University of Chicago.[24] In a way, Just viewed himself as a broker in American academia. Though he could never land a job for himself outside of Howard, he possessed a certain amount of influence when it came to placing whites. He was well aware of this and was none too shy in letting others, like Keppel, know it too.

Just met with Keppel on 9 November in New York. Somehow the Carnegie president was different, no longer a "forceful and alert" individual but a "weak and soft and flabby" one;[25] he seemed to have undergone "a sort of fatty degeneration in a spiritual and intellectual way." At least this was how Just, tired and depressed and not overly optimistic about his prospects, perceived Keppel. The meeting was rushed. Just only had time to outline his proposal briefly and to ask Keppel to request a letter of support from Oscar Riddle, a Carnegie scientist who had heard Just's Padua lecture.[26]

After the meeting, Just contacted Riddle to explain the situation. Riddle, who claimed to know Just's work and the problems he was laboring under, responded with a supportive but not glowing letter, stating that Just was "competent" but not an "advancer" of knowledge.[27] He was determined to offer support primarily out of sympathy for Just because of the racial slights Just had suffered in America, mostly at Woods Hole. Riddle himself had never gone to Woods Hole, and almost certainly felt no loyalty to that community.

A little later, Just sent a formal proposal to Keppel requesting money to continue his research in Europe.[28] It was a conflation of the letters he had written Weaver, stressing the revolutionary nature of his work for the future of biology. In a way this proposal was even more passionate than the ones he had sent to the Rockefeller. He revealed that his new theory had "ridden" him "like a witch" for three years. He had to get on with his work, no two ways about it. Marvell's verse, "But at my back I always hear/Time's wingèd chariot hurrying near" might have been written for him.

Just was asking for $10,000, to be spent solely by him for his expenses: not a penny was to be spent by or put into Howard. Keppel was moved. He felt compelled to help, but first he wanted to get another opinion on the merit of the request. He asked W. M. Gilbert, the secretary of the Carnegie Institution, who in turn asked the head of Carnegie's embryology department, George Streeter. If there was one person who was consistent in his opinion of Just, it was Streeter. A decade earlier he had been negative, and now he was no different, except that his criticisms were pitched an octave higher. He thought Just arrogant, or as he put it, "not suffering from any inferiority complex."[29] Europe had done nothing to increase Just's humility. Streeter could imagine how Just would attempt to explain his boastfulness, to justify it as "necessary to overcome the prejudices" he sensed on all sides because he was "a mulatto." Streeter knew better. In his view, Just had been judged fairly on the basis of his work alone. As a matter of fact, if any error had been made it was "on the other side," and Just's "colored blood" had "resulted in giving him greater distinction than he would have, were he an ordinary worker." But Streeter saw himself as a scientist, honest and objective, believing in merit and merit alone. He conceded that Just was "a competent investigator," though "certainly not one of the great leaders of biology." He could not lie. In his professional estimation, funds given to Just would be "well invested" and would "meet with the approval of biologists generally."

That was all Keppel needed to hear. In December, after clearing the way with the board, he handed down a positive decision to the Carnegie Institution of Washington, which was charged with the responsibility of administering and overseeing the grant.[30]

John C. Merriam, president of the Carnegie Institution, took over the matter from there. He and Just got together early in January 1934 to work out an acceptable research program and a detailed budget for the $10,000 grant. Merriam took a firm hand and was none too delicate in laying out the arrangement. He wanted either Riddle or Streeter to "help focus Just's work on particular questions," to keep in touch with him and "see that he holds to the track."[31] The bursar at the Carnegie Institution would disburse the funds and keep a careful record of receipts and expenditures.

Just disliked Merriam from the start, referring to him often as a "stuff-shirt" and a "stuck-up owl."[32] But he would say these things to himself or in private; publicly he was polite and respectful. The money was too crucial for him to risk making a social blunder with Merriam, who, after all, was simply a go-between. At most, he figured, he would have to endure Merriam's behavior once or twice, and then he would be free to go to Europe and get to work. The best thing he could do now was dress properly, put on clean underclothes as a means of protection, almost like armor, and keep his soul secretive. And this he did.

When Streeter heard about the grant and his expected role as a kind of supervisor, he responded in his usual style. A casual reader might, he said, read a recent paper by Just as "a gracefully written philosophical treastise," but Streeter knew it to be "full of holes."[33] Even worse, Just had presented his ideas as "a thesis for serious scientific consideration," not what Streeter saw them as: "a literary effort to lightly entertain a social group." (The reference was to one of Just's theoretical articles, either "On the Origin of Mutations" or "Cortical Cytoplasm and Evolution.") Somehow Streeter had not expected that the grant would really come through, especially in light of the negative picture he had painted of Just. He had conceded some merit to Just's work, but not much, and only to seem objective. Perhaps he felt in some sense betrayed when the Carnegie Corporation failed to act on his real wish instead of his lukewarm recommendation. He implied that he was both amused and annoyed by the corporation's generous decision to give Just "a joy ride to Europe." He was a man and would accept how things had turned out. He would not bother Merriam any more about it—and he never did.

Since he had obtained the Carnegie grant on his own, and since Johnson had meanwhile failed to get a renewal of the Rosenwald grant or obtain a new one from the Rockefeller, Just had little difficulty getting permission to leave Howard at the end of the first semester of the 1933–34 academic year. Johnson was not in a harassing mood and could put up no real opposition; he was fully occupied with problems of his own—hostility on the part of not only many of his faculty but also the Congress. Just was pleased, but deep down he must have felt slightly disappointed. He liked nothing better than a fight, an opportunity to plan strategy and summon up courage and, hopefully, come out the victor. Squabbles, if there were not too many and they were not too serious, spurred him on to new creative levels.

Just boarded the *New York* on 31 January 1934, going tourist this time to save money.[34] The ship arrived in Cherbourg on 8 February. Hedwig was on the tender to meet him. She had come from Germany through Paris, also on a second-class ticket. They were ecstatic. For the first time they would be together for a whole year.

In Paris they stayed for a few days at the Hotel Metropolitain, did some shopping, attended the opera and the symphony, and visited the museums. Then it was off to Naples, with stopovers in Basel, Saint Moritz, Celerina, and Rome. How different this was from his first trip to Europe five years earlier. Instead of his daughter Margaret he had Hedwig as his companion. And there was no need for Dohrn or anyone to meet him at the train station; Hedwig was an experienced traveler and knew Europe as a clever child knows the alphabet. She was fluent in German, French, Italian, and Russian as well as English. While Just could not boast such fluency, he no longer felt like a stranger. He could get around day to day about as well as he did in America. And in one sense he got around better. In France or Italy he did not have to think about whether a shop or a restaurant would serve a black man or whether a hotel might refuse accommodations to a mixed couple or whether passersby might yell obscenities when he and Hedwig strolled together hand in hand, laughing and enjoying themselves. It was certainly a change from America.

Italy gave Just the freedom to get on with his experiments, to think, to write. As soon as he arrived in Naples, on 20 February, he started work on his book. It was to be a synthesis of nearly thirty years of biological research, a disquisition with powerful theoretical and philosophical implications for biology, and, Just hoped, a means for him to find the kind of backing he needed to continue his work.

❧

The book went through several working titles, from "The Fundamental Manifestations of Life as Revealed by the Animal Egg" to "Life Reveals Itself" to, finally, *The Biology of the Cell Surface*.[35] It was a synthesis of Just's scientific achievements and a justification of his past scientific efforts, but, more than that, it was a statement of a new theory of biology. The theory concerned the role of the ectoplasm, the cell surface, in development—hence the title—and was put forward in the context of a philosophy of biology which Just outlined at the beginning of the book. Throughout the book, the philosophy and the theory are carefully intertwined.

Few books written by a biologist are so concerned with philosophy. This is not to say that biologists in the 1930s did not work from philosophical bases, only that hardly any of them were so explicit in articulating a particular philosophy. Few scientists bothered to set out their methods beyond what was necessary to justify their conclusions; insight and elucidation were not a great concern—certainly not as regards the lay public, and often not even as regards other scientists. Heilbrunn's book, *An Outline of Problems of General Physiology*,[36] was an example, an advanced textbook that made no claim to provide either a new theory or a grand overview. For the most part, scientists were content to publish their results only in the form of separate articles. Bringing individual articles together, with or without connective statements, is a notion of more recent times, with perhaps the exception of one or two famous scientists such as Jacques Loeb. For Just, however, the overview—the synthesis—was important, at a time when others did not consider it so. It was a way for him to insure that he would be remembered. Included in his book is an almost complete bibliography of his works, another unusual feature. *The Biology of the Cell Surface*, the vehicle for his overview, allowed him to reveal his theory of biology through the experience that came from years of hands-on work in the field. Though Just was taking the role of a philosopher, he was primarily a working biologist. His philosophy is grounded in methodology and technique; his general pronouncements are supported with examples from his experimental work. He sought a nice balance.

Before Just argued his thesis, he tried to clarify the "backdrop of general biological theories." He wanted to show "serious men in engineering, practical physics and chemistry and laymen outside of these professions" where biology stood and what it had to offer. He wanted to get at the important question: What is life? In contrast to many scientists of the time, he felt that biology and biological method were different from the physical sciences and their methods. But he was not a vitalist. He recognized that biology relies to an extent on physics and chemistry and mathematics. Life was special, however, and as far as he was concerned analyzing compounds by destroying life could not get at the basic question, What is life?

Just did not in any way believe in the conception of a hypothetical "life molecule." Life, he argued, is characterized by "the harmonious organization of events, the resultant of a communion of structures and reactions."[37] All attention in biology was focused on the protoplasm, consisting of two components, nuclear and cytoplasmic. Most biologists saw the nuclear component as the "kernel of life." Just wanted to shift this emphasis. In his book he would demonstrate "how far life-processes are related to the dual and reciprocal components, nuclear and cyto-plasmic structure."[38] More than that, he would spell out the role of the surface cytoplasm, the "ectoplasm," in vital manifestations. No other biologist had tried this, although many had detailed the role of the nucleus. The reason was simple enough: compared to the nucleus, the "ectoplasm" was difficult to observe with the light microscope. But though it was eye-straining work, Just had devoted himself for years to studying the "ectoplasm" closely—years before the advent of the electron microscope.

Just had a clear conception that two methods of investigation—description and experimentation—are indispensable to biology. The organization of living matter demands more description than, say, the objects of study in physics or chemistry do, but experimentation, certainly by the twentieth century, had also become integral to the science. Just wanted to explore the role of experiment in biology, since it was the newer and probably less understood of the two methods.

The first thing he wanted to do was separate his own mode of experimentation from the so-called physico-chemical school of biology, the mechanists, headed by Jacques Loeb. Just objected to the point of view propounded by this school, that the goal of biology was to reduce living matter to the physicist's "ultimate particles." The mechanistic biologists, moreover, were "more Royal than the King," pushing the notion much further than a physicist would. They misused the word "mechanist"—or rather, they misused its opposite. For them, the opposite of a mechanist was a vitalist. This was not true, Just argued. The physicists were correct in their view, namely, that the opposite of mechanist is nonmechanist. Just knew that vitalism and nonmechanism are hardly synonymous: "Not every physicist who opposes the mechanistic conception deems it necessary to support a non-physical, supernatural concept, rather he holds that the behavior of the ultimate particles of matter is not rigidly determined, perfectly predictable."[39] The same should be true of the biologist.

Somehow the role of experiment in biology had to take into account the difference between a living thing and a nonliving thing. This is not to say, Just asserted, that the experimenter had to know the precise nature of the difference, only that he had to acknowledge it as such and devise methods sensitive to it. Most biologists, especially the mechanists, focused on the similarities between the living and the nonliving, adopting the methods, and using mostly those methods, of pure physics and pure chemistry. To Just's mind this approach could threaten the integrity of the living state. He wanted biologists to continue to use the methods of chemistry and physics, to be sure, but only as a supplement to their own methods. Biologists should "count, measure and weigh and seek to detect cause and effect" through the use of physico-chemical means. At the same time, however, they should try to understand how those very methods affect, even radically change, the living state. This could be done only if biologists understood and recognized the *normal* living state, a point Just had stressed as far back as the early 1910s, and especially in the 1920s at Woods Hole.[40]

According to Just, experimentation in biology falls into three categories: experiments on nonliving systems, on killed living systems, and on living systems.[41] The first category provides comparisons between living and nonliving systems. Physical models and artificial systems are examples: semipermeable membranes, suspensions of soap in water used to imitate the cleavage patterns of the egg. Chemical analyses of substances produced by the living cell, hormones and urine, for example, also come under this heading. The second category includes experiments on the chemical composition of living matter killed in the process of analysis. Histological and cytological methods of creating fixed tissues and cells are good examples, even though they are not in and of themselves strictly experiments. The third category, perhaps the most important, Just broke down into two further groups. The first consists of experiments in which the normality of the living system remains unchanged (in situ), and the second, those in which the normality of the living system is altered. In both types of experiments in this crucial category the question of the normal living organism comes into play. Again Just emphasized that a knowledge of the living organism is as fundamental as a knowledge of physics and chemistry. He was astounded at how physicists, chemists, and biologists would go to great lengths to insure that their apparatus was up to par and their chemicals were pure, but not seem to care about the condition of the living material with which they worked. Many supply houses routinely sent investigators living material that was in appalling condition, "debased and degraded." Scientists accepted it simply because they neither knew nor cared what was normal and what was not.

Perhaps no one knew more about normal marine eggs than Just. He had studied them in their natural habitat ever since 1909. The criterion for whether eggs are normal, he stressed, is the number of them that actually develop. One hundred percent development means optimum condition; more refined criteria were associated with stages of development, the period of the mating season, and so on, and in some sense these superseded the basic criterion. In addition Just considered it important to give detailed physical descriptions of normal and abnormal eggs.

The use of quantification also concerned Just. Whenever possible he performed experiments over and over again to make sure that the scale of experimentation was sufficient to justify his conclusions. Though Just does not give a precise definition of what is statistically meaningful in biology, he certainly knew what was not. The work of Theodor Boveri, one of the fathers of modern embryology (and Margret Boveri's father as well), stood out in his mind as lacking "exactness and completeness," as requiring repetition on "a more extensive scale." Among contemporaries, Just cited the work of T. H. Morgan, the renowned geneticist and embryologist, to illustrate misleading if not incorrect uses of quantification.[42] Morgan had published the results of eleven experiments using the following number of eggs: 1, 10, 4, 8, 8, 3, 3, 1, 2, 2, 4. His error, Just maintained, was to speak in terms of *percentage* of development, when the numbers were simply too small to allow for such terminology. Where living cells are treated statistically, "the larger the number used, the more valuable . . . the conclusions drawn."

Perhaps the most profound and far-reaching belief Just subscribed to involved the aim of experimentation in biology. Basically the question was, Why do it at all? According to Just, "the main purpose of an experiment in biology should be the explanation of the naturally occurring phenomena."[43] This puts biology as a science apart from physics and chemistry. That Just held this view in no way makes him a vitalist or a mystic; it does, however, indicate the extent to which he believed in the integrity of living systems and the depth of his understanding of biological investigation.

After the prefatory exposition of his views on biological research and experiment, the discussion proceeds in a general way to the structure and composition of protoplasmic systems. In addition to the classical cell with a single nucleus, Just recognized organisms consisting of a mass of cytoplasm and no nucleus that could be observed with conventional methods, and organisms with two or more nuclei. These three types of protoplasmic systems showed the extent to which the nuclear substance is differentiated. Just wanted to focus on the most frequently occurring of the three: a simple membrane-enclosed cell with a single nucleus.

Cells vary in size "from one micron in length, discernible only under the highest microscopical magnification, to the sporozoon parasite found in the digestive tract of crustacea which measures 16mm. in length, or to the unfertilized egg of the ostrich, 105mm. in diameter or that of a shark, 220mm. in diameter, or to the nerve cells in the human spinal cord, which may be one meter in length."[44] The egg cell is particularly interesting, for out of it emerges the complex organization found in the adult human being or in any other multicellular organism, in animals and in plants. The cell is the biologist's crucial unit of observation, and the egg cell is the special domain of the embryologist.

Since 1910 Just had studied egg cell morphology—form and changes in form—as a first order of business. Though he was interested in the physiology of cells as a means toward understanding the problem of vital manifestations, the "old-fashioned" descriptions of visible form continued to excite him. He would mount a living unfertilized egg of the sea worm *Platynereis* in a few drops of sea water and place the slide under the low power of the microscope (see figure).[45] At first glance he could see "a greenish-yellow sphere everywhere crowded with smaller spheres except near the centre and at the periphery beneath the well marked egg membrane." The picture as a whole portrayed "a pebble or shagreen effect." After he brought the egg into sharper focus, he could more clearly discern an optical section of it showing the regions just described. The large clear area near the center is the nucleus, sharply set apart from the remainder of the cell by its transparency and homogeneity. Closely crowded along the nuclear boundary on the outside of the nucleus are many spherules with "some twenty or more refringent globules (oil drops) interspersed." Tiny bright granules are scattered among these spherules and globules. Under still higher magnification, Just could see even smaller granules. The spherules, globules, and granules lie in the endoplasm or inner cytoplasmic region. The outer region of the cytoplasm is a clear band extending beyond the inner. It is called the ectoplasm. In *Platynereis* "the ectoplasm is crossed by fine radial lines" which "appear to reach the egg-membrane called the vitelline membrane, though actually they end in the plasma membrane." The plasma membrane was difficult for Just to discern, as it lay underneath the vitelline membrane.

After observing it in this way, Just would fix the egg quickly with some chemical reagent to preserve it in a state closely resembling the living. He had come to appreciate and even argue for fixation, as long as it was properly done. Throughout his career he had had to observe "processes too fleeting to be followed exactly in the living state."[46] By fixing the cells he could follow many processes in their simple steps.

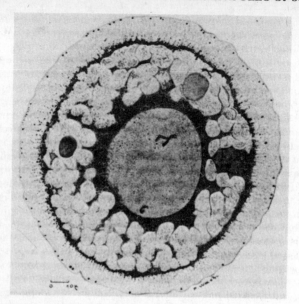

Section of an unfertilized egg of *Platynereis megalops*.

He always kept in mind, of course, that the fixed cell was dead, that its proteins were changed—"coagulated, gelated, precipitated"—and that other changes had taken place.

After fixation, Just would cut the cell into thin slices—a process called sectioning—to discern those structures difficult to make out in the living cell. A stained section from a fixed *Platynereis* egg makes a pretty picture (see figure).[47] In the nucleus there are granules of various sizes and two chromosomes, all of which "lie in a faintly granular grayish-blue field enclosed by the nuclear membrane." The cytoplasm has two regions, sharply differentiated and unmistakable: "the endoplasm, crowded with yolk spheres, oil drops and granules, the smaller of which are mitochondria; and the ectoplasm with its radially disposed lines which extend to the plasma membrane beneath the vitelline membrane."

The cell membrane, which was at the core of Just's theory, was a controversial subject. Some scientists even questioned whether it existed at all. Just held that there is a membrane around animal cells and that it is a part of the living system. The most unfortunate aspect of the controversy, he felt, was the confusion surrounding the nature and properties of egg membranes. Many scientists, embryologists included, confused the plasma membrane with the vitelline membrane. The plasma membrane is continuous with the egg cytoplasm; the vitelline membrane is separated from the egg cytoplasm by the perivitelline space. The difficulty arises because many vitelline membranes after fertilization look like plasma membranes before fertilization.

Just's aim was to define a region of the cytoplasm that he thought possessed unusual characteristics in living cells. Using a decription of the two regions of cytoplasm in a study by Haeckel in 1872, he formulated his definition of ectoplasm as "the superficial region of protoplasmic ground-substance set off from the remainder (both nuclear and cytoplasmic ground-substance) by differences in physical properties, in structure and in behavior."[48] He added that "as part of the cytoplasm, as differentiated ground-substance, it is definitely a part of the living cell." The inner region of the cytoplasmic ground-substance—with its cytoplasmic inclusions, granules and the like—is as defined as the endoplasm. Though earlier investigators were familiar with both regions, they did not usually identify them or give them equal weight in their studies.

For Just the ectoplasm was real, and it was important. It expresses itself particularly in the muscle cell, where it is preponderant. The nerve fiber also was considered by Ross Harrison to be ectoplasm, as a result of his tissue culture work.[49] Just clarified aspects of tissue culture on the basis of his knowledge of ectoplasmic behavior.

Going methodically through the history of science, Just cited embryologists who observed ectoplasmic behavior—Metschnikoff and Gatenby in the sponge egg; Kowalewsky and Agassiz in ctenophore ("comb-jelly") eggs; Fol in coelenterate eggs; Chun and Yatsu in the *Beroë* egg; Metschnikoff, Selenka, Lang, Pereyaslewzewa and Janicki in cestodes (tapeworms); and Van Beneden and Meves in the unfertilized egg of *Ascaris*. He marshaled evidence from an extensive review of the literature. His first order of business was to prove that eggs normally display an ectoplasmic structure; then he would go on and show that this structure embodied a "significance for the grand problem of biology, the revelation of phenomena." More specifically, indeed poetically, he planned to demonstrate that ectoplasmic behavior reveals the "activities that set apart the living thing from the non-living, that mark how life maintains itself ever in harmonious tempo with the ceaseless changes in its surroundings."[50]

In Just's view, the three fundamental properties of living matter—respiration, conduction, and contraction—are properties of the ectoplasm. He went through them one by one.[51] With regard to respiration, the ectoplasm "constitutes . . . an oxygen uptaking structure." In fact a number of surface changes can be correlated to the respiration of the cell—oxygen consumption and carbon dioxide output. These changes are observable in the process of membrane separation, which was tied in with his very own theory of the wave of negativity.

By adding to eggs a drop of seawater containing a few spermatozoa, Just was able to observe a most fascinating phenomenon.[52] Within two seconds the spermatozoa attach themselves to the *Arbacia* eggs, for example. The egg surface shudders back and forth from the impact of the spermatozoon, and the egg membrane goes in and out underneath the darting spermatozoon for a second or two. Then the spermatozoon stops, and buries its tip in the egg surface, and the ectoplasm begins to take on a cloudy appearance. Within twenty seconds the ectoplasm is completely cloudy. Then, "like a flash," beginning at the point of sperm attachment, a wave sweeps over the surface of the egg, clearing up the ectoplasm as it passes. This is all easily observable, especially in *Echinarachnius* eggs, which Just used for this purpose. The egg now rejects other spermatozoa along the wave, beginning at the point of sperm entry; the pole opposite sperm entry is the last point affected. The site of sperm entry is said to be a "point of injury"; it is "negative" to all other spermatozoa arriving at this point. All other points around the egg are said to be "positive." The "wave of negativity" that Just observed in slowly reacting eggs moves over the egg at a rate which varies with that at which the sperm head disappears within the ectoplasm. The egg can still take another sperm, but only for a while and only at the pole opposite that at which the original spermatozoon enters the egg. The concept of the "wave of negativity" was Just's; he had observed it around 1920, and embryologists immediately took up the observation as fundamental to work on fertilization.

The "wave of negativity," Just demonstrated, preceded membrane separation. When the wave sweeps over the surface of the egg, it clears up the ectoplasm as it goes. About twenty-five seconds after insemination, the ectoplasmic threads that have formed break at the point of sperm attachment and release a membrane. The membrane separates along a wave from the point of sperm entry, and follows (and is related to) the "wave of negativity." Just argued that this membrane is already present in the egg, that it is the vitelline membrane, not a newly made "fertilization membrane," as many biologists thought. When the membrane is fully separated from the egg, the ectoplasm builds a new structure, the hyaline plasma layer. Many ectoplasmic changes are associated with this process. Most significant perhaps is that the egg becomes more susceptible to dilute seawater; cytolysis, the dissolution of the cell, always occurs when the membrane is lifting. This is the point at which the ectoplasm is weakest—hence the importance of the ectoplasm.

Another surface phenomenon is conduction. Take nerve cells, for example. When the fiber from one nerve cell comes in delicate contact with another, the nerve impulse goes from one nerve fiber to another. Conduction, for the most part, is carried on by the fiber. From the lower vertebrates to man, "all modes of integration" are dependent on intercellular connections which, Just argued, are ectoplasmic.

Like conduction, contraction is a property of the cell surface. Muscle cells in particular illustrate this phenomenon. They, like nerve cells, are very long, and they come together to form contractile tissues with a relatively large surface area. Just argued that this elasticity is a function of the cell surface.

Just continued his synthesis by considering the question of nutrition.[53] As no cell can live without water, the way water is regulated is a crucial consideration. Just observed that water equilibrium is, for the most part, regulated by and in the ectoplasm. In fact, all nutritional processes in the cells of higher animals—the taking in of food and the secretion of wastes—can only be performed through the large surface areas of the ectoplasm. The ectoplasm is "not only a series of mouths, gateways," it is also constantly being shaped and shaped again. In a poetic moment Just concludes that "the environment plays upon the ectoplasm and its delicate filaments as a player upon the strings of a harp, giving them new forms and calling forth new melodies." Scientists could barely see these things, things that are "too nice for the indiscriminating ear of man."

Just's strongest evidence for the primary role of the ectoplasm came from his work on fertilization.[54] According to him, "the main event in fertilization is a reaction between egg-cytoplasm and spermatozoon." He detailed the visible changes in the ectoplasm and associated them with the reactions exhibited by all animal eggs in response to sperm attachment—this time going further than membrane separation and the "wave of negativity." He offered the hypothesis that underlying these changes at the egg surface is the fundamental act in the fertilization process.

For Just, the initiation of development was an ectoplasmic response to a stimulus, either an artificial one or a spermatozoon. Thus the ectoplasm is necessary for fertilization. The state of fertilization depends upon "the integrity of the ectoplasmic layer." Polyspermy, or the entrance of several spermatozoa into one egg, occurs only when the ectoplasm is slow in reacting.

Just pointed out that fertilization in the egg is independent of any stage of maturation. The egg, depending on the species, can be fertilized before, during, or after maturation. The fertilizability of eggs is therefore "independent of its nuclear state, as germinal vesicle, as first or second maturation nucleus or as completely matured nucleus." By this means Just showed the unimportance of the nucleus and by implication the relative importance of the cytoplasm, especially the ectoplasm.

The fertilization reaction is complicated, and according to Just it is confused with membrane separation. Membrane separation is the result of the fertilization reaction, but it is not the thing itself. Just's description of the fertilization process, which he saw as chemical, makes this point:

> The spermatozoon sets off a first explosion in the narrow area of the egg-surface which it touches, and kindles the spark which leads to a chain of explosions. The fertilization-reaction is thus a trigger-reaction. A large portion of the surface is shattered by the explosive effect with gas exchange and heat-liberation. One can in many eggs see this breakdown of the surface by which the egg loses substance and the membrane is separated. The pressure between egg and membrane in the perivitelline space, probably is considerable as the ectoplasmic colloids disintegrate and go into solution. Water rushes in and further distends the still ductile membrane, which then sets as a stiff structure. The fully separated membrane then becomes brittle. . . . But these changes in the membrane are due to its separation from the egg, for it no longer forms part of the living system. What therefore are important for the fertilization-reaction are the underlying surface-changes that result in membrane-separation, and neither the consequent physical act of separation nor the changes in the membrane itself. . . .
> Development embraces a series of surface-changes which vary as the surface area increases with the march of development. Those occurring at the very outset are most striking; they mark out the course, direct the way towards the final outcome of fertilization—the formation of the complex organism out of a single cell, the egg.[55]

Fertilization is one form of the initiation of development; parthenogenesis, in which eggs develop without spermatozoa, is another.[56] In parthenogenesis ectoplasmic changes are also observed. Though the changes alone do not determine the ensuing development, they give some indication of the quality it will attain. Just brought together the work of Loeb and earlier investigators and joined it with his own, especially his work in the 1920s, to hammer home his point about the role of the ectoplasm in the initiation of development.

The ectoplasm is involved with every stage of development. The pluripotent egg, one capable of producing more than one embryo, when fertilized normally develops into only one embryo because of the ectoplasmic interaction between the cells produced during cleavage, known as blastomeres; if they were separated they would develop into as many embryos. "With cleavage," Just maintained, "the original egg-substance is separated into blastomeres and the blastomeres are integrated by means of intercellular connections, that is, the ectoplasmic prolongations." In fact Just believed that in the whole animal kingdom, with the possible exception of the mammalian egg, the embryo arises from the egg surface.

In 1934 Just was still working out the influence he felt the ectoplasm played on nuclear and chromosomal behavior.[57] All attention at the time was focused on the work of T. H. Morgan, who had just received the Nobel Prize for his work on *Drosophila*. The theories of genetics and physiology of development were at odds, though Morgan had written an article trying to discuss them side by side. As Just drafted his book, he reserved some space in which he could take up the problem of genetics and development,[58] and even to venture some thoughts on evolution.

Already he had ideas on the subject.[59] Just posed the questions: How did the first living thing arise? What was the cause of evolution? He believed that the first living thing arose from a nonliving thing. Two factors were important here, the environment and the ectoplasm. The original living substance possessed a "peculiar and complex organization." At the very moment it became life it was set apart from its environment, but at the same time it was a part of environmental changes. Factors in the environment, temperature of gases, for example, affected the cytoplasmic surface first. According to Just, "The first step in the evolutionary process . . . was a differentiation of the cytoplasm into ectoplasm," and only after that was the nucleus differentiated. From then on the ectoplasm played a major role in further differentiation.

The differentiation of ectoplasm out of the ground-substance was not only the basis of the primordial living thing, it was also, for Just, the cause of evolution. This was speculation, to be sure, perhaps even an exaggeration, but Just wanted to turn scientists' minds, however briefly, away from the role of the nucleus to the possible roles played by the ectoplasm. After all, it is the ectoplasm that is the boundary of the cell, that communicates with the outside environment.

The living thing and the outside world are interdependent. Life could not, in Just's mind, be "only a struggle against the surroundings from which life came." It must be a form of cooperation. Just adopted the theory of the Russian anarchist philosopher Prince Peter Kropotkin, who believed mutual aid and cooperation to be the fundamental fact of the biological world.[60] Kropotkin's theory, with its concept of harmony, was in Just's view preferable to Darwin's idea of the struggle for existence. Just hoped to go further one day, to elaborate on the biological principles he had put forth—especially those on mutual cooperation—in order to "seek the roots of man's ethical behavior."[61]

These flights into philosophy and anthropology were new expressions of old interests. Who would have thought that the scientific question of the ectoplasm would lead so far? Just was determined to push his observations on the cell surface to the limit, to let others know that he alone had treated the subject in a comprehensive and far-reaching way. He was synthesizing, pooling others' work, highlighting cases where they had missed the significance of the ectoplasm, elucidating this significance himself.

In drafting the book Just was very conscious of himself as a scientist and an author. The pronoun "I" is used in a way that would have made most scientists of the time uncomfortable. The observer, the narrator is always present, to lead, to guide, to shape opinion. Science as a discipline is portrayed as objective and powerful, and Just as a scientist comes across as perceptive and brilliant. No reader could miss this. But Just was more than an observer, a scientist; he was a poet as well. Metaphor after metaphor is used to captivate the reader and entice him into looking more closely at the wonders of biology. Just is one of the few biologists to take such literary liberties. It was not a way of showing off; it was genuine enthusiasm, joy in his experimental work and pride in the larger synthesis that he had managed to put together after so many years of hard work and struggle.

"We feel the beauty of Nature because we are part of Nature and because we know that however much in our separate domains we abstract from the unity of Nature, this unity remains," Just wrote in the last paragraph of his book. "Although we may deal with particulars, we return finally to the whole pattern woven out of these. So in our study of the animal egg: though we resolve it into constituent parts the better to understand it, we hold it as an integrated thing, as a unified system: in it life resides and in its moving surface life manifests itself."

Only in Europe, only with Hedwig at his side, could Just have written with such passion. The year 1934 brought them closer together. He explored his mind and his heart: What was he, a black man, doing loving this white woman? Was it because she would have been forbidden him in the States? Was it his way of putting himself on a level with Sturtevant, Heilbrunn, and Schrader? Or was it simply that she had touched his heart with her kindness, her strength, her intellect? The questions were too many and too involved to be answered straight out. He had to find other ways to get at them.

To him, Hedwig was a goddess. She was Venus, Aphrodite, Pallas Athena—an embodiment of all the religious and classical literature he had imbibed as a child.[62] His fantasy world was now becoming real. Hedwig was in no way beautiful, was even gawky in appearance. But to Just she was an angel, and her mind seemed almost mystical, with its delicate flights of creativity. Unlike Margret Boveri, she took a deep interest in his work and his problems, and she spent a great deal of time consoling him, sympathizing with him, even flattering him.[63] In Naples they had lunch together every day in a little café near the Stazione Zoologica. They talked about the book, about literature, about their future life together. Before returning to the laboratory in the midafternoon, they strolled along the bay and enjoyed the cool breeze coming in off the Mediterranean. How glad Just was to be working on marine eggs rather than chick eggs! Johnson had once wanted him to ship eggs to Howard from Naples or Woods Hole.[64] The suggestion, though misinformed (by the time they arrived the eggs would be useless) and perhaps vindictively motivated, had some basis in reason. Marine eggs were the subject of choice for embryology, but they could not be had year round. Some scientists were even beginning to move away from that medium, to look at salamander eggs, frog eggs, and, yes, chick eggs instead. But this change was not for Just, nor did he ever contemplate it. He had spent too much time amassing detailed knowledge about the life histories of *Nereis, Arbacia, Chaetopterus*, and the like. Besides, he loved the ocean. His search for marine eggs had taken him to Europe, to the Mediterranean, to Hedwig. He thought about these things as he gazed through the laboratory window between the palm trees at the shining blue water of the bay and at the quaint houses of Sorrento silhouetted in the pink distance. And gazing out with him was Hedwig.

Just worshipped Hedwig, and perhaps even that is not putting it strongly enough. The only thing he loved as much was his microscope. Both competed for his time and attention. Often he saw them as one and the same. There were times when he referred to her as "a glorious globule," to it as his lover.[65] Romance and science became intertwined in a curious fantasy. They played out that fantasy every day—lying in a field peering through his microscope at an ant or a worm, sitting by a lake catching tadpoles and telling stories of crabs and toads that carried diamonds hidden in their heads, of frogs' songs coming from age-old cisterns, of deep-sea animals turning on their little lanterns on the dot of eight o'clock, the hour prescribed by nature.

Hedwig could bring herself back to reality perhaps more rapidly than he could. She was always conscious of what a life for them together would require in practical terms: a flourishing career and money. And one without the other was hopeless. Hedwig wanted Just to make checklists of people and decide who his friends were and who could not be trusted, who would help him and who would not. Through her he became ever more critical of people like Heilbrunn, Schrader, and eventually Lillie. His letters from Europe were pointed and demanding, pitched at a higher, more strident level than those he wrote in America.

Just saw himself working for Hedwig as a slave works for his master. The analogy was his, used time and time again.[66] He thought her footprint on his neck was as natural as dew on honeysuckle in the early morning. This submissive role was one of their private fantasies, however. In their day-to-day work in the laboratory, Hedwig took the orders and obeyed them to the letter. He planned their schedule, she carried it out. She worked over and beyond the call of duty—typing, translating, compiling bibliographies, making microscopic sections. If a slave is one who works for a master and carries out his every wish, the description fits her better than him.

The will to work, the drive to produce had to come from somewhere. At State College and Kimball Union Academy he had his mother and her expectations to push him. At Dartmouth it was the memory of his mother. When that memory faded he married, joined the black community, and dedicated himself to being a model to the race. Founding Omega Psi Phi, winning the Spingarn Medal—these things were part of a drive he had to both serve and somehow be a leader among blacks. And then there was his passion for science. It was deep, and being virtually the first black in the field made it deeper. But drive lasts only so long, and even less when obstacles are put in one's way. The temptation is to give up. By the 1930s Just had no cause—none to work for, none to live for. Hedwig became his new cause. He could work for her. She would direct his every move. Her will was his.

They worked together through the summer in Naples, with its dirt and wretched heat. Aside from the book, there was a lecture to be prepared for the International Congress on Radiobiology to be held in Venice. Just went alone. He spoke on the effects of ultraviolet rays on animal eggs and was listed on the program as "Research Associate of the Carnegie Institution of Washington."[67] Because he was late in finding out about the congress, he was not among the main speakers and not listed as such. This nagged at him. All during the conference he complained about the accommodations and the lack of attention to his work.[68] His success at the Padua conference and all the attention his new theory drew there were fresh in his mind. The Venice conference was disappointing by comparison.

On the whole, Europe was a little unnerving this time around. The political situation in Germany was sad and growing worse, and in changing his base of operation from Berlin to Naples Just had done the only feasible thing. But there were problems in Naples too. When he arrived there in March he had been assigned a laboratory next to the aquarium, and the smell of methyl alcohol was unbearable. Some days he could hardly work. But he was so grateful to have the space at all that he chose, initially, not to complain to Dohrn. In addition he had trouble keeping his time for himself. The American scientist Ethel Browne Harvey, knowing his old role at Woods Hole, began badgering him for help with her experiments. Not only did he dislike the woman personally, but he thought her ignorant of biology and was convinced that she would not recognize a sea urchin egg if it were as big as the Stazione itself.[69] He had to avoid her somehow. Also, the collecting crew was not doing its job well. Many marine animals were arriving dead or half dead. What could Just do? He was too much of a purist to work on substandard specimens, but he did not want to jeopardize his place at the Stazione by offending Dohrn. At one point he even considered changing his work—something he had never been willing to do before. He could not muster the courage to discuss the problems with Dohrn face to face; instead, he wrote out his criticisms, and only then with great trepidation and many apologies.[70]

Problems or no problems, Italy was still better than America. Just felt more at ease at the Stazione than he would have at Woods Hole. And he had Hedwig to help. They continued working hard almost through November. Then Hedwig fell ill. The work was too much of a strain, and so was the climate.[71] Over several weeks, well into December, Just nursed her back to health, at the same time keeping up a rigorous schedule of experiments and writing. They took some time off around Christmas to visit Capri and a few of the other coastal islands. The first of the new year saw them again working hard, to finish up details before their departure at the end of the month. They said good-bye on 23 January. Hedwig was off to Zurich to continue bibliographical research for the book; Just had to get back to teaching at Howard. It was a difficult farewell. Just watched Hedwig grow smaller and smaller as the ship pulled out to sea. He "broke down," but soon pulled himself together thinking about "life and all that it promises."[72]

On 31 January 1935 Just disembarked from the *Rex* in cold, desolate, ice-clogged New York Harbor.[73] He had been away a year. He wasted no time getting back to Washington. Before boarding the train at Penn Station around noon, he sent two telegrams: one to Maribel announcing his arrival in Washington, one to Hedwig announcing his arrival in New York. His train pulled into Union Station at 4:05 P.M. Margaret and Maribel met him there.[74] After gathering his bags and finding a taxi, the three began to catch up on their year apart. For Just the most shocking news was the death of Bamba, the children's grandmother, on 17 December. She had had a cold that developed into pneumonia, and her heart was too weak to carry her through. She would have been seventy-five soon. Just was hurt. Bamba meant much to him, and he would have liked to see her again, or at least to have been told sooner about her illness and death. This reminded him too much of the suffering he had undergone when he journeyed to Maryville from Kimball Union Academy, only to find that his mother was dead. Now, as then, he nonetheless had to move forward, continue toward his goal. Suddenly he became angry, even bitter, as he realized that his plans for the future might be ruined by this turn of events. Somehow it had not been like this when his mother had passed away.

Ethel was at home to meet them. She was cordial, and stressed the inevitability of death for an old woman with heart trouble. Just felt uncomfortable, sick even, as Ethel denied him the opportunity to express his condolences. He did not know how to take his wife's behavior, her calm and almost gracious resolve in the face of tragedy.

Ethel had been trained too well—by Bamba herself—to betray her emotions in public, even before her husband. The next day, when Highwarden and Maribel were at school and Margaret had gone to the station to collect the rest of Just's luggage, she accounted for her silence. She had not written Just about Bamba because she wanted him to reach a decision, quite apart from extenuating circumstances such as a death in the family. Without going into the details of his life, he informed her that he had in fact come to the decision that he wanted a divorce. When Ethel agreed and demanded only support for herself and the children, Just could not believe his ears. In less than twenty-four hours they had reached an agreement. But Just felt that something was wrong. He was sure Ethel was up to something, probably something not so good. All he could do was wait and see.

All three of the children were subdued, and Margaret and Maribel were physically unhealthy as well. Their situation weighed heavily on Just's conscience as he continued to chart out his future plans—ones that did not include them. But it was either him or them, and his choice was clear. He saw himself engaged in a fight for his own soul. He could not change his course of action for anything or anyone. He was determined to be hardhearted if that was what was necessary for him to win life with Hedwig.

Another surprise awaited him at Howard. Johnson was most cordial. But by this time Just had undergone too many battles to be deflected by the president's changes in mood and behavior, or to worry about motives and methods and the best course of action to take with the Howard administration. He was determined to keep a clear mind and go about his business as he saw best. First, he would arrange his teaching schedule and start work. Then he would bring up the matter of money and status with the president. Then he would go back to the Rockefeller Foundation to make them come to a decision one way or the other. Next, he would seek out private individuals and try to interest them in supporting his research. And he would contact the Carnegie people again, and also the Commonwealth Fund. Meanwhile, he would continue working on the book. It was only February, and for the next few months he had a heavy schedule in front of him. But there was June to look forward to. By then he would be away from Howard and back in Europe with Hedwig.

∿

The Howard campus was in an uproar the week following Just's return. Behind library and office and classroom doors and around laboratory benches, everybody was whispering about the impending investigation of Johnson and the school's administration.[75] Many in the Howard community, including the Executive Committee of the

General Alumni, had accused the administration of misusing funds. In the fall of 1934, while Just was in Naples, Secretary of the Interior Harold Ickes had impounded approximately $100,000 of Public Works Administration (PWA) funds, part of an original sum of $800,000 earmarked for building a library at Howard. An investigation was pending into the record of disbursals to date. Abraham Flexner, chairman of the Howard trustees since 1932, was called in immediately. The issue centered on money that had been appropriated out of PWA funds to equip the proposed chemistry building. In December Flexner explained that the original appropriation for the library had been reduced from $800,000 to $600,000 so that the remainder could be used for the chemistry building—a move that did not follow the original recommendation of the Buildings and Grounds Committee and one that Flexner took it upon himself to authorize. But nobody had taken up the matter of reallocation with the Department of the Interior. Nobody could trace the reallocated funds. What had happened to them and how had they been spent? Flexner promised to get to the bottom of the matter and straighten out all difficulties. The investigation continued. Secretary Ickes asked Flexner to meet with him on 4 February. Four days later Flexner called a special meeting of the board of trustees at the Hotel New Yorker.

Everyone knew about the special meeting. The opinion on campus, including Just's, was that Johnson would be dismissed for malfeasance and maladministration involving fiscal affairs. Just was pleased. For some time before the PWA scandal broke he had believed such things were going on. He looked happily and hopefully to the future. If Johnson were fired, Just's past contentions would be vindicated and he would be able to "profit" by the president's downfall;[76] if by an unfortunate turn of events Johnson was allowed to stay on, Just would watch eagerly as a noisy and recalcitrant person became "a very quiet and a very much more amenable, reasonable being." He could not lose, whatever the outcome. Things seemed to be working in his favor.

The Howard campus was rife with confusion, with rumor upon rumor. The newspapers gave front-page coverage to the investigation during February and March.[77] The faculty was in a frenzy about the future of the university. Just found it difficult to work. But he needed to get the book into final shape so that he could leave in June. He used to come back to the university late at night, when he could find a little peace and quiet, and write and rewrite until early morning. If ever he needed to get away from Howard it was then, during perhaps the worst turmoil in the history of the institution. The hectic time at Dartmouth during his sophomore year had not been nearly as disturbing. The Durkee years, put back to back and compressed into one, still could not match the turmoil Just had to face now. He was constantly depressed and lost seven pounds as a result. His intestines became inflamed. He saw a doctor and was told he had colitis.[78] He found himself going to the bathroom eight times between breakfast and lunch, and work became almost impossible. He put himself on castor oil and went back to the doctor. He was told that his condition was due purely to "nerves." His life was putting him under potentially deadly strain. But there was no shell to close into.

Flexner pulled out of Howard on the day after the special meeting in New York. Unable to deal with the troubles or unwilling to risk further failure, or both, he tendered his resignation and submitted his explanation in writing at a meeting of the board on 18 March:

> After two and one-half years spent in the effort to improve the conditions, the special meeting (of February 8) showed clearly that the fundamental conditions which existed at the time I took office had really not changed. Under these circumstances, it is idle for me to continue in office, and no effort on the part of the Board or any member will change my decision to retire. I have given notice to the Department of the Interior to this effect.
>
> With all good wishes for the welfare of the University, I am
> Sincerely yours,
>
> A F (Abraham Flexner)[79]

No more would he have to direct and chide and encourage the Howard administration, faculty, and student body. By anybody's standards he had done his share, pitched his inning. He was due a rest. It did not occur to him that Just and others had been in the ballgame—the same one—much longer, and had had to face these difficult players all along. Even if it had crossed his mind, he would have thought that after all it was their game, win or lose. He was not in it to the same extent, only insofar as he could help to win. Anything other than success—well, he did not have the misfortune to have to deal with that. Besides, the research institution he had founded, the Institute for Advanced Study at Princeton, was on the upswing. It had recruited Einstein and other intellectual giants. Why then, Flexner wondered, should he waste his time in Washington? The decision to leave Howard was his, and he had only to decide to go. He did not need to think about *where* he would go. For him, unlike for Just, that simply was not a problem.

The investigation continued without Flexner. Johnson, about to be fired, fought back, and denied to the end that he was "aware at all that the Treasurer had not taken up with the Department of the Interior the matter of the reallocations. . . ."[80] The blame was in the end put on the treasurer, V. D. Johnston, and Johnson was absolved of all guilt. The final report on the investigation also concluded that "no dishonesty" was involved.[81] The problem arose, it said, from Johnston's difficult personality. But the board of trustees chose not to blame Johnston in full. At the annual meeting of 9 April 1935 Johnston was partially vindicated and the board unanimously reelected him treasurer—contrary to the recommendation of the Department of the Interior. How Johnson influenced the outcome is unclear. Naturally, Just thought the president had cleverly wiggled out of a tight hole that Just wished he had remained stuck in.

Not that Just should be blamed for wishing ill on his archenemy. What cornered and scared person does not? From all sides life seemed to be closing in—especially on the domestic end. Just and Ethel were having ugly sessions over the divorce question. He wanted her to get the divorce by June, hopefully on noncriminal charges, in order to protect his job at Howard and free him from alimony payments and child support.[82] Determined to give up as little as possible, he pushed her to get a job, and thought about how nice it would be if Margaret were married and out of the way.[83] But Ethel would not budge. She held her own, determined to protect herself and the children. Bamba had left her a little money, around two thousand dollars, to be divided between them, but it would take some time for the money to come through as her will had not been written up in proper legal form.[84] Aside from the financial issue, Ethel also wanted to protect the children from the disgrace of a divorce. She had to protect herself too. But from what? Just had not told her about Hedwig, and all she could do was suspect.

Just was temperamentally unfit for the ugly sessions with Ethel, and each one made him physically sick.[85] Deep down he knew that he was shirking a lifetime commitment, but he also realized that he had himself to think about, his life—whatever was left of it—to fight for. The children were young and had their whole lives in front of them. Though initially hurt perhaps, they would get over it in the years to come. Just thought ahead to June. By then the divorce, if not over completely, would be under way. Should Ethel decide not to go ahead with it, he was prepared to go to Nevada and do it himself.[86] He would return to Europe—to Hedwig—immediately after.

Problems that had settled into the background for a while surfaced again in the spring of 1935. Roger Arliner Young began to haunt Just. After siding with Johnson over Just's affair with Boveri in 1930, she went about her teaching duties at Howard, and continued her research at Woods Hole during the summers while Just was in Europe. She made many friends at Woods Hole—mostly students of Heilbrunn and members of the Chicago crowd. In a sense she was the link between Just and the Woods Hole community. When was Just coming back? what was he doing in Europe?—these questions were whispered to her over and over again. Dissatisfied with her own relationship with Just, Young revealed what she thought the Woods Hole community wanted, but was shocked, to hear: Just, she whispered back, was having affairs

with white women in Europe. Such a revelation was sure death professionally, and often literally, for an American black man—and Young
knew it. Her motive was vindictiveness pure and simple. The true
feelings of the Woods Hole people did not surface until 1936, but
hostility toward Just was already brewing earlier. It was all right, or at
least permissible, for the white geneticist Calvin Bridges, known for his
theories on free love, to try to have an affair with Roger Young, but it
was not all right for Just to fall in love with a white woman.

Young's own fortunes went progressively downhill. She and Johnson
drifted apart after she ran to him with the letter proving true the
rumors of Just's love affairs in Europe. Just ignored her as much as
possible, giving his attention instead to the two other junior professors
in the department, Hansborough and Chase, but he suspected she was
spreading rumors. Also, he held it against her that as a scholar she had
turned out to be a disappointment. She had had her chance at Chicago
in 1930 and failed. No doubt Just had started planning to get rid of
Young in 1933, possibly even earlier, but he preferred to ease her out
rather than cause a stir by firing her outright. Besides, he was too often
in Europe to attend to the matter in a concerted way.

Young was anything but docile. Considering Just's other problems,
she could not have chosen a better time than the spring of 1935 to begin
her counterattack. She confronted Just, in no uncertain terms, about
his behavior toward her. She accused him of deliberately disrupting her
work as a researcher and a teacher ever since 1932. He would not let her
use equipment such as centrifuges for her research, she charged, and
worse, he would not grant requisitions for her classroom work either.
She characterized his behavior as "averse to a true scientific or real
university spirit."[87] Young did not stop there. She accused him of
ignoring her requests for appointments and of not reading and commenting on her latest article—on a subject he himself had suggested.
She felt abandoned, especially since at one time she had been "so
heartily encouraged" by him in "seeking knowledge of the Truth."

The last person Just wanted to trouble himself with was Young. In
no way was he about to feel sorry or guilty, especially in light of the
other problems on his mind. Young had been "a sore trial" to the other
department members.[88] Her work as a teacher was "far below standards"
even for Howard. She often missed her classes, and according to Just
the problem became so bad that students went to the administration on
their own to complain. She was careless with scientific equipment and
would not return it promptly for the use of others, so Just felt justified
in locking it away from her. Though he had "no respect" for her "as a
human being," in the past he had leaned over backwards for her
professionally, and, as a result, had endured much "nonsense." His
experience with Young was "the most disastrous" of his life. It had
caused him more "lacerations of the spirit" than he cared to recall.

Why could she not have been like Chase and Hansborough—very highly respected "both as teachers and as gentlemen"? Just did not know what would become of her, nor did he particularly care.

Ethel was uppermost in his mind throughout the spring. He wondered if she would go through with the divorce as promised, and if so, how much the whole thing would cost. His finances were shaky. The tax people were after him to pay income tax for the previous year.[89] The Carnegie grant had run out in January, except for a couple of hundred dollars for secretarial services.[90] Just's prospects for more money were uncertain. If he was indeed going to marry Hedwig in the summer, he would need a clearer financial future. He had put some money in a Swiss bank, and Hedwig had some saved up from the $5,000 she had earned from the Carnegie grant for services to Just as "translator, assistant, and secretary."[91] But none of this was substantial enough to get them started on a life together. Just had to find money somewhere. Since the foundations did not seem interested, he began to think more seriously about locating a wealthy individual who might serve as his patron.

His life was all worry and work, work and worry. To get away from it for a while he had someone drive him (he never knew how to drive himself) through Washington to see the new federal buildings. Afterwards they crossed the Potomac over the white stone bridge behind the Lincoln Memorial, and stopped and looked back at the city, with the river in the foreground and the parks and the tops of the city buildings beyond. There was a beautiful sunset, "long bands or ribbons of soft fleecy clouds of pink that glowed with fire and then became like smoke and disappeared."[92] But Just stood there without much feeling, without really being touched by the beauty of the scene. He wondered why. Was it because he was tired, anxious? He did not think so. Sunsets in Naples, no more beautiful on the surface, had "filled" him to "bursting," given him a sense of oneness with the world. The problem with Washington was the place itself, the people in particular. They were mechanical, lacking warmth, very much like "the wax figures which display hats and necklaces" in the shops. Never would they have had an urge to stick their "bare feet in sand on the sea shore"; never would they have thought of sex or love or the earth around them as anything but "dirty." Just wondered why they bothered to live at all. To him they were strangers, and he longed to get away from "their world and their city."

By 5 June, the day Just sailed for Europe again, nothing had been settled. There was no money and no divorce. Yet another summer Just and Hedwig had to live, as her family put it, "a criminal existence," as husband and wife without a marriage license.[93] They stayed in Switzerland for the most part—Saint Moritz in June, Zurich in July and August, and Basel in September—and worked on a revision of the book. They did not go to Naples. It is unclear why they stayed away: Naples had always been one of their favorite haunts, and they were still on good terms with Dohrn. Perhaps scientists at the Stazione had begun to talk about them; perhaps the militaristic atmosphere created by the Fascists precluded the peace and quiet they relished.

Back in America in the fall of 1935, Just set his mind to finding money. He tried to get himself permanently attached to the Carnegie Institution as a research associate. The effort failed.[94] Next he sent out his usual round of appeals to funding agencies, to the Rockefeller Foundation and the Carnegie Corporation in particular but also to the Davis Foundation, the Chemical Foundation, the Guggenheim Foundation, and others.[95] All of the responses were negative—except for one. The Oberlaender Trust in Philadelphia seemed interested.[96] When Just found out about this new prospect some time around mid-December, he was in the midst of final preparations for the annual conference of the American Society of Zoologists, which was to be held in Princeton right after Christmas. He had no time just then to pursue the Oberlaender Trust. He would wait until after the conference, or better yet, maybe he would pass through Philadelphia on the way to Princeton and present his proposal in person.

Just was looking forward to the conference. It was the first he had attended on American soil since 1930. He would have a chance to reveal his new line of scientific thinking, to quiet Heilbrunn and Schrader and others who had been insistently complaining about his absence from the American biological scene. The conference would give him an opportunity to renew friendships and contacts that might be useful in helping him find money to relocate. He was determined to make his presence felt as never before. He was not scheduled to speak in Princeton, but he took it upon himself to get on the program—late and therefore without prior publicity. His paper, he hoped, would have as much if not greater impact than the one he had delivered in Padua. He planned to speak on genetics, one of the top scientific issues of the day. T. H. Morgan, known to some as the father of modern genetics, had received the Nobel Prize for his work in the field just two years earlier.

Throughout the fall, Just checked out the content of his paper with A. P. Mathews, Frank Lillie, and others.[97] The problem was a difficult and controversial one, combining the approaches of genetics, cytology, and embryology. The interpretation put on genetics by the professionals in each field differed, and it was this difference that captivated Just. He wanted some kind of unified approach, showing the interdependence between chromosomes and the inner cytoplasm and cortical cytoplasm of the egg. Lillie was cautious, and advised Just to tone the paper down, to avoid being nasty to the big names in American biology.[98] Just was annoyed by Lillie's attitude. He would have appreciated serious criticism of the substance of his talk, but the question

of whether this or that bigwig might be offended was of little consequence to him. Mathews offered more encouragement, advising Just
that if he had the evidence to overturn the current gene theory he
should publish it, regardless of the consequences—much as Lavoisier
had revolutionized chemistry in the eighteenth century.[99] If further
evidence was needed, however, Just should wait, as Darwin had done
with his *Origin of Species*, until there was certainty of "demolishing"
old ideas. The only goal was scientific truth.

Just was anxious to lift his work from the experimental plane to the
theoretical. In the past he had not had a chance to theorize, partly
because of the nature of his work at Woods Hole, partly because he did
not have the time, partly because he had lacked self-confidence. But
Hedwig, the Germans at the Institut, and Europe had showed him that
it could be done. He became more and more interested in leaving a
scientific legacy, and as a result he grew more and more intrigued by,
perhaps driven to, the notion of what he called a "revolution" in
science. His experimental work had long been praised for its precision
and creativity, its unusually consistent attention to detail. Just was
proud of this aspect of his work, but he was determined to show that as
a biologist he went beyond technique and method. Besides, he was
aware that many who complimented him on his experimental work
were being condescending, were giving him a sort of backhanded
compliment, for they thought of theory as a higher mode of science
than experiment. He had to change their image of him. If he was lucky,
he would win recognition within his own lifetime.

Just thought he had grasped his destiny. His letters took on the
character of documents written to justify his life. He wrote regularly to
Hedwig, sometimes as many as four or five letters a day, as if he were
compiling a diary or putting together a loose autobiography, rewriting
his struggles and his goals for the historical record. His letters to
foundations were comprehensive, and not always strategic, repeating
again and again the same pleas for help, the same descriptions of the
struggle he was engaged in to keep alive the work that formed his very
soul. He wanted these things put down on paper for posterity. Similarly,
his lectures had to be memorable, dramatic. He wanted to make an
indelible impression on a suspicious and hostile audience, who would
then go home and talk about him, perhaps even write their feelings
down. He had begun to live life on a theatrical level. As he boarded the
train for Philadelphia he had no idea whether he would be received
with cheers or boos at the Princeton conference, but he was going to
make sure that his performance did not allow his audience to sit back
indifferent.

Princeton was a kind of reunion. Many people there had not seen
him since 1930, when he gave his farewell speech at the Lillie birthday
celebration; some had seen or corresponded with him in Europe, but

none had kept up with his work. The rumor at Woods Hole was that he had stopped work altogether. Just set the story straight, stressing that while he had indeed stopped publishing regularly in American journals, his recent work, if anyone cared to look, could be found in prominent European journals such as *Protoplasma, Wilhelm Roux' Archiv für Entwicklungsmechanik der Organismen,* and *Zeitschrift für Zellforschung und mikroskopische Anatomie.* Furthermore, the *focus* of his work had changed, and so had his image of himself as a scientist. His presentations, he promised, would show how and why he had changed.

❧

Just gave two papers. The first, "Alcoholic Solutions as Gastric Secretogogues," was not the one he had prepared long and hard for, but it was important to him too.[100] In it he demonstrated some of the possible medical relevance of his work. Foundations were likely to support research that could be shown to have practical applications in fields such as medicine and engineering. Perhaps, Just must have thought, he could bring this up with the Oberlaender Trust; perhaps he could even try to interest a specialized organization such as the Distilled Spirits Institute in Washington, D.C.[101] Ten years earlier he had had no success in getting the National Research Council to take seriously his idea for protecting wharves from shipworm, but there was no harm in trying the practical route again, from a different angle and with a different organization. Also, it was good to show his Princeton audience that he was more than a tinkerer with marine eggs. He had never published or even talked about his shipworm work with the Woods Hole people, so for them this was a completely new angle.

The second paper, the big one, "Nuclear Increase during Development as a Factor in Differentiation and in Heredity," explained his theory of the gene and attempted to reconcile two opposing factions in the gene controversy, the embryologists and the geneticists.[102] It was part of a larger paper, "A Single Theory for the Physiology of Development and Genetics," which was to be published a few months later at Just's own expense.[103]

The talk at Princeton and the subsequent paper were an attack on the geneticist T. H. Morgan, who, though he had recently won the Nobel prize for his work on *Drosophila,* had not come close to offering a gene theory that would satisfy an embryologist, especially Just. Just had spent time earlier with the Morgan group at Columbia and Woods Hole,[104] but he had lost contact after his trips to Europe and no longer felt any loyalty in that direction.

At the time a scientific controversy was raging about genetics and the physiology of development, or more accurately, about how those two fields could best be brought together. The embryologists were concerned with the development of the fertilized egg into differentiated adult characters, geneticists with the transmission of heredity elements from generation to generation. Both groups were dealing with concepts of heredity which related closely to the known distinction of genotype and phenotype. "Genotype" is the term used with respect to the genetic composition of an organism, that is, what genes it carries and passes on to the next generation. "Phenotype" is used to talk about how the organism looks, that is, its visible adult characters. The distinction had been made in 1911 by Wilhelm Johannsen in an article published in the *American Naturalist*.[105] Just, though he did not use those terms, was attempting to show the interdependence of both phenomena. In other words, Just thought that "the same mode by which the grosser parts become differentiated calls forth the Mendelian differences."[106]

Just admitted that there was no *a priori* reason to see development and heredity as the same, but he was prepared to show them both as different sides of the same coin, "merely two aspects of life-history."[107] Morgan, the leader of the genetics group and himself a one-time embryologist, had tried to reconcile the two fields. But his book on the subject, published in 1934,[108] treated the two disciplines separately, and by his own admission there was no synthesis to speak of. Curiously, Just never mentioned Morgan's book in either his talk or his article. As far as he was concerned, the problem remained—embryologists and geneticists were still not talking the same language.

The issue of differentiation was at the heart of the problem. The gene theorists, according to Just, postulated that "the genes order the progressive differentiation of development by liberating to the cytoplasm at different stages a different something which brings about differentiation."[109] Just dismissed this as an explanation of the physiology of development. He also argued against the then popular theory known as embryonic segregation, held by embryologists such as Lillie and Conklin. He raised a number of objections to the theory,[110] and thereby put himself between the geneticists and the embryologists. He would join neither.

Just offered his own theory—a theory of restriction. He had arrived at it by observing the egg cell during cleavage and, by careful experimental work, noting the increase of the nucleus during this stage. He assumed that his observations of differentiation during cleavage were true for all other stages of the embryological process. The nuclear increase had to come from somewhere, he reasoned, since the nucleus could not build something out of nothing. Before identifying the source of the nuclear increase, Just further assumed that what is true of the nucleus is true of its constituents, the chromosomes. He cited direct indications, however, work that had been done on determining the presence of nucleic acid in eggs, that the chromosomes increased in mass. This building up of nuclei and chromosomes, which always went with differentiation, was done at the expense of the cytoplasm.[111]

Just used his theory to explain numerous embryological concepts, from polyembryony to asexual reproduction. But he was not willing to stop there. He forged on, using his theory to place, somewhat incorrectly as it turned out, the basis of inheritance in the egg cytoplasm. His conception differs from the gene theory in two chief respects: by locating the factors for heredity in the cytoplasm instead of in the genes, and by offering "an interpretation of 'how the genes in the chromosomes produce the effects in and through the cytoplasm' of which Morgan confesses ignorance."[112]

Biochemistry was becoming more and more important for Just and especially for his work on the nuclear increase during cleavage. In the mid-1930s no one knew the precise composition of the genetic substance, although by then it was becoming generally accepted that the nucleus consists essentially of nucleoprotein. But Just's mind was open to other possibilities. He called for further biochemical research into "the role of each of the three organic compounds in protoplasm."[113] His work on nuclear increase pointed to that need. Only once (later, in 1940) did Just actually speculate rather offhandedly that "the genes are nucleic acids."[114] He was never tied down to a belief in the primacy of the nucleus, as he knew some protoplasmic systems that had no discrete nucleus as such. Some bacteria and blue-green algae, he remembered, were nonnuclear organisms, containing mostly nucleic acid. But what he wanted was a theory of heredity applicable to all animals and all plants. He asked, then, how "upon the thread of the nucleic acid common to plants and that to animals, the specific gene entities are imposed."[115] Biochemists would need to work overtime to provide the answer.

Just's theory of genetic restriction was an attempt to provide an explanation of the mechanism of differentiation. Never mind that the problem of differentiation was, and still is, a difficult one; exactly how it occurs continues to elude embryologists. In Just's view his theory had several advantages; perhaps most important, it gave the nucleus, cytoplasm, and ectoplasm, definite roles in development—something that other theories had not done.

His "interpretation," as he called it, was recognized as imaginative and provocative. Younger scientists, Lester Barth among them, walked away from the lecture in awe. They wrote for reprints of the paper and read it with deep interest. Still, these scientists did not carry on and develop this work in any real way, even though they saw Just as one of the most creative men in zoology in the United States, someone who had dared to advance a new idea, someone who was different from the older scientists but showed no fear of them, someone who should be taken seriously and given equal respect. Biology was perhaps changing, Just thought, thanks to this new younger breed—which would eventually discover the real magic of the gene. Just hoped so, and if he did nothing else, he wanted to spur them on.

ᘔ

Many things happened in the three days at the conference in 1935, and Just did not have time to sort them out immediately. Most important was that he had had a chance to air his new work. Once again he had faced his old antagonists, Plough, Chambers, and the Morgan crowd, and once again he had shown that his was a mind to be reckoned with. E. G. Conklin made an "outburst" at the conclusion of Just's paper, but Just maintained his composure, even though he had not expected an attack from that quarter.[116] Conklin had never been hostile before; no doubt the rumors about Just had reached him too. Also, he may have been annoyed by Just's presumption in criticizing the work of men like Morgan, and it is certain that he was still piqued that a few years earlier Just had refused an invitation from him to act as commentator on a presentation he was scheduled to deliver at a Society of Zoologists conference.[117]

Just was trying to carve out a new role for himself in American science, a more assertive, self-assured one. But he could see, perhaps more clearly than ever before, how closely intertwined his private life— his character, his relationships, his personal goals—had been and always would be with his status as a professional scientist. At the conference things had been said, or had been thought, even if left unspoken. This depressed him at first, then jarred him into thinking about his mixed-up life, then depressed him all over again. He needed somebody's ear.

On his return to Washington he became even more depressed, to the point of being incapacitated. Letters came from Hedwig accusing him of being passive, lazy, no good. It was, she said, no accident that they were not married, that he had no laboratory in Europe, no position, and no future, that he would probably live and die a run-of-the-mill scientist rather than the creator of revolutions. Her accusation jolted him. The cold air in the laboratory that Saturday night in January 1936 grew colder. Just sat down to think. He could not let Hedwig's words or the experience at Princeton go unchewed-over. He poured himself a drink. His slides and microscope sat appealingly in front of him, but they would have to wait.

He was worried about Hedwig and her health, especially now that she was undergoing a severe bout with influenza, brought on most likely by the stress of living apart from him. They were separated. The poison of society had done this, "a poison in the soul" that was eating at the marrow of their bones like a dreadful cancer waiting to spread.[118] It was affecting their work, their outlook on life, the very basis of their relationship. They would not be together again until June. In the meantime, he was trying to settle problems on the home front. Or was he? Could Hedwig be right that he was not doing everything in his power?

Just hated himself for having these problems. He blamed, not pitied, himself. He was living a double life, moving between two households, without really doing as much as he should to sort out the confusion. In the past he had lashed out at Johnson, Ethel, Howard, Lillie, and the MBL community for having put him in this bind, but it was dawning on him now that the responsibility lay in large part with him. He was no different from other human beings, every one of whom, he was convinced, carries around within him his own undoing, his own poison. Just's was fear. He had tried to run away from his problems by going to Europe. He could not cope. He was weak. In a sense it was good that he had met a strong woman like Hedwig, someone who could help him be strong too. But the problem was that he could not live up to her expectations. Perhaps he should deny himself Hedwig, since he knew that he was and always would be passive, tender, weak. He could disappear from her life, he could give up his science. But something inside made him push on, stay with Hedwig and continue his work, trust to the future. Hedwig had given him the breath of life, hope, something to live for. Perhaps she could change him.

But the problem went beyond him; it involved other people, society, the world. On his way to the conference in Princeton two weeks before, an incident on the train had prompted him to think about the other reasons for his passivity. As the train pulled out, he went straight to the smoking compartment, hoping to have it all to himself. Unfortunately, a young white man, about thirty years old, was already there. The man tried to engage Just in conversation, first about the weather, then about other things. At first Just responded in monosyllables. He was not used to talking socially to white strangers in America. On his earlier trips to Chicago, New York, and Boston, he had spoken to whites only to ask the time of departure or arrival—and not even that if he could help it. But since going to Europe he had begun to feel a bit more comfortable. He no longer feared the very thought of speaking to a white passenger next to him, or of having one speak to him for that matter. After a few minutes, he opened up to the white man sitting opposite. They talked for over two hours, until the train reached Philadelphia.

An architect and a native of Washington, D.C., the stranger was a liberal who had studied and traveled abroad. He expressed admiration

for blacks and what they had accomplished despite their severe social
and political handicaps. But at the same time he was angry with blacks,
and he chided Just for the "timidity" he sensed in them. Over and over
again he stressed his view that blacks in America were "all too fright-
fully fearful." He was angry about it.

Just tried to explain. "What if blacks were not timid?" he asked.
"What if they did not submit to whites and forgot even for a moment
their role in society?" Just knew what would happen. He had grown
up in the South and could never forget the vulnerable position blacks
were in. They had less protection than rabbits; at least there was always
a season when rabbits could not be shot and slaughtered at will. How
could this architect appreciate the lot of a black man, even begin to feel
the fear and the anguish? Just himself had never really thought
about the psychologically torturous experiences blacks were forced to
undergo. He had only felt them, never grasped them intellectually.
Now he was beginning to understand why so many blacks, himself
included, had passive personalities.

When the train reached Philadelphia, Just was in a frame of mind to
do some more thinking about his life from the racial angle. At the
conference he was painfully conscious of people's behavior—the chilli-
ness on the part of some, the hostility from others. Afterwards he went
with the Schraders to their home in Tenafly, New Jersey. On one
occasion, when Franz and Sally teased him about his ability to assess a
new face very quickly, to determine a friend from an enemy, he reflected
on how he had trained himself to do this from early childhood, as a means
of protection against insults. It was no gift; it was technique pure and
simple. As a black he was aware that every new face was a possible foe.
He cultivated a fear that grew and grew—a fear that shielded him from
enemies. He had been reserved, cautious, wary in each new experience
—at Kimball, Dartmouth, and Woods Hole. Only with time did he
loosen up, and then only slowly and never completely.

At Woods Hole he had given so much, to get just a little. He had paid
a high price there for his work. For every three things he could call his
own, he had had to give away seven—discoveries, ideas, techniques.
This was the only way he had been allowed to work at Woods Hole.
Again this was passivity, even weakness on his part. But it was also
what kept him going. If he had been otherwise, pushy and aggressive,
he would have gone into medicine. In medicine, as a "plodding, money-
grubbing" man, he could have made a success among blacks, a success
based on greed and "predatiousness." He saw it around him every day,
and was always repulsed. Besides, if he had chosen that route, there
would have been no Europe, no Hedwig. His life would have been
different. His outlook on life would have been small and narrow. He
would have been a "Negro" physician, not a scientist seeking "a wide
point of view." The likes of Loeb would have won out. As a physician,

of course, he would never have been allowed to work at a white hospital, while as a scientist he had been accepted, to a small degree at least, by the white scientific community. Woods Hole was bad, but it was better than nothing.

Just had cringed at Woods Hole for many years, but not because he loved it. He hated it. "If I fought for crumbs from the Masters' table," he stressed, "I did not do so to give them pleasure and enjoyment." He took what he could get before "dogs could get it." During his first years at Woods Hole he had started out as a menial, serving food in the mess and scrubbing Lillie's laboratory. He could not have gotten a start at Woods Hole otherwise. He hated getting on his hands and knees to clean up "Lillie's dirt." He detested the laborers and janitors, whom he had to serve meals. But it was all so that he would be allowed to do science at Woods Hole, in America. The price had been high. He was just seeing this now, nearly thirty years later. All the bits and pieces were falling together, showing up in sharp relief after his experience on the train, at the conference, with the Schraders. Whether or not we believe him about his early days at Woods Hole, whether or not we choose to think that his European experience may have colored his views too strongly, we must take him as full and indeed sole authority for what his feelings were in 1936. It is clear that he felt that he had been cheated out of a meaningful life and career by the Woods Hole community.

Just could see now that he paid a high price in other ways as well. He had suffered even in his own home. He had had to "buy" freedom for his work. He had given up everything to keep a portion of his mind clear. Each year, as the time to go to Woods Hole came around, he and Ethel would have bitter quarrels. She was hostile to the place and his work. But he never left her or suggested a divorce; he simply endured. Passivity again. A separation would have been different then, ten or fifteen years earlier; by 1936 it was a frightening problem. There was, however, no other way out. Just had come to the point of living in his home like a stranger, sitting down at the evening meal, hearing a language he did not understand, having "no physical, spiritual contact." He felt like a "walking dead man—a ghost," suffering through "nights of torture—physical suffering, heart-breaking longing, spiritual loneliness" as he dreamed of Hedwig. It was as if hell were closing in on him. Could he summon up the willpower to do something about it?

His professional position was no less disturbing. And in a sense it was the reverse of passivity that was the problem here. As a scientist he had forgotten to be passive. He had dared to theorize in biology and he had decided not to depend on America for intellectual guidance. He sensed that for this he was being punished, even persecuted, by Conklin, Schrader, other Woods Hole people. He was growing to suspect that even Lillie was not behind him. He believed that his former mentor could help him improve his bad situation, could help him obtain some form of permanent support for relocation abroad. Lillie, who had recently been elected president of the National Academy of Sciences, wielded power. Just called him the "Pope of Science." All Lillie had to do, Just thought, was put in the right word at the right time. But that did not happen.

"Come back and do the dirty work and all will be forgiven" is what he interpreted the Woods Hole crowd as saying with their strange glances, awkward attempts at conversation, in some cases outright declarations. "We need you, can use you" were the words that hurt him most. He was forming new opinions about everyone, especially his old friends. While he had come to expect a measure of ill-will, even animosity, from the Woods Hole community at large, he had always put Heilbrunn, Schrader, and Lancefield in a special category. But the visit to Schrader's home after the conference was revealing. He found out that even his friends had not acted right by him. Schrader and Heilbrunn, he discovered, had advised Johnson to keep him in America, and had discussed with each other how to get him to stop theorizing and return to experimental work at Woods Hole. They too, it seemed, had turned against him.

Just stood firm. He had decided that he would not be forced back to Woods Hole, that he would return only on his own terms, if ever, that he would continue to fight for his theoretical positions in science and plan a life with Hedwig in Europe. Alone, he would continue to canvass foundations for money. His one fear now was that his so-called friends would "kill off" his chances with the funding agencies. But he was determined to remain calm, confident, self-composed. He would keep his "fixed purpose" clearly in mind. He would not be afraid to act, to do what had to be done, at the appropriate time. The only thing he had to fear was time. But he was certain that one day, somehow, he and Hedwig would talk together and work together and live together. Half-drunk, in the cold of his laboratory late that mid-January night, he might indeed have felt as if he were hallucinating. But he saw a bright, clear future shining through the misty, confusing present. He got up and walked home with new resolve. The sun was just coming up.

The problems did not go away, and dealing with them grew no easier. Ethel was no more ready to actually give Just a divorce in the spring of 1936 than she had been the previous year, and he was too busy, perhaps too nervous, to go off and get it done on his own in Nevada. He had a new plan. He would wait till he got back to France, where conditions for divorce and remarriage were apparently less stringent than in America.[119] But those were technicalities. More importantly, Just still had no definite prospects for the future. His attempts to secure funding for a permanent setup in Europe were all futile. Discouraged, he came more and more to consider life in America as a necessary, if undesirable, option. At least he could hold on to his job at Howard. Bad though it was, he had a steady income from it. The question, though, was where he and Hedwig could stay. Washington was not a place for a mixed couple, nor would it be a good idea to live so close to Howard. Naively, Just thought that life for them would be better in one of the suburbs, perhaps even Baltimore, than in Washington. At least they would be detached from the Howard community. The best place for them would be New York, but Just had gotten nowhere in his search for a job there—at the Rockefeller Institute for Medical Research or elsewhere.[120] He was resigned to Howard for the time being. There was nowhere else to go.

The longer Just stayed on at Howard, the more disgusted he became. Each new incident was worse than the previous one, depressing him to the point of physical illness. In the spring of 1936 he was forced to fire Roger Arliner Young.[121] Johnson had demanded that Just name one of the three junior professors—Young, Chase, or Hansborough—to be dropped from the zoology department. Though Just had grown to despise Young, he did not want her to lose her job, the economic situation being what it was and she with an invalid mother to support. Just claimed, rightly, that the whole thing was a setup by the president and the dean. For some reason, Young had lost favor with them, and Just was being brought in to rubber-stamp a decision that had already been made. He did not disagree with the decision—in his view if anyone had to be let go, it should be Young—but he had neither initiated nor made it. He complied with Johnson's request that he rank the three faculty members in order of preference, and he chose Hansborough and Chase over Young; both had been loyal and cooperative in a way she had not.

Young blamed Just—and Johnson too, but less so—for what had happened. By this time she had come to see that Howard was no place for her, anyway, and she was ready to leave. She did not know exactly what she would do, but she was determined to get out. At Woods Hole Heilbrunn had become a kind of mentor to her, and she considered doing graduate work under him at Pennsylvania. Her contract at Howard had a year to go, and she served it out in a halfhearted way. She was annoyed and bitter. In June 1937 she left the place for good.

At this time Just's mind was on money and position—two factors that kept magnifying themselves a thousand times over in his mind. After the Carnegie grant ran out in July 1935 the only hope he had was the Oberlaender Trust. He had visited with trust officials on the way to the Princeton conference in December 1935 and impressed them not only with the substance of his work but also with his close ties to German scientists.

The Oberlaender Trust was part of the Carl Schurz Memorial Foundation, which had been set up in Philadelphia in 1930 "to cultivate and promote closer intellectual relations between the United States of America and Germany" by maintaining "an interchange . . . of students, teachers, scholars, lecturers, artists, and men of affairs."[122] Two of the founding members were Julius Rosenwald and Oswald Garrison Villard, and Rosenwald served as a member of the first board of directors, elected 7 May 1930.[123] A year later Gustav Oberlaender, a German national residing in the United States, set up a principal fund of a million dollars with the foundation to contribute to the general goal in a specific way—sending American citizens to German-speaking countries for "research, studies and surveys."[124] The money was distributed through a special office set up as the Oberlaender Trust under the directorship of Wilbur K. Thomas, a Schurz Foundation official. The grantees included university professors, social workers, journalists, and physicians. Later, during the war, the purpose of the trust was modified to focus primarily on finding positions in the United States for displaced German and Austrian scholars.[125] But in the early and middle 1930s the notion of cultural exchange was paramount. An office was set up in Berlin under the management of Georg Kratzke, who, along with his able American secretary, Elizabeth May, looked after the needs of visiting Americans during their tenure as Oberlaender Fellows.

Just's work drew heavily on that of German scientists and vice versa, and he saw himself as fostering a link between the American and German intellectual traditions. Trust officials agreed. The grant was awarded to him to "complete his work on the German contribution to biological thought and its significance for a social theory of human behavior."[126] His work extended beyond this, of course. He used German scientific thought only as one facet in a larger synthesis involving his own work and the work of Russians, Americans, scientists of every nationality. But to get the money he set out a narrower perspective in his proposal to the trust.

The grant was a pittance, $1,800, nothing on the order of what he had received from the Rosenwald Fund and the Carnegie Corporation.[127] It was travel fare and not much more. For the first time since 1919 Just was dependent on the Howard payroll. He had no money, but it was not for want of trying. He was still sending updated versions of his letters to the foundations: the Julius Rosenwald Fund, the Rockefeller Foundation, the Commonwealth Fund, the Carnegie Corporation, and the Carnegie Institution. Frederick Keppel of the Carnegie Corporation came in for special pleas.[128] So did Lessing Rosenwald, Julius's son. But

Keppel had helped before and could go no further; Lessing Rosenwald, who had already declined Just once, had not changed his mind.[129]

Just had too much at stake to end his campaign now. For some years prior to 1936 he had sought out millionaires, hoping that he could find one who would take him under his wing and nourish and sustain him. He met them on his transatlantic voyages—newspaper executives, Park Avenue doctors and lawyers, Hollywood actors, heiresses. There was Mrs. Arnold Gottlieb, a close friend of John D. Rockefeller, Jr.; Lewis Gannett, publisher of the New York *Herald Tribune*; John Barrymore, the foremost matinee idol of his day; Henry Blackman Sell, a prominent lawyer; J. A. Lazarus, a New York physician who promised to arrange an interview with Filene, the multimillionaire businessman; Robert H. Davis, a member of the executive board of the *New York Sun*; Francis P. Garvan, president of the Chemical Foundation; and Dorothy Norris, an heiress of some means.[130] With such acquaintances, it is not surprising that he once appeared on the society page of *Fortune*, the prominent business magazine, alongside Joseph P. Kennedy and Alice Longworth Roosevelt.[131] He kept up his contact with millionaires and wrote them begging for help. An enormous amount of time and effort went into setting up appointments and pouring out his story. Though that story was sad, and in many ways dramatic, his contacts were not for the most part moved to the point of making any monetary commitment. Some did make promises, but they never made good on them.

Dorothy Norris was an exception. She had met Just on board the *Roma* some time in the early 1930s.[132] She was genuinely interested in him, his work, his philosophical outlook. Though either unwilling or unable to help out herself, she seemed to have connections all over New York. Through her Just became friendly with Max Wolf, a New York physician on Park Avenue who had a flourishing clientele that included some of the wealthiest people in the country. Wolf and Just grew close, and before long they had mapped out a strategy for finding money for Just's research.[133] Through Wolf—his practice, his social ties—they would find sympathetic rich people for Just to meet. Just would step in and sell himself and his work. The goal was $50,000 or more. Among the possibilities were Mrs. Evelyn Walsh McLean and Mrs. Eleanor A. Patterson, prominent Washington socialites.[134] A potential problem was that Wolf's contacts would have no scientific background or training, and would be unable to understand the intricacies and significance of Just's theoretical work. A potential advantage was that, unlike foundation officials, they were not by and large connected with the scientific community and so had no easy way of finding out how his work was judged by his colleagues. What interested them most were practical advances in medicine and biology. Just understood this well, and decided to stress that side of his work.

Just had always kept up his connections with the medical world, serving on committees for the Harlem Hospital and giving lectures to medical groups from time to time.[135] Though cell physiology is different from mammalian physiology, he had always tried to understand the problems of doctors and put his mind to drawing a link between their concerns and his work; he had chosen to do medically oriented courses in his doctoral program at Chicago and he had taught for ten years at the Howard Medical School. Much of his work shed light on medical problems, even though its primary concern was usually cell biology and his primary goal was usually tied to pure biology. In the late 1920s and early '30s he had published some good ideas on water movement, water balance, and hydration and dehydration in the living cell.[136] None of these had ever been seriously followed up by medical scientists.[137] Just was convinced that the work on water, if pursued, could ferret out leads on diseases such as edema and nephritis, both of which involved an imbalance in the water retention of certain cells.[138] Most important, the work seemed to him to be relevant to cancer studies. There was the question of the reaction-producing agent versus the reacting system. Because so many different agents were known or suspected to cause cancer, Just felt that a successful solution to the problem lay in a study of the reacting system itself. He lifted this principle, perhaps subconsciously, from his work on parthenogenesis, in which the egg as a reacting system took preeminence over the sperm and other activating agents. A study of the behavior of water within a system, he hypothesized, would give a clue as to what happens when an agent is applied as a stimulus. How far he could go toward finding a cure for cancer remained to be seen. He only wanted money to try, $50,000 or so.

Any biologist who could link his work with cancer or any other medical problem did so: money could be had far more easily from individuals and foundations if there was a chance that the proposed research would produce a breakthrough, or an advance of any kind. In the 1920s and '30s Frank Lillie secured funds from the National Research Council for his research on fertilization by using just such tactics. The money he received was part of a larger fund slated for a study of the social problems of sex, and Lillie managed to convince the council that his laboratory work was germane. To a great extent, in fact, it was not. (What eventually came of the council's program was the 1948 Kinsey report.)[139] There was nothing particularly strange about Lillie's claim that his work had practical consequences. He was just one of many scientists who tried to highlight this side of what he was doing; he was merely more successful than most.

Just's approach was different, in that his target audience was to a large degree made up of private donors, and therefore a particularly subservient attitude was required on his part. But he had a mutually respectful relationship with Max Wolf, and was positive that he could go far with Wolf acting as a buffer. He had a talent for begging, but hated to do it. If he had not been struggling for a new life with Hedwig, and if she had not pushed him so hard, he probably would not have taken this course. But he had no choice. The summer of 1936 marked a peak in his campaign. While he was traveling in Switzerland and Austria as an Oberlaender fellow, he and Hedwig composed letter after letter.

Just had appealed to Wolf over and over again for an introduction to John D. Rockefeller, Jr., whom Wolf claimed to know intimately. Nothing happened. In Lanersbach he decided to write Rockefeller on his own, introducing himself and portraying his struggle in the most vivid terms:

In June, 1916, when I took my doctor's degree at the University of Chicago, I first saw you and heard your short address. I do not expect that you remember the moment when, as the file of candidates passed before the Dean, I, the one Negro in line, had to halt before your seat and you gave me a quick glance. I on my side recall this incident and wish now that the experience had been more personal; it thus could serve as a means of re-introducing myself to you. Four years ago persons who know you offered to state my case to you. It seemed to me, however, more in accordance with ethical procedure to place my case before the Foundation directly. That I do now bring my situation to you personally grows out of several reasons.

First, the limitations which the policy of the Foundation sets prohibit grants for scientific research to the individual worker directly. Thus I can hope for no aid as long as this policy is strictly adhered to because my experience of now almost four years clearly proves that I can not obtain the Foundation's aid through the only channel possible, namely, that of the institution of which I am a member. My case being beyond the scope of the Foundation, my only recourse is a personal appeal. This must be made to one who has shown by his wide and generous aid a broad and tolerant spirit and who need not be bound by those rules necessary for the sustained benevolence of an organization as large as the Foundation.

Second, although a Negro, I sincerely hope that the badge of color will not too greatly color the appraisal of my scientific work. When in 1914 I received the Spingarn medal, the first awarded, Mr. Whitman, then governor of the State of N.Y., said in his speech of presentation that he hoped that what one of those most influential American scientists in recommending the award had said of me was not true, for it would be a sad reflection on American civilization. The scientist's words to which he referred were to the effect that my scientific work was then of such value that were I a white man I would receive instant and universal recognition. This point of view, as we both know, based on pre-judgment (prejudice) is far too prevalent in America even among many who would quickly disclaim having it. Blackness in the minds of most connotes a distinct inferiority and renders easy the continuation of a policy which aims to support and to prove the "sound reason" for this policy. If, as is sometimes said, the Rockefeller Foundation (and others as well) has a set philosophy with respect to the place of the Negro in American civilization, this might hang together with conditions in America; it might be difficult in America to do other than subscribe to the prevailing and majority opinion. Moreover, it is surely the privilege of any organization to dispense its funds as it sees fit. If then the Foundation has a formula for aid to Negroes, again I must appeal to one who in tolerance and individuality can pierce beyond skin-color and can deal with an individual whom attributes and aspirations place beyond the chance circumstance of race.

In the third place, it might be that my scientific work which has developed an independent point of view and has shown the need of a new basis for the attack of biological and medical problems—indeed indicating both a much needed synthesis and a philosophy—because it is so revolutionary may fail to attract the Foundation's interest especially if it assays my work by the measure of those who hold to ideas and theories which I had to attack. A group is often slower to see the flame of truth than a single open-minded individual. A group must be conservative and being so may run the danger of attempting to hold with both the old and the new—an impossibility when the one is incompatible with the other. But I must not over-emphasize either the novelty or the revolutionary character of my work—its fundamental basis is far too simple, too well-grounded in experience and in observation both of others and of myself.

But the strongest reason for this personal appeal to you remains.

Scientific work must of necessity be narrowly prescribed; it advances the more it is particularistic, clearly defines minutiae. As such it abstracts, sets apart, the infinitely small from its relatives. But the time comes when these parts must be gathered again to make the science, as physics, chemistry or biology. These natural sciences must in turn be brought together in relation to all life, stand in some juxtaposition to art and religion. One meaning of science—the larger, in my judgment—is that to human life and spiritual being. Out of my work, without any will or desire on my part, has come a clear conception of man's spiritual nature, a sound and simple philosophy far more potent than that, for example, derived from the physics of Eddington, Jeans and others. After all electrons and atoms are farther from man than the living cellular units which make up man's body.

The world to-day needs a spiritual regeneration. If this can come from biology as biology it has a power in this age of science over those who see science as the end-all of life, and gives strength to others who know by faith alone. But spiritually poverty-stricken as the world is to-day, how few realize this plight! Where is there given adequate emphasis on things of the spirit? And yet, what is man and all his science but the reflected light of this inner flame? In this situation any, even the slightest spiritual contribution merits support.

As an "ambassador in chains" I approach you. That I have been limited and yet have been able to carry on should not be an argument to keep me limited. As one who despite handicaps has been able to accomplish works of great value—the records sent herewith attest this—I ask for the opportunity to go forward with my work. If you would grant me a personal interview at your convenience in October next, I should then tell you more explicitly my aims.[140]

The only response Just got to this appeal was a cold and abrupt note from Rockefeller's secretary.[141] Desperate, he continued to push Wolf to follow up on the letter. Wolf finally had to admit that there was no chance that Rockefeller would help. He suggested that it would be best for Just to wait until the next generation inherited the empire; Just's chances would be better then. John D. Rockefeller III was, in Wolf's view, as liberal as his father and grandfather were conservative.[142]

During this time, Just and Hedwig must have sensed the brutality overtaking the political arena of Europe. Nazism's power was on the increase. Hedwig was part Jewish, a fact that was becoming more and more difficult to ignore, even though Switzerland was relatively safe. It was that summer that Jesse Owens, the black hundred-meter-dash runner, captured four gold medals at the Berlin Olympics and was rebuffed by Hitler. Just must have felt the growing race hatred, especially in Germany. But unlike many people, he could not bring himself to denounce the Germans outright. He made excuses for them and came up with rationalizations, and substituted his personal worries for any broad political perspective. He was aware, however, that Germany, and perhaps German-speaking countries in general, might not be a pleasant place for him and Hedwig to live.

Because he had been treated nicely by the Italians he came to see their country as a place of possible refuge. Never mind that under the Fascists Italy was becoming in many respects another Germany. The Fascists' invasion of Abyssinia was sad, Just felt, but he still hoped some good would come of it.[143] If he could not live in America—and now that he was in Europe again he was returning more and more to this view—then he and Hedwig would consider taking out Italian citizenship. On the whole the Stazione was a nice place to work; he would also apply for a post at some university or scientific institute. As an expert on the rearing of marine animals, he thought he might be able to increase the practical value of the fisheries industry in Italy—a country that, in his words, "because of its geographical location, presents the most attractive possibilities of success to her sons, animated, as they are now, with an indomitable will to work."[144]

Just began to make appeals to the lower ranks of the Italian government, but got no response. He was not permitted to see the minister of education, despite numerous attempts to do so.[145] But as he had done in the case of Rockefeller, the financial king in America, Just set his proposal before the highest power in the land:

Celerina, August 13, 1936

Il Capo del Governo d'Italia
S. E, Signore Mussolini
Roma

Your Excellency:

Although I have, in accordance with regular procedure, presented a request to the office of your government to which such requests should be addressed, I nevertheless take the liberty to appeal to you personally because I am motivated by the thought that mine is such an unusual case that it merits your individual attention. As an American negro who for more than twenty-five years has contributed to the progress of biological science without having attained a place in America which such service deserves, I desire earnestly the opportunity to continue my labors in Italy and thereby co-operate in the ardent activity with which your energetic leadership accelerates Italy to a magnificent destiny. My scientific labors fructify in this moment and promise a new basis for both the investigation and the solution of fundamental problems of vital phenomena; they reveal a significance for medicine, for example, the elucidation of cancer, and they relate to the practical improvement of a natural food-resource in Italy—i.e., the piscatorial industry.

I request that you grant me an audience that I may have the possibility to explain my purposes. I maintain myself at your disposal until August 29 when I must return to the U.S.A. Enclosed are my vita, a bibliography, and a separate of a recent publication of mine. I further enclose copies of two statements, translated into Italian, addressed to your Minister of Education, which present my case in detail.

I have the honor to remain, Your Excellency, with expressions of highest esteem,

most sincerely Yours,

E. E. Just[146]

But Mussolini was too busy building his Fascist machinery to be concerned with the problems of a lone black American scientist, though Just seemed to have little understanding of this fact.

From day to day the problem, already enormous, became magnified many times over in his mind. He became obsessed with the future, with Hedwig, with the contributions to science he was still hoping to make. Photographs taken at the time show him worn and haggard, graying fast and losing weight rapidly. But he kept pushing on, determined not to give up, ready to fight to the end. When he returned to Washington in the fall of 1936 he renewed his efforts on the home front. He contacted Max Wolf and Dorothy Norris, urging them to attend the annual conference of the American Society of Zoologists, to be held in December in Atlantic City.[147] He wanted them to come and see for themselves—to find out more about his work and to feel what it was like to be treated as he was by the scientific community. He planned to take up where he had left off the year before. If the geneticists were upset then, they would be livid this time. Just was determined not to spare Morgan or anyone, especially now that he had two outside supporters as witnesses.

Just looked forward to the conference all through the fall. When the time came he was ready, except for some slides that Hedwig was supposed to send him from Switzerland. They did not arrive until late on the day he was scheduled to leave. Breathing a sigh of relief, he rushed to catch the last train at Union Station. When he reached Atlantic City, around midnight, he took a taxi straight to the hotel.[148] The room clerk was shocked to be approached by a black man carrying luggage and asking for his reserved room as a member of the annual conference of the Society of Zoologists. Wherever he went in America (he never traveled in the South), Just always was certain to make advance reservations in order to give himself a better chance of not being refused a room. But even then clerks were likely to hesitate. This one, on second glance, saw Just determined and ready "to raise hell," so he ordered that Just be taken to his room. It was a good suite with two beds, a large bath, and a lovely view of the ocean.

Just had breakfast the next morning with the man who had been president of the American Society of Zoologists the previous year. This man, "though of German descent," was unpleasant and poorly behaved, and scolded Just for being half an hour late without stopping to think that there might have been a valid reason. The dining hall was busy, but Just received special attention from the waiters, while many of the other scientists grumbled about poor service. He attributed this to the fact that the waiters were European—Italian and Swiss. He knew how to get along with them, speaking to them from time to time in Italian, German, and French. The Italians were especially gracious, and made him feel as if he were "really someone of importance."

After breakfast Just gave a paper on "Ultra-violet Radiations as Experimental Means for the Investigation of Protoplasmic Behavior." He received many compliments—something he was sure to note—and considered the paper "a distinct success." Wolf and Norris both attended and went away impressed by his performance. Afterwards Just took Heilbrunn to lunch and caught up on old times. He spent the afternoon talking to people and received an invitation to a formal dinner—the celebration of S. O. Mast's twenty-five years at Johns Hopkins. That evening, at the hotel where the dinner was being held, Just noticed how shocked the doorman was that a black was entering. But he proceeded unflustered to the manager and asked for the organizer of the dinner by name. Though he was politely conducted to the dining hall, everyone in the lobby, prim ladies sipping tea and gentlemen reading newspapers, glared at him as though he were "a man from Mars or . . . demigod." They refused to believe what they saw—a black man attending the Mast dinner; they shook their heads and assured themselves that they must have been mistaken. After dinner Just went to the "smoker." There he saw many of his old Woods Hole colleagues. Schrader acted embarrassed, as if he felt guilty about clandestinely not supporting his friend's cause. Just responded in a "cool" and "impersonal" way, keeping him at a distance. This awkward interchange ended, fortunately, when Just was taken off into a corner by Ross Harrison, the embryologist from Yale. Harrison had become more and more interested in Just's ideas on genetics, and they talked until two o'clock in the morning.

The next day was the big one. Just was scheduled to face the geneticists head on, as he had done the previous year. His talk, "Phenomena of Embryogenesis and Their Significance for a Theory of Development and Heredity," drew an enormous crowd and, at the conclusion, "a tremendous burst of applause." Most of the audience left after hearing his lecture; apparently, they were only interested in that part of the program. The thrust once again was against Morgan and his crowd. Marcella Boveri, mother of Just's old lover Margret, swore it to be the best talk of the entire conference. As a kind of "barometer," a "reflecting pool for others' minds," she could be trusted to give a good judgment on matters of that sort. The best compliment, however, came from Harrison at the closing session later that evening. He praised Just for stressing "larger changes in the whole organism" rather than "the lesser qualities known to be associated with genic action," on being interested "more in the back than in the bristles on the back and more in eyes than in eye color."[149] This too was a slap at Morgan. In Just's opinion the published version of the talk lost "not only flavor but color"; it lacked "the fire and dash of the spoken word, . . . the caustic and sarcastic passages, the irony."[150]

The conference was a success, but Just was no better off than before. Wolf and Norris had come and gone and promised nothing in the way of tangible support. Just faced the new year, 1937, with no more chance for funding than the previous year, no possibility of a permanent position, and perhaps even less certainty of a divorce. He had fewer supporters than ever in the American scientific community. He had become friendly with Harrison, but only because Harrison's work on tissue culture dovetailed with much of his own on the cortex. Just did not care. He began to take pleasure in seeing the scientific community in a truer light. He was finally coming to question not "the cold-blooded scientific method," but rather its practitioners. After all these years he was finally realizing that scientists were human beings, no

294 BLACK APOLLO OF SCIENCE

more, no less, no better, no worse. They had the virtues and failings of society at large. If the Woods Hole experience had not taught him that in the 1920s, the experiences of the 1930s brought it home full force. He faced up to facts in the spring of 1937. America was no place for him. He wanted to be in Europe, with Hedwig, to try and plan their destiny there. He was off again in June.

Hedwig had never given up hope that she might help Just. But by and large she had no leverage and few connections that would have provided permanent support for his work. And there was no guarantee that any pleas she might make would elicit a favorable response, especially under the circumstances. Her affair with Just had caused consternation in the family, especially on her father's part, and her relationship with them was up and down. She had no financial independence or moral support. All she had was Just—and only during the summers. Her one source of income, for the most part, was Just. She lived off his salary, or, that one time, off his Carnegie grant, on which she was listed as his assistant.[151] She needed other people to help her think through her own difficulties, and perhaps to help Just with his. She might have been able to talk to her sister about the moral questions, but Ursula was too caught up in her own love affairs to help Hedwig attend to hers. Just and Hedwig spent many hours worrying about Ursula's romantic problems when what they needed was for someone—almost anyone—to worry about them.

Only Europeans knew the details of their relationship, and few were sympathetic. By 1937 Naples had lost some of its charm for them, because people had begun to talk. Helen Hartmann, Margret Boveri's replacement as secretary at the Stazione, was the lead gossip.[152] Hedwig and Just tried to ignore her and the others, but it was not easy. In the end they had to forgo Naples altogether.

Aside from Just, Hedwig had one source of sympathy and companionship—Nancy Astor. Lady Astor knew everyone and everyone knew her. She was a perfect connection for Hedwig. She had power, fame, and wealth, as well as a bubbling, honest, incisive personality. American by birth, a native Virginian, she had found her way to England and married Waldorf Astor, one of the wealthiest men in the world. He was a lord and she became his lady, but not content to be a lady in the traditionally passive and often frivolous English style, she took a deep interest in politics and social reform, and eventually became the first woman to occupy a seat in the British Parliament. By 1930 she had become internationally known, entertaining diplomats and leaders from throughout the world at her country estate, Cliveden, beautifully situated on the Thames.[153]

Hedwig had met Lady Astor in 1933. Her brothers, Karl and Otto, were friendly with Lady Astor's son, David, who used to go to Switzerland to ski each winter.[154] The Astors and Schnetzlers got to know and like each other, and Lady Astor became a surrogate mother to Hedwig. Hedwig often went to Cliveden. She found the setting restful, a retreat from her problems and from the realities of the world. She used to spend as much time talking and strolling with Lady Astor as her hostess's busy schedule would permit. In a sense she was a part of the notorious "Cliveden set," coming and going just like George Bernard Shaw and Philip Lothian. She spent Christmas of 1933 there, but did not mention her affair with Just. The truth came out at last, three years later, not from Hedwig but from her mother, Elisabeth, who, while visiting Cliveden with her husband, appealed to Lady Astor out of fear that Hedwig would abandon Europe for America.[155] It was a shock all around.

Soon afterwards Hedwig confided in Lady Astor about her black lover in America, and expressed hope that she would help them realize their ultimate destiny—to live together as husband and wife in Europe.[156] Things were even further complicated by the possibility that there may have been some romantic goings-on between Hedwig and David, Lady Astor's son. But Lady Astor kept an open mind.

In the spring of 1937 Just had still not arranged his divorce, and Hedwig was distraught. She had not revealed to Lady Astor the murky side of Just's family situation, but Lady Astor discerned its broad outlines by roundabout means, and Hedwig, somewhat embarrassed, confessed all the details. She needed and wanted help.

Just did not want Hedwig to put her confidence in Lady Astor, whom he suspected of still harboring traces of a Southern upbringing, even if she did happen to be in another country and going back and forth to Parliament fighting for legislation on social reform. She had to be prejudiced, perhaps overtly and certainly deep down.[157] Hedwig listened to Just, but continued to seek out Lady Astor's advice. More than anything, she wanted Lady Astor and Just to meet.

Lady Astor thought about Hedwig's situation and came up with a practical solution right away. Hedwig and Just should get married and live in America, where his job was.[158] The immediate problem was for Just to get a divorce, and this was best done in France, not in America.

On 11 June 1937 Just boarded the *St. Louis* for Southampton. On arrival he proceeded straight to Paris, where he planned to establish residency and arrange the divorce. He employed a lawyer in Paris, Guillaume Valensi, to handle the matter.[159] Then he was advised that the most expedient divorce could be obtained in Riga, Latvia. In a few months, after the case was prepared, Just and Hedwig would go there for a week or two and get married as soon as the divorce papers were granted.

Planning for the divorce and their marriage left Just and Hedwig little time for science per se, but they used every minute at their disposal to arrange a way for him to carry on his work in the future.

There was no use hoping about Italy, as government officials there seemed not the least bit interested in him or his proposal. Paris was a possibility. Just was beginning to feel comfortable there. He had come to know many members of the Parisian scientific community, and they had made him feel welcome. A special friend was Raoul May, an embryologist at the Sorbonne. May looked into a number of possibilities for Just, including a position at the Institut Pasteur. The problem at the Institut was that things had changed a great deal since the death of Pierre Paul Emile Roux, the previous director. Now the focus was on applied bacteriology, and pure research was being discouraged; the trend was to attract young medical men, not established scientists; the prejudice was in favor of hiring French scientists rather than scientists from other countries. May suggested, however, that there was much hope if Just wanted simply a laboratory and an intellectual atmosphere without salary. In that case, any number of institutes would open their doors to Just, including the Institut de Biologie Physicochimique, the Sorbonne, and the Collège de France. Just was well known and highly respected in the French scientific community, and to them that was all that mattered. They prided themselves on being "as free from caste and race prejudice as is humanly possible."[160] The only problem was money.

And that was the problem that kept returning to haunt Just. It was good that he was welcome in France, but he had to have money to stay there. Both the Paris and New York offices of the Rockefeller Foundation were opposed to his working anywhere but Howard. Earlier the foundation's director of the division of natural sciences, Warren Weaver, had given Just the impression that if he could get institutional backing he could get Rockefeller funding. In 1937 Just had offers, without salary, from institutions in America and France. The Jackson Memorial Laboratory, a marine institute in Maine specializing in cancer research, was willing to take Just on. But now Weaver said no to the proposed arrangement.[161] In France a number of institutes offered positions. Again Weaver said no.[162] By institutional backing he had meant Howard, nothing else. Just's future looked bleak as the summer rolled around. But Hedwig had hopes across the Channel. They packed their bags and headed for London.

ॽ

Who would not eagerly await luncheon with Nancy Astor? The outspoken member of Parliament who capitivated England with her charm, wit, and at times unorthodox behavior was sure to be in top form. Just watched her car pull up to the apartment building in Empire Court, Wembley Park, where Hedwig's brother Karl had lived since he moved to England to work with General Electric. It was the middle of August, Friday the 13th to be exact. A petite, energetic woman of about fifty-five got out of the car and made her way up to the third floor. The group—Just, Hedwig, Karl, and Lady Astor—lunched on cold cuts, mainly cucumber sandwiches garnished with watercress, and a pitcher of ice water in the Southern style. There was no wine, not even tea.

Afterwards Hedwig and Karl moved to the other side of the apartment and left Just and Lady Astor, "the two Southerners," to find their social bearings.[163] Not surprisingly, they took to each other right away. Lady Astor lost no time in getting down to business. What was the future for Hedwig? How would Just support her? Was their love strong and durable enough to bear the weight of an oppressive society—one that ostracized mixing of the races, especially through marriage? When would he know about his divorce from his wife? Would his wife go along or would she make trouble? Lady Astor always stated what was on her mind, not always with tact and grace. It was for Just to decide how he would take this inquisition.

Of course he loved Hedwig. As to the future—well, it was unclear. He was determined to get a divorce, but money, money, money was the eternal problem. What he needed was someone in the right place to say the right word. Lady Astor perhaps? Just wanted her help and begged her to put her influence to work. It was not just for him but for the larger cause of science, and religion too. Just put himself forward as "an instrument of the Higher Intelligence."[164] Knowing that Lady Astor was a devotee of Christian Science, he described how he had returned to religion after meeting Hedwig, and pulled himself up from the "lower level" he had sunk to after leaving South Carolina. His science was spiritual, not just materialistic. He wanted to develop that aspect further. All he needed was "a place in the physical sense" to do his work—somewhere to set up a microscope and simple apparatus, to study and write and publish. He would be glad to dispense with position or title in a university or research institute if he could get something as simple as "a table in a cottage." Not once did he mention the Cliveden estate as a possibility, nor did he or Hedwig ever make a direct appeal for help from the Astor fortune. Lady Astor was, after all, a *friend*, and friends are different from patrons.

The afternoon ended pleasantly. Lady Astor promised to do her utmost to help these two desperate people.

The following day was Just's birthday. He and Hedwig celebrated before departing for a rest in Riccione, a peaceful resort in nothern Italy. They spent the last hot days of August in the Grand Hotel there.[165] In September they stopped briefly in Zurich before going on to Paris.[166] Just had already made up his mind to resign his post at Howard, request the back pay he felt was due him, and remain in Europe with Hedwig. Divorce and remarriage would come in time.

After receiving a sharp telegram from Lady Astor, however, he changed his mind.[167] She urged him to keep his post at Howard. Only in that way, she thought, would he have a chance to go somewhere else. Just knew that in general this was the right advice, but that in his own case such things no longer mattered one way or the other. Somehow he would have to get this message across to Lady Astor. But there was no time now. He agreed to do as she suggested. Sadly, almost broken-heartedly, he made plans to return to Washington.

Lady Astor began her campaign almost immediately. She contacted her old friend Philip Lothian, the British ambassador to the United States. Was there, she asked, any position open, and if so, were there any chances for Just to occupy it? Lothian knew no science but had good contacts. He made inquiries of Frank Aydelotte, secretary of the Rhodes Scholarship Trust in America. Aydelotte "researched" Just's case, or rather he asked Abraham Flexner for an opinion. He reported back to Lothian that while Just's scientific investigations were "credit-able, especially for a man who has had the disadvantages which a Negro inevitably must meet, nevertheless they are not considered by scientific men of sufficient importance to make it probable that any foundation would, under the circumstances, give him . . . an appoint-ment."[168] Flexner's opinion was based on earlier inquiries made of Carnegie scientists. Just was "very well qualified to be a teacher in a colored university," and he would have received all the help he needed had he been interested in "that kind of career." At least Flexner's assessment, as reported to Lothian through Aydelotte, was more favor-able than the one offered almost two decades earlier by Jacques Loeb, who had suggested that high school teaching was the appropriate level for Just. But it was still negative. Lothian scribbled in pencil at the bottom of Aydelotte's memorandum that Just was "very good for a negro but not for a white."

There could be no help for Just with reports like this. When he heard about it from Lady Astor, he was not surprised. In fact he had fore-warned Lady Astor about what would happen.[169] He had come to regard Flexner as an enemy on three counts. First, he wrote, in 1920 Flexner had persuaded Rosenwald to decrease the amount of his original grant. (It is unclear what Just was referring to here, but more than likely it was Flexner's refusal to endorse his research trip to Jamaica.) Second, Flexner had advised Rosenwald against permanently endowing a Rosenwald Institute of Zoology at Howard. Third, he had made no effort as president of Howard's board of trustees to "correct the injustices" of Just's position at the university. This last certainly was true, but it would perhaps be more accurate to say that Flexner never admitted, perhaps never perceived any "injustices." On Flexner's part it was not lack of action on a commitment; there simply was no commitment.

Curiously, Just linked his troubles with Flexner to his troubles with Loeb, and both with what he saw as his troubles with Jews. His controversy with Loeb over the theory of parthenogenesis, he had found out, somehow had had serious repercussions for him, though he did not know the details of his consideration and subsequent rejection for a post by the Rockefeller Institute for Medical Research. In explaining to Lady Astor the forces acting against him, he exaggerated the extent of the controversy with Loeb, but not the hostility associated with it. Just portrayed himself as having overthrown Loeb's life work and of having supplanted it with a theory of his own. By no stretch of the imagination was this the case. Perhaps out of desperation, Just asserted his due and more. Lady Astor knew no science, and Just must have felt that there was no need to be subtle about the particulars. He told her that he had become an outcast because of his controversial ideas and actions. Further, he had been "put in the bad book by Jews generally."[170] This remark was made for her to interpret as she pleased. Just was familiar with the popular opinion of Lady Astor and the "Cliveden set" as anti-Semitic. The remark had strategic intent, but in some sense Just believed it was true.

ᘯ

Just spent the fall of 1937 at Howard attending to the zoology department. Young was gone; Hansborough and Chase were the only junior faculty. Just was determined not to have a graduate program for the following year. He wrote Charles Wesley, dean of the college, to this effect, and Wesley relayed Just's intention to President Johnson. Johnson did not agree, but instructed Just to direct correspondence about the matter to Wesley. [171]

What Just really wanted was a leave of absence, so that he could get himself established in Europe without giving up the security of his job. In another letter to Wesley he requested leave beginning February 1938, at the end of the first semester.[172] He outlined the resources in the department and suggested that the educational program would not be adversely affected by his absence. He spoke fatalistically about his diminishing chances of leaving an intellectual legacy to science. He was fifty-five years old; he had graying hair, intestinal ailments that were bad and growing worse, and bad eyes from years of straining microscopic work. There was little time left. How long he could last physically he did not know. But Howard was the place he suffered most, undergoing "tremendous nervous and spiritual let-down" each year.

In mid-January Just made another appeal to Johnson, this time not for leave but for retirement and back pay.[173] He was broken in body and spirit, and the university was "no place for broken down workers." He wanted to bow out gracefully with a $15,000 lump-sum payment on date of retirement, a sort of alimony settlement after a not-so-good marriage between him and Howard. Johnson ignored the request. Two weeks later, by which time Just had hoped to be already in Europe, he made an even stronger appeal.[174] Again Johnson ignored him. Desperate, he asked Dorothy Norris straight out for money, even if it was only a short-term loan. If she really wanted to help him, she would have to be more than a go-between for him and her rich friends, none of whom had come up with anything but empty promises. Norris refused to make the loan, explaining that since her mother's estate was still unsettled she had no money.[175] Just had nowhere else to turn.

By March Just could take his life in America no longer. The last seven years had been a constant uphill struggle. Weak, almost feeble, he was fighting battles on many different fronts: at home, at Howard, at the doorsteps of the rich, on the thresholds of foundations, in the arena of the American biological community. Hedwig had kept him going. She had devised strategies, urged him to take stands, increased his self-confidence. She had made him become more aware of the world and his place in it. To him she was a goddess, his life and soul, his symbol of Europe. He could endure racist America no longer. Now he was leaving, never, he hoped, to return.

CHAPTER 8

The Exile,
1938-40

Paris is especially beautiful in spring, and the spring of 1938 blossomed with narcissuses and tulips and hyacinths—and new growth everywhere. Just had perhaps never seen anything like it before, not even in the lush gardens of Charleston and its environs. Even if he had, there is no doubt that he had never been able to appreciate the annual rebirth in quite the same way, because never before had he too been bursting alive in such perfect resonance with nature.

He was free. He had no money, but he had Hedwig and the prospect of doing science in a congenial atmosphere. The details of income and divorce would be settled in time. But first things first. He found a flat at 1 Avenue du Général Balfourier and staked out a laboratory space at the Sorbonne.

His work required extensive use of the library. He was studying general philosophical problems important for the whole of biology, and there was no time or need for him to spend all day in the laboratory as he used to. Now was the time for synthesis of his past work. He felt that he could give important answers to difficult questions. Any laboratory work he put his mind to involved mainly studying previously embedded slides and sections. An avid experimentalist, he was of course hoping to get back to live animals in the summer. He planned to go to Roscoff, a marine station connected to the Sorbonne and located in Finistère on the northwest coast of France.

After settling into his life in Paris, he began to think again about finding secure funding. He was not on salary at the Sorbonne and there were no prospects of getting a salary anywhere in France, or anywhere in Europe for that matter. He had almost given up on the foundations, so Howard was the only place to turn to. His efforts to deal with the question of his university status and salary from Europe were the culmination of the long line of struggles with Mordecai Johnson.

Just wanted to retire from the university with a lump sum of $15,000, payable immediately.[1] This was not much to ask, he thought, as according to his calculations Howard actually owed him $25,000. He arrived at this figure by totaling the amount he had "saved" Howard by having obtained part of his salary from outside sources for nearly twenty years.

Johnson did not deny the appeal outright, but asked Just to modify his demands. He suggested that rather than insisting on immediate retirement Just should take a sick leave in the second semester at full pay. This would pave the way for the larger request, slowly, more reasonably. Just agreed. He was willing to do anything to bow out gracefully. He felt neither bitterness nor hostility nor a sense of persecution, he claimed, only "an overwhelming feeling of failure, . . . collapse, . . . complete frustration, utter defeat."[2] At some later date he could probably be counted on to fight again, but at the moment he was too tired, too worn out. He would submit a request for sick leave as Johnson suggested. Anything so that he could have some income while the larger request was being sorted out.

The plan was that Johnson would present the request to the Howard board of trustees at their spring meeting in April. From past experience Just doubted that Johnson would carry through, so he reported the details of the plan to Peter M. Murray, an old friend from the National Medical Association who was a member of the board.[3] That way if Johnson did not present the matter properly, Murray would be able to do so.

As Just suspected, Johnson did not bring the matter to the attention of the board, and Murray had to present it. It was highly controversial and the board members were divided. The issue remained undecided. Still, some members, including Murray and Charles H. Garvin, were determined to continue fighting on Just's behalf.[4]

Neither the board nor Johnson contacted Just about the status of the case. The matter was put in the hands of the Committee on Instruction and Research, which was supposed to produce a verdict over the course of the summer. The committee, composed of Guy B. Johnson, Mordecai Johnson, Peter Murray, and T. J. Jones, essentially rejected the proposal in mid-August, but at the same time urged Just to come home and appear before the board on 25 October.[5] But Just was then in Roscoff, happily working away on his experiments, and he did not even want to think about returning to America.

Instead of getting full pay for the spring semester, as Johnson had promised, Just received only half pay. The retirement proposal went nowhere. The meeting of the board on 25 October focused instead on the question of Just's salary for the 1938–39 academic year. It was decided that Just should be granted sick leave for a year, at $250 per month.[6] His status beyond that point would be decided over the course of the winter and spring. Though this was not the decision Just had hoped for, he was relieved that he would be receiving some income. He had to continue to fulfill his family responsibilities. In February 1938 he had arranged for his paycheck to be turned directly over to Ethel, and he was glad that that would continue. Because of his own financial problems, however, he was eventually obliged to reduce the amount sent to Ethel from $250 to $166.75, and have the balance sent to him in Paris.[7]

❧

Around the time Just began his exile, in the winter of 1938, Howard was not the only place he tried to obtain substantial funding from. He also organized a new plan to get backing from the Julius Rosenwald Fund. Knowing that Edwin Embree was not favorably disposed toward him, he decided not to make his own appeal and planned instead to have a third party push the case. As in 1920, the National Research Council was called in to serve as a buffer, to act as matchmaker of scientist and patron.

By a fortunate coincidence, Ross Harrison was president of the council at the time. An avid supportor of Just's new work on the ectoplasm, since it supported his own on tissue culture, he was also an intimate friend of Marcella Boveri, Margret's mother, who was teaching biology at Albertus Magnus College in New Haven just down the road from Yale. Just appealed directly to him.[8] He wanted Harrison to go to bat for him with the Julius Rosenwald Fund. At some point, Just knew, Harrison would probably touch base with Abraham Flexner, as Rosenwald had done in 1920 and Lothian in 1937. For blacks, Flexner was perennially a pivotal figure. After so many years in the game himself, Just wanted to make sure Harrison understood Flexner's position before attempting any plays. Flexner controlled the game. He was "the dictator in all matters concerning Negro education," something that Just had known all along but was only now, now that he was in France and had no intention of returning to America, gaining the courage to say outright.[9] Flexner had never been directly involved in black education, but he continued to be "the arbiter of many" who needed "the stamp of approval in order to approach any body whose aim is to further Negro scholarship." He often handed down "judgements" and issued "fiats" in matters about which he had no knowledge.

This was a common criticism of Flexner, one made not only by Just but by white scientists and administrators as well. Out of the apparent objectivity of the Flexner report had come strong policy changes. Yet how could a layman with no medical training make grand pronouncements about medical education? Of course, it was one thing for whites to raise such questions, quite another thing for a black to do so. Besides, any white who took an interest in black affairs was considered an expert; no special training or experience was needed. Also, had not Flexner been fair with Just, soliciting expert opinion from the highest ranks of American science? Was not Flexner a sincere friend, trying to help? On the surface this was so, but Just was no fool: he warned he could recognize "paternalism" under "the guise of friendship."

Now that he was settled in at Roscoff, Just wanted Harrison to help him get on with his work. He was anxious to continue his study of inclusion-free cytoplasm (ground-substance) in eggs; extend his past work on waterdrops in normal and treated but still viable eggs; and further his investigations into the synthesis of nucleoprotein during cleavage. These three areas of research would form a significant part of his unified program: "the biological analysis of embryogenesis as a series of chemical reactions."[10] Just was beginning to focus more on the chemistry of the living cell, but it was new to him, and he was fifty-five years old. To maximize his chances of accomplishing his goal, he needed time, a variety of animal forms, and funds. The first two required that he continue his extended exile; the lack of the third, it seemed, might disrupt his whole plan.

Perhaps teaching and research do not mix, and in Just's case they certainly did not. The "teaching load," the "poor quality of students," and the "small number of advanced students with inclination or the financial background" made it impossible for him to give time to research at Howard.[11] For the most part his work was done abroad, and not at all when he was involved in teaching.

As regards animal forms, Just had in the past handicapped himself by focusing mainly on one species at a time. He needed desperately at least one species from each of the following classes: eggs fertilizable in the germinal vesicle stage; eggs fertilizable in the first maturation stage; eggs fertilizable in the second maturation stage; and eggs fertilizable when completely matured. No one knew better than he that all four forms were readily available only in Europe. But he must have realized that this reason would not get him much sympathy from American embryologists, who themselves were working under the same disadvantage.

The question of funds was ticklish. Here was Just, a black man, asking for support—in Europe. The request could only have seemed arrogant to people who had always felt proud of the support that was already being given to blacks. Besides, funding agencies in America, as Just well knew, were inclined to support things that were American. Many were helping immigrants escape Nazism in Europe, find jobs, and settle down in America. They could not see that Just was a political escapee too, though fleeing in the opposite direction. The immigrant to America did not look back to Germany or Austria or Poland for help once he escaped. The sad thing about Just was that he had to look back; only America—the place he was running from—had the financial means to support his asylum. He was being denied the independence though not the pride of a noble exile.

Harrison knew little about the particulars of Just's case, and he asked Flexner for advice.[12] Flexner declined to get involved, explaining that he no longer had contact with the foundations. He did offer a brief suggestion on the route Harrison might take to procure funding for Just, but he was perfunctory and hardly insightful.[13] Harrison also contacted Frank Lillie. Apparently he did not know how close Just and Lillie had been or what role Lillie had played in previous efforts to secure funding. Harrison told Lillie of the "importance" of Just's work and his own desire to help Just achieve "what . . . is necessary for him to carry on his research with equanimity."[14] He did not have in mind the kind of permanent exile Just was trying to carry out. Essentially he agreed with funding agencies and others who thought Just should "remain in this country and share the burden" of improving the social conditions of his race. Harrison was really no different from Lillie or Flexner or Heilbrunn. True, he was entering the scene several years later, but like them he was only willing to seek "help to enable Just to have, say, one more year in Europe."

As he had done earlier in his letter to Weaver, Lillie tried to clarify his relationship with Just. He told Harrison that Just had been his student, and that he, Lillie, had been influential in getting him funding in the past.[15] Lillie had hoped that Just "would reconcile himself to the inevitable and take some interest and pride in working for his own race." Though proud of Just's research, he was sad that he had never been able to convince him on the question of racial duty. To Lillie, Just was "a very sensitive, intellectual person" who had in recent years acquired a "rather exaggerated self-appreciation" mainly because of his enthusiastic reception by the Europeans. No wonder he could not be happy in America. Heilbrunn and Schrader also knew Just personally, but Lillie prided himself on being Just's "best and most devoted friend." This, of course, was a matter of opinion. Something of a distance had grown between them. But the fact remained that they had come a long way together. In 1938, with this in mind, they both thought back on the relationship—how it had grown, what it had meant, where it might go.

ℜ

When Just took Patten's advice to write Lillie at the University of Chicago in 1909, he had no idea what kind of man Lillie was or what kind of response to expect. All he knew was that Lillie was a top American biologist at a first-rate university, who might be interested in having him as a student for the doctorate.

Born in 1870, Lillie had grown up in Toronto, Canada, in the household of very respectable parents.[16] His father, George Waddell Lillie, was a successful businessman, a partner in a wholesale drug company; his mother was the daughter of a Scottish tobacco merchant who later became a Congregational minister and amateur astrologer. On both sides of his family were long traditions of strong morality and upright character.

In high school Lillie became intensely interested in religion and joined the Church of England. When he entered the University of Toronto he planned to study for the ministry. But in his senior year he switched his interests to science, concentrating heavily on the problems of embryology. After graduation in 1891 he went to work with the embryologist C. O. Whitman at the Marine Biological Laboratory in Woods Hole. There he found E. B. Wilson, E. G. Conklin, A. D. Mead, and L. Treadwell, all aspiring scientists conducting studies in cell lineage. On Whitman's suggestion he too began work on the cell lineage of lamellibranchs. In 1892 Lillie followed Whitman to the zoology department at the University of Chicago, where he became Whitman's prize student and wrote an outstanding thesis for the Ph.D. in 1894.

The summer after earning his doctorate Lillie taught embryology at Woods Hole. He married one of his students, Frances Crane, a year later, and she joined him at the University of Michigan, Ann Arbor, where he took up an instructorship. He stayed at Michigan until 1899, then moved on to Vassar. After a year there he returned to his alma mater as an assistant professor of embryology, ready to work himself up the academic ladder.

Very early Lillie showed himself to be an adept science administrator as well as a far-reaching researcher. Soon he began taking over many responsibilities from his teacher and mentor, Whitman, both at Chicago and Woods Hole. In 1900 he was appointed assistant director of the MBL under Whitman, while keeping his post as assistant professor at Chicago. Both Chicago and the MBL were young institutions that needed nourishing and developing—and much of this work fell to Lillie. It is not clear exactly what Whitman suffered from in the early 1900s, but his problems were psychological and often affected his capacity to operate efficiently. More and more responsibility fell on Lillie, who was at the same time always careful to deal with Whitman quietly and sensitively. He remained loyal to Whitman to the end. And he was loyal to the MBL's existence as an autonomous institution; despite many problems, he helped keep it from being absorbed by the Carnegie Institution of Washington. By 1909 he was well on the way to a brilliant career as an administrator. In 1910 he took over from Whitman as director of the MBL and also as chairman of the zoology department at Chicago.

When Lillie received his first letter from Just, he was about to take a new direction in his own career. He had a feeling of fresh beginnings. He wanted to be sympathetic to this new student of a different race, but he also wanted to be open and honest. He set out his advice in a letter. Though the letter is no longer extant, Just kept it always and in the 1930s referred to it often, describing it as "cold, if not actually hostile."[17]

At the time, Lillie did not think Just should get a Ph.D. or strive for a career as a research scientist. A doctorate was useless for a black. After he obtained the degree, no white university or research institute or industry would hire him, so there was no chance that an advanced degree would give him any employment advantage. Lillie was explicit on these points. The only possibility for a black, he stressed, was to work at a black college. Therefore there was no need for Just to work for a doctorate, since black students were usually underprepared and far from being Ph.D. material themselves.

Lillie was a practical man—one who always looked toward reasonable goals. He was especially attuned to Just's problem because another black scientist, C. H. Turner, had done his doctorate in zoology at Chicago and afterwards had nowhere to go. But science attracts people who are not necessarily looking for a straightforward career path. There are other considerations—of interest, of talent, of dedication. Lillie viewed Just's request only from the career perspective. It was not until later that he was able to recognize Just as a person who, despite the very real difficulties he encountered along the way, was captivated by the joy of his work, and fascinated by the wonders of nature. But even in 1909 Lillie was not totally discouraging. He offered to give Just a try as an assistant working in his laboratory at Woods Hole.

From that first summer in 1909, Lillie was impressed with his new assistant's brilliant, alert mind and appetite for hard work. Just used to be on the job in the morning before Lillie arrived, and would finish up long after Lillie had gone home for the evening. He became friendly very quickly with the collecting crew, going out with them at all hours and bringing back specimens (usually the best) for the next day's work. By the time Lillie arrived, usually a little before noon, after attending to his administrative duties, Just had set up the laboratory and experiments were often already under way. The men worked together for hours in complete silence and deep concentration. They were scientist and assistant, master and pupil, devising experiments and techniques for the futherance of their scientific aims. Time was always left toward the end of the afternoon for discussions of the theoretical implications of the practical work.

Masters are often referred to in music, art, and other fields, but rarely is the notion associated with a career in science. As a profession science is supposedly more public than private, resting on objective and repeatable techniques rather than subjective and idiosyncratic ones. This attitude has set science apart from fields in which a technique or skill or trade is handed down from generation to generation, from master to pupil. But to some degree science is not so different. A form of apprenticeship takes place, usually just before and right after the Ph.D. And this was certainly the case with Lillie and Just. They had a deep master-pupil relationship, one that was both complicated and enriched by their separate roles in society, Just as a black, Lillie as a white.

The two men fitted together well in their work habits, personalities, and outlook on life. Just was clean and orderly and methodical. Lillie was also, traits that helped him become the top-notch administrator that he was. Both men were strong-willed and determined, but at the same time outwardly shy. By and large this shyness was no handicap, for scientists were permitted to be as reserved and withdrawn as they pleased; in fact, a kind of separateness was often regarded as an essential characteristic of the serious workers in the field. Further, an image of quiet and reserve was a real advantage for a black. Just's shyness was viewed as a good thing by his colleagues; it kept him out of trouble in the isolated, all-white town of Woods Hole and made his presence around the MBL less obtrusive. Just later said that his shyness was developed mainly as a defense mechanism against the white world. But perhaps he was shy long before he encountered the outside world. Whatever its origin, his shyness helped deepen his relationship with Lillie.

Just was unusually devoted to his mentor. The word "loyalty" comes up over and over again in their correspondence. Though Lillie was only twelve years older than Just, he seemed more like a father than a brother. It was a paternal kind of relationship, and those overtones were intensified by Lillie's relationship with others at the MBL. He maintained his position of authority as director by being firm and wise, by giving himself the image of a sage. This air put a distance even between him and others his own age. His wealth only widened the gap.

Lillie served as the director of the MBL throughout the 1910s and '20s and on numerous national science boards. He was a solid citizen. Unlike Loeb, he did not have to push to get things accomplished. Often he would carry matters through in such a smooth and graceful manner that it was hardly apparent that he was the guiding force. Typically, after a meeting everyone would realize that somehow Lillie had accomplished what he wanted, often without saying a word.[18] He had a gift for nuance, and for using even silence to great effect. His power might have been a bad thing were it not for the fact that he usually made the right decisions.

Lillie carried diplomacy and finesse into all parts of his life, especially into his science. He never put forward his work in an extreme or controversial way. Though his fertilizin theory was inconsistent with Loeb's work, he let the exact discrepancies go unstressed. This reticence is interesting, since, like any scientist, he wanted his theory to gain acceptance. How did he expect his work to survive if he did not boost it more, become controversial? Perhaps Just did it for him. Just recognized the need to discredit one theory in order to establish another, and he pushed Lillie's work. This loyalty was no different from that shown by other students to their mentors. Loeb had his followers, Morgan had his. It was natural for Just to work on the same problems as Lillie, and to take up arms on his behalf. What was unnatural was the degree to which he felt compelled to express his feelings, and the extent to which he was willing, almost anxious, to antagonize adversaries. In part it was his sense, as a black, of being on the fringes of the scientific community; in part it was a response to Lillie's refusal to get involved in scientific squabbles. Lillie always tried to get Just to "tone down" his work and be more diplomatic. Curiously, though, as a scientist he benefited from Just's stance in the long run.

For about a decade Lillie and Just worked side by side on the problem of fertilization in marine invertebrates. At the same time Lillie was pursuing other research interests, in particular the development of the freemartin, a sexually imperfect, usually sterile, female calf twin-born with a male. Added to this were his numerous administrative responsibilities. Much of the work on his fertilizin theory, then, was left to his students, mainly Just. The theory was innovative and had an immediate impact on the world of embryology, but it needed experimental proof to back it up; in short, it was nothing without experimental proof, and Lillie was aware of this when he conceived it. It was the kind of theory that opened up several avenues of research, each requiring many hard hours of laboratory work, collaborative and otherwise. Just got involved. From the beginning of his career in 1912, and especially from 1919 on, he piled one beautiful piece of research on another, making ever more refined observations and developing ever more ingenious experiments to find the truth about the fertilization process. Among his many contributions were the detailed observation and definition of the process whereby the vitelline membrane separates from the egg cytoplasm, the movement he termed "the wave of negativity." This "wave of negativity" was referred to time and time again by others.[19] Observations like these were valuable to embryologists whether or not they accepted Lillie's fertilizin theory. To some extent, Just saw himself as working on problems and producing results that were cumulative in their support of Lillie's theory. Still, he was an independent investigator, no two ways about it. The article on fertilization in Cowdry's text names Just as a collaborator with Lillie; his name falls second, though alphabetically the order would be inverted, so the impression given is that he was a junior author. The fact is that Just wrote an equal share, if not more, of the article, a point Lillie made to Rosenwald. Subsequent investigators in the field have, however, usually credited Lillie alone with the work. Few people understood, as Lillie did, that Just made an enormous contribution by "keeping alive and extending" the fertilizin idea throughout the 1920s and '30s. By his own admission, Lillie was "not the least bit of a propagandist" for his own work and in that regard was "somewhat of a fatalist."[20] Just was the one who kept the work going.

During the 1920s Just never pushed hard to be independent from Lillie. In many ways he was nourished by the connection. Lillie wrote numerous letters on his behalf and worked behind the scenes to keep him going in his career. He wrote to Just too, more than to anyone else, to offer advice, boost morale, and simply express friendship. Just wrote as many, if not more, letters in return. There is no evidence that Lillie ever gave Just money out of his own pocket—he was much too wise and discreet for that—but he constantly maneuvered to get support for his prize pupil from a variety of sources. Just was a favorite. He got better-than-usual laboratory space at Woods Hole, even though and perhaps because he did not enjoy complete social acceptance there. For his own part, he felt deeply loyal to Lillie, as some servants feel toward their masters, and he had a sense of being able to produce best while working for a superior. Also, his work meant "infinitely more" to him, he told Lillie, when he was allowed to express his "feeling of loyalty."[21] The drive, the motivation to push forward had to come from somewhere, and for Just it was Lillie who towered in the background as a measure of the success of E. E. Just. To be sure, he had had other godheads—his mother, Rosenwald, von Harnack, Dohrn, Hedwig, Keppel, and even Mordecai Johnson for a brief time—but his most constant source of motivation, moral and scientific, was Lillie, the embodiment of a paternal white master.

Society reinforced this tendency in Just. Everywhere he looked he could see blacks seeking appreciation from the white world and being handicapped if they did not. Howard had always sought the approval of whites; it had to. Even the Negro Renaissance, often thought of as a black movement through and through, had transposed white standards onto black culture. The movement's leading figure, Just's close friend Alain Locke, was proud of his Rhodes Scholarship and his Harvard degree, which suggests a degree of acquiescence to, if not absolute admiration of, a white value system that was in many ways founded on black subservience. Given the circumstances, it is hardly surprising that Just too needed to look up to something—something that embodied white culture in all its pervasiveness. For him, it was Lillie.

But the attraction extended beyond social issues, beyond the relative position of blacks and whites as reflected in the society of the day. Loeb, Morgan, and Conklin were also white, but Just did not feel the same way about them. Lillie was special. He was honest. The word "integrity" might have been invented for him. He was decent, and he was fair. He was generous to people of all races and ethnic backgrounds. He tried to protect Heilbrunn from anti-Semitism in much the same way he tried to protect Just from racism. He took special interest in Oriental students, and made an effort to insure that their adjustment to life in America was trouble-free.[22] Just knew Lillie as a sensitive man as well as a first-rate scientist—one who would do a favor and never mention it again. He was helpful and straightforward. When he warned Just about racism in the academic community, he was not expressing a personal feeling but dealing with the issue openly and with understanding. Not that he was free from racial feelings himself: he had his prejudices, but he also had the good sense to articulate them and to try and overcome them. For example, he defined democracy as "The conviction that in spite of external differences all men are in the deepest sense equal. This belief was not trained into me and was not native to me, but has been a gradual acquisition, and held pragmatically now as the only sound basis for social organization. I am in fact still full of prejudices, both racial and personal."[23] It was this kind of honesty and willingness to change that made Just truly loyal to Lillie.

Lillie was of two minds about Just's loyalty. He appreciated it as "quite a bright spot" in his life, and let Just know this time and time again.[24] He was a loyal individual himself. He had been devoted to Whitman in the early years, and he was consistently supportive of his wife, Frances, who underwent one conversion after another in politics and in religion. Just's loyalty was not dissimilar from Lillie's. But Lillie was concerned that Just might take this loyalty too far. What he seems to have meant by this, but never said directly, was that he feared that Just would begin to expect too much in return. Lillie had taken Just on and expressed an interest in him and benefited from his work, but he alone could not deal with the numerous problems that Just faced. He did not want Just to have any high hopes about what could be done to help him. Though he was wealthy and well-connected, Lillie recognized the boundaries of his influence in Just's case. There was a larger society to contend with.

Intense loyalty produces other problems as well. Often it is associated with lack of originality. An individual devoted to the ideas of someone else has difficulty coming up with ideas of his own. Often people lacking in talent and creativity fall back on loyalty as a way of making good in the world. George Streeter as much as accused Just of this. But Just's loyalty pointed not to any lack of creativity on his part, but to the circumstances in which he worked. It was not so much an intellectual problem as a response to social and political obstacles.

Besides, the issue of Just's creativity, raised by Streeter and others in the context of funding, was little more than a red herring. Science moves forward on the hard work of men and women making substantive cumulative contributions to a collective knowledge. Funding agencies, even in the mid-1920s, had begun to support the notion of steady advance in science, instead of continuing to make wild gambles on one or two gifted individuals who might or might not produce breakthrough discoveries. Now the aim was to build a sizable group of competent working scientists. There was no need for these scientists to be great advancers of the frontiers of knowledge; Lavoisiers, Darwins, and Einsteins, after all, are rare. Just was sometimes compared to Lavoisier by his old teacher at Chicago, A. P. Mathews, but his work never approached the stature of Lavoisier's. Nor did anyone else's at Woods Hole or the Carnegie or Harvard. But many scientists were brilliant and had contributions to make. They deserved support, and so did Just.

In the 1930s Just's loyalty to Lillie changed. In Europe he had found a more congenial working atmosphere and gained self-confidence. He began to feel cramped working in Lillie's shadow. He developed his own theory, using his work on fertilization to stress the importance of the cell surface in development. The European experience gave him the courage to question Lillie. In "A Cytological Study of Effects of Ultraviolet Light on the Eggs of *Nereis limbata*," he was quite deliberate and pointed:

> In my paper (1915) on normal fertilization in the egg of *Platynereis* I took Lillie's point of view concerning the origin of the sperm aster. More recent observations convince me that I must now abandon it. And a farther study of Lillie's work on the size of centrosomes and of asters associated with sperm-fractions strengthens my conviction; for his interpretation needs at least some modification.[25]

At the same time, Just thanked Max Hartmann of the Kaiser-Wilhelm-Institut profusely "for his kindnesses, his interest and his stimulating criticisms." This was the kind of acknowledgment he had given Lillie in the past. It was not that easy, however, to break free. Before submitting the article for publication he checked back and forth with Lillie, nominally on the correctness of his point about the sperm aster but also about the tenor of the criticism.[26]

Lillie handled the new Just with deliberate care. Remembering their deep and long-standing friendship, he continued to be helpful to Just's students—Young, Chase, and Hansborough—at Woods Hole. But he realized that something new had entered Just's life, something foreign. He wanted Just back at Woods Hole and Just would not go. How different this was from the time twenty years earlier when Just had given up his honeymoon in order to follow Lillie's advice. Lillie had not changed. Never once did he waver in his opinion of Just's work or back down from his commitment to help Just in, as he put it, "every concrete situation."[27] The point is that they could not reach agreement on what was best for Just. For the most part, it was their tacit assumption even into the 1930s that Lillie should have a say in the matter: Just would ask for advice and Lillie would give it. Up until 1935 Just was still telling Lillie that he wanted to dedicate his new book to him; while Rosenwald was alive he said the same to him. He actually dedicated it to neither. Quietly, beginning in the early 1930s, he had begun to question whether Lillie should have any say in his life at all. By 1936 he had decided in the negative. But he never exactly said so to Lillie. He needed Lillie's support more than ever, and so he maintained his "unwavering loyalty," at least on the surface. He did begin, however, to make more urgent pleas and demand payment of debts that he considered overdue:

It is now a little more than three years since I wrote you from Naples requesting your help in releasing me from a situation which then bad enough has steadily become worse. In your answer to my letter you quoted me the old adage, "out of sight, out of mind." That I am again in Europe may again elicit from you the same or a similar response. I sincerely trust that this will not be the case. This trust I build upon the hope that I can reveal to you my situation with sufficient clearness that you envisage it in a moment. First permit me to endeavor to separate from this domain of mine all the clouds that obscure the landscape.

That I have always felt out of place at Howard University you know, for you more than anyone else lightened as far as possible the more onerous burdens that I there carried. Out of a sense of duty, I sought to stay on knowing full well that the University is such only in name and as far as I can see never will be one in reality. The reasons for this belief I have often stated to you: the too small number of untrained students scattered among all too many colleges, the scandalously profligate waste of time and money, the lack of energy, of intellectual and spiritual enthusiasm, on the part of both students and faculty, the maladroit, more truly, maleficent, administration.

As you may recall, beginning 1920 I had the privilege voted by the trustees of Howard University to be in residence as a teacher only one half of each school year. It was your own idea that in this way I should have more time for research. Against the advice of Dr. Angell, then president of the National Research Council, I remained in residence two thirds of the teaching year with freedom for research beginning the middle of March. This treaty with the trustees persisted during the years 1928–1933 when I worked under the grant of the Rosenwald Fund with the difference that I was not allowed to teach undergraduates but only such qualified graduates who for one half year were to carry on research. The teaching of these all too few students proved more difficult than that of freshmen. The prevailing gossip had it that I sacrificed the University for my own personal advancement by absenting myself both from Howard University and from America. This was not true. I lived up strictly to the five year program made in 1928. When first Mr. Embree asked what I would do had I means, I answered at once—go to Naples to work out problems which I had in mind because the animals needed were not to be had in America. True, I made the initial mistake in electing to remain at Howard. In the first negotiations with Embree it was clearly understood by both of us that my remaing [sic] at Howard was not a prerequisite for the award of the grant. I never would have accepted the grant on terms other than those which stipulated a limited residence at Howard during each school year.

Since you had suggested that I be given less teaching in order to have more time for research—and this suggestion led to the plan of my being in residence during one half the school year only—the aspersive criticism of my absenteeism could not have infected you and so changed your attitude to me or have altered your opinion of the quality of my research. Nevertheless, the whispering campaign might have put you on the defensive; so I interpret your suggestion written in 1933 that I return at once that summer to Woods Hole. In a childish manner the gossips dissociated my absence—complaining more about my being so often in Europe than about my being away from Howard University: my suddenly so important corporeal substance was still in Washington if only I was in Woods Hole.

Whilst it is true that my situation in Howard University has yearly grown more intolerable, especially because of the attitude of the administration, it might be said that escape from the ugly circumstances did not of necessity mean leaving America; spending my time at Woods Hole could have sufficed. This would be true were it not for the fact that in the European atmosphere I have grown intellectually, have grasped more firmly conceptions which now underlie my work, and have obtained a clear point of view from which I envisage my work as the basis for a new biology. If I have felt freer in Europe and have won there a sense of dignity that I never knew before, this is owing less to myself than to the American milieu into which the chance of birth thrust me. This, however, someone may counter by saying: Good! Having won this feeling of respect for self you ought now remain in America. I on my side say that the value of my ideas for biology is such that it puts me out of place in a make-believe university. And since there appears to be no chance that I get a place in a real university or research institution in America, I must use the years before me in every way possible to keep aglow the flame within me which in my particular circumstances and as a Negro I am not allowed to nourish in America. There are several most important pieces of work that I can do only at Woods Hole and I look forward to working there again.

But I want if possible first to be established in Europe. To this end I strive; for it I ask your support.

You yourself once wrote me very clearly that whereas my work would be remembered long after you and I were both dead, I need never expect to obtain a place in a white American University. Although it is true that a Negro has occupied a place in the department of pathology of the University of Chicago and another in biology at McGill, that since you wrote the above quoted statement I had tentative offers from white institutions and that Loeb in 1914 and Carlson in 1916 offered me positions, nevertheless I very well know that in writing thus you but expressed the prevailing American point of view. And I do not now bring it up as a charge against you for I remember how honestly you wrote me when first I sought the privilege of becoming a student of yours. After all, the world is full of prejudices. At the same time, I cannot easily forget my long association with you, so easily overthrow my feeling for our cooperative work. Nor can I believe that this association and my loyalty to you have not in some measure overcome your prejudice. Moreover, you yourself, even as late as last fall when we talked, have given me belief in the value of my research and in the fundamental significance of my ideas for the progress of biological thought. It having been ordained that I should not find a place in America where I may have the opportunity to develop these ideas, I seek a place where more likely I can find one—in Europe. In 1930, in 1931 and again in 1932 I had offers to remain in Germany—offers which out of a sense of patriotism for America I declined. I still prefer to work as an American citizen. I believe that one strong and vigorous individual could win for me from the Rockefeller Foundation support which would enable me to carry on my work in Europe. No one to my knowledge in American science has more power than you have. European scientists are often shocked to learn that I have no real place in America. And I have often been embarrassed when asked why you do not support me because your position and prestige are known far beyond the boundaries of the U.S.A. It is not only to my former chef, therefore, but also to the head of American science that I write: to the former who realizes that a student carries on the professor's work and program, to the latter whose duty it is to see that every scientific idea of value in whatever man, black or white, flourishes, is nourished and brought to fruition.[28]

In response, Lillie claimed to have little influence. He had been retired from active duty at Chicago since 1930; he continued to go to Woods Hole, but more out of habit and for vacations, since he was no longer director there. His opinion was rarely sought in filling positions and putting together recommendations.[29] These disclaimers struck Just as insincere, considering that his former mentor, as president of the National Academy of Sciences, was in one of the highest positions of authority in science administration. Lillie promised, in any case, to continue to endorse Just's research proposal. It is curious that never once, now or earlier, did he propose Chicago as a place for Just to work; nor, for that matter, did Just ever ask for a position outright. Lillie wanted them to remain friends nonetheless. But in 1936 he was still preaching patience and still urging on Just the need to make the best of life's circumstances.

It was hard for Lillie to accept Just's need to give up America and live in Europe. Lillie had a sense of what he called national pride, of devotion to one's native country—ironic, since he had given up his own country, Canada, for the United States: on learning that he was a British subject, his wife had insisted that he become naturalized as an American, and he had done so. Just was always quick to point out this inconsistency, not to Lillie but to others.[30] What Lillie really meant by national pride was pride in America. But there was no way Lillie could expect Just to feel the same degree of loyalty to America as he did; the country had bestowed its bounties so unevenly.

In some sense, Just's desertion of America reflected poorly on Lillie and his efforts to make Woods Hole a refuge. Lillie had solved problems for Heilbrunn and others, but he was never able to even approach finding a solution for Just. This nagged at him. As someone who liked order, who extended enormous energy and forethought to arranging neat solutions, he must have been a little upset—frustrated is perhaps too strong a word—that things did not come together for Just. He kept thinking about the problem. By 1938 he was beginning to wonder whether any solution was possible; he was sincerely distraught to see how his old student, friend, and collaborator had to struggle to carry on against innumerable odds and with no money. He hardly knew what to think, less what to do.

Howard University was not, for the most part, in sympathy with Just's exile. In the 1938–39 academic year Just wrote each of the trustees separately, demanding retirement and a lump-sum settlement. The board was split. Peter Murray and Charles Garvin continued to support Just; the others, headed by Guy Johnson, did not. Murray and Garvin knew from firsthand experience how much true service Just had rendered Howard, and they felt that the institution owed him something in return. The precise debt they had not determined, but in their view it was substantial. Guy Johnson and the others also held Just's work in high esteem, but they could not agree with his demands. They did not think that any reward should be given to him for his initiative in getting outside grants. Johnson had been connected with a research institute (the Institute for Research in Social Science in North Carolina) where workers were salaried via their grants as a matter of course and did not expect additional pay from the institute.[31] Actually, workers at such institutes often made demands similar to Just's. But the parallel was hardly strict anyway. No doubt Johnson's institute had explicit policies along these lines, and grants were the rule rather than the exception there. At Howard, where the opposite was true—unclear policies and grant money extremely rare—perhaps a different approach was warranted. Rewards may have been needed as incentives in an environment where few existed. Faculty members at Howard must have wondered why they should struggle with foundations over funding if it was only to get the same income they would ordinarily receive. There were other benefits, of course, not the least of which was the chance for independent research, but these could only be fully appreciated and taken advantage of in a serious research institution.

The issue then turned from being one of retirement benefits to one of sick leave. Mordecai Johnson wanted to focus on the question of health, and convinced Just to do likewise. This was a tactical mistake on Just's part. Because he had always been seen as a little odd at Howard, working in his laboratory at all hours of the day and night, it was not long before the community concluded that now he was finally paying the toll. Rumors flew; he was said to be losing touch fast, if he had not already gone mad. Nothing was further from the truth. He was haggard, distressed, and worn, but he was also free and happy and optimistic about his future life in Europe with Hedwig. His letters continued to be clear, precise, and rational.

The rumors circulated beyond the Howard campus as well. Everyone was upset. Benjamin Karpman, an old friend and a psychiatrist, wrote Just inquiring about the "many disconcerting rumors" and insisting that he rest so as to overcome his "nervous tension."[32] Newspaper reporters bothered Ethel constantly, and the family was worried, especially because they had no way of knowing whether the rumors were true or false. Just wrote them few letters and his old friend Locke was left to "boost morale" at the Just homestead.[33]

Though thousands of miles away, Just soon got wind that his mental competency was being called into question. He was quick to confront Johnson, and Johnson was quick to respond.[34] He claimed to have repeated only what Just himself had said. In his letters to Johnson, Just had indeed referred to his "intestinal and nervous conditions" and spoken of being "broken . . . in body and spirit."[35] But he had simply been describing the agony and frustration Howard had caused him; he had certainly not meant to give the impression that he was losing control of his faculties.

The Howard board of trustees again took up Just's case in April 1939.[36] They decided to restore him to full-time active duty at full salary for the 1939-40 academic year. If full-time service was not possible because of total disability, he would be given an allowance of one-third of his salary of $5,000 a year for the duration of the disability by submitting to a medical examination performed by a physician of the university's choosing. The board refused his request for back salary and a lump-sum retirement payment. They determined that there were no grounds for the former, and as regards the latter, Just did not meet the age requirement and had not paid premiums to the Retirement Annuity Fund.

Just was not deterred. He accused Johnson of underhanded tactics, of prejudicing the trustees against him, and he continued to reiterate his demands, reminding Johnson of the unexpended grant from the Rosenwald Fund and Johnson's "failure to carry through the negotiations with the Rockefeller Foundation—a settlement which would have meant a large money-saving for the University."[37] Just wondered whether the board knew the whole story.

A settlement from Howard would have helped him get his life in order and carry on with his work. But by this time he had grown cynical and did not expect anything very positive to happen from that side. One thing was certain: he was not under any circumstances going back to Howard for the 1939–40 academic year, or ever. He was determined to stay in France—even if it meant being absolutely penniless.

Not everything was bad about the first months of 1939. One thing, in particular, was good: Just's magnum opus finally came off the press. *The Biology of the Cell Surface* was printed in an elegant, distinctive typeface and enclosed in a handsome dark maroon leather binding. Judged by its cover, the book was tasteful, even arresting. And the contents measured up to the cover.

Just dedicated the work to his mother. Only three others were singled out for appreciation: Julius Rosenwald, Adolf von Harnack, and Frederick P. Keppel. Discretion precluded any mention of Hedwig. The same reason cannot, however, be posited for the absence of Lillie's name. It seems that Just simply no longer felt indebted to his mentor, let alone to colleagues like Heilbrunn and Schrader. In late January he sent out a number of announcements from his Paris apartment. Lillie, Embree, Heilbrunn, Flexner, and the Schraders were all on the list.[38] To each he explained that owing to his destitute financial circumstances he could not send them complimentary copies, but that he hoped that they would nonetheless buy the book and read it. Aside from the pride he took in announcing his achievement, he was evidently anxious to remind these people that they had all in one way or another been responsible for his hard times. The book was proof that his days in Europe had been well spent. Those who had criticized and in some cases tried to hamper his work had been proven wrong. He sent out only three complimentary copies, one to the MBL library, one to his youngest daughter, Maribel, and one to Keppel.

Just's regard for Keppel went far back. Keppel had been sympathetic all along. When he was approached by Harrison in the fall of 1938, he expressed interest in supplying funds to Just for another year of work in Europe.[39] *The Biology of the Cell Surface* reached his desk in February 1939. He was touched by the acknowledgment in the preface, and promised Harrison to deal sympathetically with any application Just might make.[40]

Harrison did indeed receive a proposal from Just, but he thought it "in some respects a little vague and in others too comprehensive" to interest the foundations.[41] He suspected that part of the proposal was a follow-up to *The Biology of the Cell Surface*. He had not seen the book, so he went to Lillie for advice again. Just's former mentor and almost-abandoned friend still had faith in the value of Just and his work. This time he was more positive than ever. He offered a fresh look at Just's professional problems, and made an urgent plea for something to be done:

> I am glad to have your letter about Just and to know that Doctor Keppel is taking an interest in his case. A little over two years ago Just wrote me that the Carnegie Corporation at that time did not see its way clear to make a grant for his research, so I am particularly pleased to know that the situation now appears to be somewhat different. Indeed, I feel that Just's case is precisely in accordance with Doctor Keppel's announced plan, in that it would enable him to proceed with his program of research in which he has been terribly handicapped by circumstances.
>
> Just has qualities of genius; nothing whatever turns him aside from his purpose. I have attempted over and over again to get him to conform more to the conditions which his race and the nature of university life in America impose. I think now that this attempt was unwise; certainly it was futile. Earlier in his scientific career I aided him to secure support from the National Research Council and by personal appeal to Julius Rosenwald, but I thought that after that he should settle down and devote himself to promoting the interests of his own race at Howard University. As I have just said, I think that this opinion was probably not well considered.
>
> His program of work would perhaps sound vague and too comprehensive if one did not know as I do, and of course many others, the many years of hard, consistent and objective investigation that form the foundation for it. I have just been through his new book and am quite impressed with his point of view with reference to the principles of scientific analysis in biology stated in the introduction, and with which I agree. He has stated this situation very well.
>
> Just needs support, and I think it would be a serious blot on American science if he should come to grief. I am convinced that he will continue to give a good account of himself in his research, and so I hope that Doctor Keppel will make some provision for him.[42]

It was a little late, but Lillie had finally come close, the closest yet, to understanding in a profound way Just's severe problems. But was it too late to rescue the remaining years of a tortured life?

Unknown to Lillie, Just had already settled into a new life. At the Sorbonne he found everyone "extremely appreciative" and enjoyed the "warm and friendly atmosphere."[43] He was taking lessons in French from a young scientist, Geneviève Bobin, who became a close friend to him and Hedwig. His French was good enough for him to lecture in that language, so he held seminars from time to time for members of the laboratory. The seminars were well attended and received, and everyone was impressed by the way Just managed to reveal interesting ideas about his work in clear and idiomatic French. He had adapted remarkably well, and was working with renewed spirit, with the knowledge that he had found a new lease on life.

Spring rolled around and the date for his divorce grew closer. Just was happy, like an adolescent experiencing first love. For eight years he and Hedwig had been living as lovers rather than as a married couple; for eight years they had caused consternation all around. Now they would be married. Not that this would remove all their anxieties; in many ways it would be just the beginning of their problems. But the problems would be different. In the eyes of God, they would at least be clean and decent—an urge that had dominated Just's life going back to his early childhood. According to the laws of man, they would at last be technically above-board. There were people who would no doubt moralize on the pros and cons of racial intermarriage, but that was sure to be less of a problem in France than in America.

In late April Just and Hedwig took a trip to Riga, Latvia, the place to go for the quickest divorce with the least amount of trouble. It is uncertain what problems travel to this Baltic country posed for them. A black man and a part-Jewish woman would have been prime targets for Nazi terrorism. But somehow Just and Hedwig made the trip without too much difficulty. They filed the papers and were told that it would only be a few months before they were processed. They returned to France, spent May in Paris, and headed for Roscoff in early June.

Ethel was not informed of the divorce. Whether it would have been recognized in America is difficult to say—probably not. Besides, Just was not thinking about the United States at all. He had made his move to France. The Riga divorce and marriage satisfied his needs there. He planned to become a French citizen.

Just and Hedwig continued their work at the Station Biologique on the outskirts of the little town of Roscoff in the Finistère district of France overlooking the English Channel. Brittany had a bleak but interesting landscape, and picturesque peasants decked out in high headdresses and speaking a language entirely their own. The living conditions, however, were far below those Just was used to, even at Woods Hole or in Naples. There was no running water; buckets had to be drawn at the wells. Just found the sanitary conditions appalling, and it is a wonder, considering his unusual concern for cleanliness, that he could work there at all.[44] But Roscoff was his home away from home, and for the most part he enjoyed it. Still, he complained bitterly about the unsanitary conditions and wondered how Louis Pasteur, whom he had always admired, could have come from a culture so unclean and prone to the transmission of germs on a large scale.

The Station Biologique was founded in 1868 by Henri Lacaze-Duthiers, a professor of zoology at the Sorbonne.[45] On a trip to the area he had been impressed by the abundance of marine fauna and flora, and so had set up a small marine laboratory. The beginnings were modest, almost primitive, with just one little house for conducting experiments and housing equipment. The laboratory functioned in this way for several years, but in time other buildings were added to house a library and several classrooms, and a boat was purchased. Lacaze-Duthiers held the post of director until 1901, when Yves Delage took over. Under Delage the laboratory became part of the Sorbonne and took on the name Station Biologique de Roscoff. As at Woods Hole, teaching and research were carried on together. Though smaller than the MBL or the Stazione Zoologica, the Station Biologique was well on its way to becoming an international laboratory by the 1910s. In 1921 Charles Pérez assumed the directorship, and was still in that position when Just went to Roscoff in the summer of 1939.

The big advantage for Just was that Roscoff was a peaceful little town isolated from the usual worries of the world. Only a small number of scientists were ever there at a given time. Those that were there he knew well. He had become a celebrity of sorts and was treated with deference. He and Hedwig stayed to themselves for the most part, but from time to time they got together with Geneviève Bobin and the Huttrers, a Jewish couple from Austria. No one knew that Just and Hedwig were not married.[46]

In early August they left for Riga to achieve their goal. The marriage took place on 11 August.[47] They stayed in Riga for a few days, honeymooning at the Hotel Bellevue, before returning to Roscoff on 26 August. At last they were husband and wife; their life together was no longer a lie. They wrote the Schnetzler family, and Hedwig announced the news to Lady Astor.[48] But it would be some time before Just would say anything to Ethel.

Roscoff continued to be a place of refuge for the Justs even when fighting broke out in France toward the end of 1939. They were determined to stay through the fall and winter. They were working on a rather long article on philosophical problems in biology, and they needed time and isolation. In the meantime concern about Just was growing in America. Newspaper reporters were beginning to inquire about his whereabouts, no doubt because they wanted a story on Hitler's treatment of blacks. Ethel knew little. When reporters called her in early September she informed them that her husband was studying and resting in France and Germany and that, according to a letter she had received from him two weeks earlier, he did not consider the war scare serious enough to cause him to leave Europe.[49] Like everyone else, however, she was alarmed, and promised to contact the State Department if there was no further word from Just within a few days.

Just felt the growing political tension, especially when he went on occasion to conduct business in larger neighboring towns such as Quimper. These trips were frustrating. Just had to pack himself into an overcrowded bus. Delays were frequent because no one ever entered the bus with a ticket ready or moved in time to get out. In Quimper the banks were totally disorganized. Just could never get correct information, only errors followed by beautifully expressed apologies. There were lines miles long. Shopkeepers could not count, and the peasants bustled around aimlessly. A passage from *Henry V* constantly ran through Just's mind:

> 'Tis certain, he hath pass'd the river Somme.
> And if he be not fought withal, my lord,
> Let us not live in France; let us quit all,
> And give our vineyards to a barbarous people.[50]

There was an atmosphere of diffidence and confusion. Even the soldiers, lolling around cafés and slouching on street corners, did not seem to know what they should be doing.

Late in October the Station Biologique received orders from the French government to close off its facilities to foreigners. Charles Pérez sent Just a long personal letter of apology.[51] He knew that France was Just's last haven, but there was nothing he could do. The Nazis were taking hold of France, and all foreign scientists, not only Just, would have to find quarters elsewhere to conduct their research. The French had simply surrendered to the German authorities, and in Just's opinion had shown themselves to be cowards and hypocrites.

Just and Hedwig decided to stay in Roscoff anyway. The editors of *Physiological Zoology* had invited him, along with Heilbrunn, Ralph Lillie, and B. H. Willier, to contribute an article to a series of synthesizing papers. Just's topic was "Unsolved Problems of General Biology," and the deadline for completion was the spring of 1940.[52] At last Just had the chance to put others in the field on the track he considered productive; at last he was being afforded the role of a statesman of science—perhaps not yet an elder statesman, but a statesman nonetheless. He did not want to uproot himself from Roscoff in the middle of this important work.

In Just's view, the most pressing need in general biology was to find out "what in any protoplasmic system, cell or otherwise, is the living substance."[53] Biologists had focused on this for some time, but no one had yet put together a general synthesis, a reflective piece to suggest fruitful paths to travel. The larger questions surrounding a discipline have to be articulated periodically to keep the goal in mind, Just began. Questions about "how life begins, how it is continued, and how transmitted" depend on a knowledge of protoplasmic structure, and Just had come to realize that this was an issue for science generally. The truth behind protoplasmic structure could not be found by biologists alone; it had to be attacked by biologists, chemists, and physicists, all those "who appreciate the limitations of the life-state, the range of normal life-processes." This call for collaboration between scientists from different fields was new on Just's part. A decade earlier he had felt pushed out of Woods Hole by chemists, biochemists, and physicists; now he had come around, there is no doubt. But the changing state of the field, his many experiences with a wide range of scientists and philosophers in Europe, and his closeness to Hedwig and her philosophy all contributed to the new role that he saw for biology. Descriptive biology, the kind Just had spent a great deal of his career doing, was still important to him. In a manner of speaking, the descriptive biologist was the one who had to keep the rules of the game straight, "to establish, beyond question, criteria of normality, the range of normal processes, and the extent of . . . normal variability." In the final analysis, chemists and physicists were dependent on descriptive biologists. For Just, this meant that his life's work had been justified. And it was appropriate that he should make a statement to that effect in this, his last full-length published paper.

Even though Just received no salary from Howard in the 1939-40 academic year, he and Hedwig lived well in Roscoff. Hedwig's brother Otto transferred funds to Just's account at Crédit Lyonnais from a bank in Canada.[54] Most of the money, amounting to several thousand francs, was to be transferred eventually to the Chase National Bank in New York for Otto's use later, but Just and Hedwig no doubt used some of it. Otto had long been contemplating a move to America. He had been there on company business in the mid-1930s and liked what he saw. Now the growing political tension in Europe made a permanent move almost essential.

In early June 1940 the Germans began their siege of Paris. Hedwig and Just were still in Roscoff. The Huttrers had left France for America a month or so earlier. No doubt Just and Hedwig would have gone then too, but she had no immigration papers. They had to spend the summer trying to get things settled at the American embassy. On 11 June Just reserved passage for himself and Hedwig on a transatlantic liner scheduled to depart later in the summer, but on 17 July he canceled those tickets because of "the changed situation."[55] The Germans had taken over. In the ensuing confusion it was difficult to get the documents they needed. When Just went to Paris to try and get official papers for himself and Hedwig, there were long lines in the embassy and government offices. French slowness and inefficiency seemed magnified a hundred times. Outside there were military processions, and in the distance was the muffled sound of gunfire. Life was miserable.

Just had finally come to despise the Germans—a people whom he had so revered in the past, a people from whom he himself had probably descended. He had hated Hitler ever since March 1939, the time of the German invasion of Czechoslovakia. So incensed was he over that action that he expressed willingness to give his life to fight against "the agencies of destruction."[56] He offered himself for military service "the very next day." When the French government turned him down, he was deeply hurt. No longer was he the old "dyed in the wool pacifist" of World War I days; he was anxious to make whatever sacrifices were necessary to promote the cause of a just war, the cause of humanity itself. But he was too old and ill and tired to be of much use to any army.

Just was worried about whether or not he could get the necessary papers together and whether or not he and Hedwig would then be allowed to travel. To make matters more complicated, Hedwig was several months pregnant. Travel was difficult. Even the trips to and from Paris and Roscoff were trying; how much worse a long and possibly stormy Atlantic crossing would be. In early August the Nazis interned Just in a camp, probably Chateaulin. When Hedwig's father found out he mobilized his Nazi contacts on the board of the Brown, Boveri Company, and an official was dispatched to France to obtain Just's release.[57]

Americans were gravely worried about him. Lillie had been receiving constant inquiries, and so had Harrison. They got together and decided to contact the American Red Cross and the State Department to see if anything could be done to locate him and send help if he was in trouble.[58]

By this time Just and Hedwig were making their way out of France, down into Spain, and finally to Portugal. In early September they boarded the S.S. *Excambion* in Lisbon.[59] But their problems were not yet over. The captain told them that he would not let Hedwig make the crossing: he had had an unpleasant experience with a childbirth on a previous passage and did not want a repeat episode. After much pleading, Just and Hedwig changed his mind. At last they were on their way to America.

CHAPTER 9

America Again

Let America be America again
Let it be the dream it used to be.
Let it be the pioneer on the plain
Seeking a home where he himself is free.

(America never was America to me.)

Langston Hughes,
"Let America Be America Again"

Life is precious, especially to the biologist who has spent his life studying it in depth. Life is activity, consciousness; it is the will to struggle, to hang on to existence, to fulfill oneself. The urge for life is instinctual, even if the outlook is bleak and death certain.

Just pushed to stay alive and active, using every instinct he could draw on. His escape from the Nazis had been difficult, a flirtation with death, but he had done it. He was on his way back to America. He had no choice but to hope for a life there for himself and Hedwig. But earlier, while in Europe, he had been convinced that a return to America would be the end of him.[1]

His daughter Margaret met them at the dock in New York. A few reporters were there too, to find out about his experiences abroad. He put them off, promising to recount the story later.[2] He arranged for Hedwig, who was due to give birth shortly, to check into the French Hospital in New York.[3] Then he sent word to his sister Inez that he would be back in Washington before long and would like to stay with her.

Despite his poor health, Just went back to his teaching at Howard in the 1940–41 academic year. His salary was $5,000, about half what he had received under the Rosenwald and Carnegie grants. Hedwig lived with her brother Otto and sister Ursula (they too had managed to escape Europe) in a small house in East Orange, New Jersey, an out-of-the-way suburb of New York City.[4] The plan was for her to remain there and look after the baby, a girl, named Elisabeth after Hedwig's mother. Just would commute, living with Inez in Washington during the week and spending weekends with his wife and new daughter in East Orange.

Just did not know what the future held, but he was anxious to get started again on his work. Two years (three seasons) worth of work in Roscoff was down the drain. He had been working on two species of starfish, three of *Nereis*, and two of the sea worm *Phascolosoma* (curiously, his old teacher at Darmouth, John Gerould, had done the original classic study of *Phascolosoma*), but in the confusion, all his notes had been left behind. But he was not discouraged. He wanted to get back to work, this time on *Chaetopterus*, the worm that Lillie had worked on over three decades earlier.[5]

Somehow the bitterness he had felt toward Lillie was no longer there. In fact, he felt closer to Lillie then, on his return to America, than ever before. The struggle, it seems, had only served in the long run to deepen his loyalty. When he forgot to send Lillie a birthday card Just was sincerely chagrined. Again he embraced his old dream of an almost religious tie with Lillie. Though "time and experience" had "mellowed" his ties with Lillie, his image of him as a kind of surrogate father, he was glad to be back in America, near the man who had been an inspiration, a source of constant encouragement for so many years.[6]

Everyone was happy to see Just come home. Lillie and Harrison were particularly pleased; Heilbrunn and Schrader joined in the welcome too. Lillie wanted Just to begin all over again, and was certain that he would somehow find the courage to carry through. Even if Just did not complete his ambitious program, he would at least be able to bring it to a natural terminus. Lillie suspected Just was worn, with no energy and perhaps hardly the will to live. But be that as it may, Just was back in "the freest country in the world."[7] This, Lillie told him, was all that mattered.

Though thin and tired, Just had no intention of giving up. He had had to retreat from a battle, but he did not feel the war was lost. The man possessed amazing resilience, an almost defiant inner strength. Nothing could get between him and his work. Repeating Robert Browning's verse, "Grow old along with me/The best is yet to be," was one way of getting himself to press forward.[8] In mid-December 1940 he prepared three papers for a talk at the American Society of Zoologists' conference to be held in Philadelphia.[9] Even though he had lost most of his notes, he put together from memory some of the results of his work at Roscoff on certain species unavailable in America. His paper on "Egg-laying in *Nereis diversicolor* at Roscoff" was an appropriate way for him to make his reappearance on the American biological scene.

While at the Philadelphia meeting Just had a long talk with Heilbrunn at his house one evening, catching up on news and renewing their friendship. He refused an invitation to stay over, but not because he was still suspicious of Heilbrunn's true feelings. He knew he would not be a good guest; he was quite ill.[10]

It was difficult to keep up appearances in front of Hedwig and his friends. At the start of the second semester in February 1941 he suffered severe, almost incapacitating stomach trouble. At first he tried to ignore the symptoms, but finally he went to a doctor. It was a nervous stomach, the doctor said. Just did not believe this; he thought it was liver trouble.[11] His nerves, for the most part, were all right. He had stayed out of Johnson's way the entire first semester and had only met up with him once since the start of the second semester. Really there were no immediate problems bothering him.

He kept at his work. Long hours in the laboratory became the rule again. He worked hard on what was to be his last manuscript, "Ethics and the Struggle for Existence."[12] At last he was putting together a synthesis of biology and history and sociology—anthropology too—as he had wanted to do as far back as his Dartmouth days. The world was on the brink of a crisis in September 1941: could there have been a better time to think deeply about ethics and man's behavior? Unfortunately, Just's thoughts were fragmentary. He was too ill to focus clearly on the subject before him.

No matter. He had already reached his main goals. He had put out his book, *The Biology of the Cell Surface*; written an article offering speculations on the whole field of biology; married Hedwig and begun a family with her—all of this from a position of disadvantage. At fifty-seven, what man, scientist or no, could do more? He still had much he hoped to do, but deep down he did not know how much time he had to do it. He started tidying up his affairs. One outstanding obligation was a loan of $372 he had received from Dartmouth three decades earlier. When Dartmouth wrote him and asked for repayment, he explained that after living for two years in Europe he had no money left.[13] But he wanted to repay the debt. It went beyond money. In the twilight of his life he thought wistfully about the traditions he had been exposed to and the world view he had acquired in Hanover. Patten and Gerould had been formidable examples to him, as a science student and later as a scientist. He recalled with gratitude "the inspiration, kind sympathy and intellectual aid" he had received so freely at Dartmouth.[14] Incidentally, Gerould was one of the few to whom Just sent a copy of his two books, *The Biology of the Cell Surface* and *Basic Methods*. And it was not only Patten and Gerould: the members of the history and sociology departments at Dartmouth had helped him shape a broad view of the world so that in later life he was able not only to appreciate the social sciences but also to bring them closer to biology. There was no way for him to go back to those good old Dartmouth days, when he and E. E. Day—who had recently been appointed president of Cornell University—published poetry side by side in the college literary magazine.[15] The best he could do was try and continue the tradition, and do what he could to make it possible for other blacks to go through the same experience. Though penniless, he made out a check for $100 payable to Dartmouth College. Not long afterwards, he sent a smaller check, $25, to Kimball Union Academy.[16]

As the spring approached, Just was busy making arrangements for his summer work. He wanted a country house by the seaside, preferably near a small marine laboratory where he could have a table for his microscope. Places in Maine and New Hampshire came to mind—places right along the coast. He sent out inquiries to a marine station in New Hampshire, but there was no accommodation to be had.[17] He went through the *New York Times* every day looking at advertisements for cottage rentals in Maine. He sent out scores of letters, but was refused time and time again; no doubt the owners knew that Howard was a black school.[18] Finally he found a place in Portland, Maine. He, Hedwig, and little Elisabeth would spend the summer there.

They stayed only one month, June. In early July Just became so ill that he had to return to New York, where he was admitted to Mount Sinai Hospital. Clearly his intestinal trouble was not just psychological, for his doctor found a real organic problem—one that may, however, have been aggravated by stress. The structure of his spinal nerves had been altered on the right side, resulting in an abnormal neuralgia. The therapy was to block off the nerves and follow this up with electric treatment.[19]

After several operations Just was no better. The pain only increased, and he began to waste away. In early August his old friends started dropping by to see him, some on their way home from Woods Hole. Heilbrunn and Schrader came. Not only did they express their concern over his health, but they also initiated a campaign to find him a position and some money, joining forces with Lillie, Harrison, and Abram Harris, a Howard economist. They knew he was penniless.[20]

Letters of concern came from all over. Especially touching to Just was one from Lady Astor, who urged him to try Christian Science.[21] But Just did not feel so moved. His trouble with spinal nerves, it turned out, was merely a symptom of a more serious problem—cancer of the pancreas.

He was in constant pain. When he left the hospital to go home to East Orange with Hedwig some time in early September, he weighed less than one hundred and thirty pounds. Heilbrunn went to visit again.[22] In spirit Just was the same, determined to fight and get on with his science. He liked Heilbrunn, felt as though he always had. He began to wonder what Heilbrunn really thought of him, indeed how others, and the scientific community at large, viewed his warped life. Would he be remembered: If so, by whom, and how?

Just knew Lillie to be eloquent, but never passionate. He could never have guessed that his old mentor, in many ways an introvert, would make a public outcry on his behalf. In *Science*, the official journal of the American Association for the Advancement of Science, Lillie would call Just's scientific career "a constant struggle for opportunity for research, the breath of life."[23] He would tell the world of Just's science (a reversal of roles to be sure), of his notion of the importance of the ectoplasm in development, of his single theory of heredity, and of his interpretation of evolution. At last he would publicly confess that Just had been "condemned by race to remain attached to a Negro institution unfitted by means and tradition to give full opportunity to ambitions such as his." As Just lay in bed, sick, choked with pain, fading away to nothing, he might have found some consolation in Lillie's characterization:

> An element of tragedy ran through all Just's scientific career due to the limitations imposed by being a Negro in America, to which he could make no lasting psychological adjustment in spite of earnest efforts on his part. The numerous grants for research did not compensate for failure to receive an appointment in one of the large universities or research institutes. He felt this as a social stigma, and hence unjust to a scientist of his recognized standing. In Europe he was received with universal kindness, and made to feel at home in every way; he did not experience social discrimination on account of his race, and this contributed greatly to his happiness there. Hence, in part at least, his prolonged self-imposed exile on many occasions. That a man of his ability, scientific devotion, and of such strong personal loyalties as he gave and received, should have been warped in the land of his birth must remain a matter for regret.

Just had no way of knowing that Lillie cared so much, though he had always hoped he did.

Selig Hecht had not crossed Just's path for nearly two decades, not since he had defended Loeb against Just. A professor at Columbia since 1929, Hecht was at the peak of his career in the early 1940s. Years and success had changed him. He had come to a more compassionate understanding of Just's plight. He remembered Just as "an old friend," and counted himself among those who "loved his gentle nature and appreciated his scientific stature," who "saw with increasing pain the racial tragedy that slowly warped Just's life and which certainly shortened it in the end."[24] Hecht too had come to feel that racial antagonism in science, evident in small and large ways, was as malignant as cancer.

The future was difficult to see and the past was a mere glimmer. Just was living more and more in the present, each day separate, taken one by one. Time was closing in, from ahead and behind. He could hardly remember that Ethel had left Washington, taken up a teaching job at Virginia Union College in Richmond, Virginia, and started a new life for herself and Maribel; that Margaret was married and living in Maryland; that Highwarden had left Amherst College, in part due to the high cost, and was now studying at Howard.

For Just to get his paycheck, he had to go in to the university to work. Now, especially now, he feared the thought of a classroom or laboratory at Howard. Though perhaps he needed to get away, to go somewhere, he wanted to stay with Hedwig in New Jersey. However unbearable the pain of having her, young and fresh, see him in this dreadful condition, a failing and broken man, he wanted to stay. But a salary was crucial for a man with a wife and baby, so Just left East Orange to live with his sister Inez at 1853 Third Street in Washington, within a few blocks of the Howard campus. When the fall semester began he stayed alone in Inez's apartment; almost bedridden, he was too sick to move. Directives came from Johnson and other university officials ordering him to return to the classroom. He could not muster the strength even to answer; he simply did not have the will to fight this last battle. Inez, herself a nurse, attended him into the autumn, and his son Highwarden often stopped by after classes to look in. Nothing helped. His was a lost war: he died on Monday morning, 27 October. The Howard trustees, in session at the time, stood for a moment of silence.

PUBLICATIONS BY E. E. JUST

"On Government Monopolies." *Kimball Union* 11 (1903): 87–90.

"The Crucifixion." *Dartmouth Magazine* 21 (1907): 248.

"On Going Home." *Dartmouth Magazine* 21 (1907): 288–92.

"When the Light Went Out." *Dartmouth Magazine* 21 (1907): 363–65.

"The Relation of the First Cleavage Plane to the Entrance Point of the Sperm." *Biological Bulletin* 22 (1912): 239–52.

"Breeding Habits of the Heteronereis Form of *Nereis limbata* at Woods Hole, Mass." With F. R. Lillie. *Biological Bulletin* 24 (1913): 147–69.

"Breeding Habits of the Heteronereis Form of *Platynereis megalops* at Woods Hole, Mass." *Biological Bulletin* 27 (1914): 201–12.

"Initiation of Development in *Nereis*." *Biological Bulletin* 28 (1915): 1–17.

"An Experimental Analysis of Fertilization in *Platynereis magalops*." *Biological Bulletin* 28 (1915): 93–114.

"The Morphology of the Normal Fertilization in *Platynereis megalops*." *Journal of Morphology* 26 (1915): 217–33.

"The Fertilization-Reaction in *Echinarachnius parma*. I. Cortical Response of the Egg to Insemination." *Biological Bulletin* 36 (1919): 1–10.

"The Fertilization-Reaction in *Echinarachnius parma*. II. The Role of Fertilizin in Straight and Cross Fertilization." *Biological Bulletin* 36 (1919): 11–38.

"The Fertilization-Reaction in *Echinarachnius parma*. III. The Nature of the Activation of the Egg by Butyric Acid." *Biological Bulletin* 36 (1919): 39–53.

"The Challenge." *Oracle*, spring 1919. Quoted in Herman Dreer, *The History of the Omega Psi Phi Fraternity, 1911–1939*. N.p.: Omega Psi Phi Fraternity, 1940, pp. 93–94.

"The Fertilization-Reaction in *Echinarachnius parma*. IV. A Further Analysis of the Nature of Butyric Acid Activation." *Biological Bulletin* 39 (1920): 280–305.

"The Susceptibility of the Inseminated Egg to Hypotonic Sea-Water: A Contribution to the Analysis of the Fertilization-Reaction." *Anatomical Record* 20 (1921): 225–27.

"Initiation of Development in the Egg of *Arbacia*. I. Effect of Hypertonic Sea-Water in Producing Membrane Separation, Cleavage, and Top-swimming Plutei." *Biological Bulletin* 43 (1922): 384–400.

"Initiation of Development in the Egg of *Arbacia*. II. Fertilization of Eggs in Various Stages of Artificially Induced Mitosis." *Biological Bulletin* 43 (1922): 401–10.

"Initiation of Development in the Egg of *Arbacia*. III. The Effect of *Arbacia* Blood on the Fertilization-Reaction." *Biological Bulletin* 43 (1922): 411–22.

"The Effect of Sperm Boiled in Oxalated Sea-Water in Initiating Development." *Science* n.s. 56 (1922): 202-4.

"Studies of Cell Division. I. The Effect of Dilute Sea-Water on the Fertilized Egg of *Echinarachnius parma* During the Cleavage Cycle." *American Journal of Physiology* 61 (1922): 505-15.

"The Fertilization-Reaction in *Echinarachnius parma*. V. The Existence in the Inseminated Egg of a Period of Special Susceptibility to Hypotonic Sea-Water." *American Journal of Physiology* 61 (1922): 516-27.

"On Rearing Sexually Mature *Platynereis megalops* from Fertilized Eggs in the Laboratory." *American Naturalist* 56 (1922): 471-78.

"The Fertilization-Reaction in *Echinarachnius parma*. VI. The Necessity of the Egg Cortex for Fertilization." *Biological Bulletin* 44 (1923): 1-9.

"The Fertilization-Reaction in *Echinarachnius parma*. VII. The Inhibitory Action of Blood." *Biological Bulletin* 44 (1923): 10-16.

"The Fertilization-Reaction in *Echinarachnius parma*. VIII. Fertilization in Dilute Sea-Water." *Biological Bulletin* 44 (1923): 17-21.

"Fertilization." with F. R. Lillie. In *General Cytology*, edited by E. V. Cowdry. Chicago: University of Chicago Press, 1924, pp. 449-536.

"Science and Human Needs." *Howard Medical News* 1 (1925): 1.

"Dr. Patten Lectures on his Spitzbergen Trip." *Collecting Net* 1(1926), no. 3:4.

"Experimental Production of Polyploidy in the Eggs of *Nereis limbata* by Means of Ultra-violet Radiation." *Anatomical Record* 34 (1926): 108.

"Vacuole Formation in Living Cells." *Anatomical Record* 34 (1926): 109.

Review of E. G. Conklin, "Localization Phenomena in Development." *Collecting Net* 2 (1927), no. 3: 8-9.

"Effect of Ultra-violet Rays in Altering the Polarity of *Nereis* Eggs." *Anatomical Record* 37 (1927): 130.

"Cortical Effect of Ultra-violet Radiation on *Nereis* Eggs." *Anatomical Record* 37 (1927): 130-31.

"Decolorization Wave of Pigment Granules in the Jelly Hull of the Inseminated Egg of *Echinarachnius parma* Exposed to Dilute Sea-Water During the Process of Membrane Separation." *Anatomical Record* 37 (1927): 147-48.

"Bodies in the Egg of *Arbacia* Described by E. B. Wilson as Golgi." *Anatomical Record* 37 (1927): 158.

"Fertilization of *Arbacia* Eggs in Solutions of KCN in Sea-Water." *Anatomical Record* 37 (1927): 160-61.

"Mitochondria and Golgi Bodies in Mayonnaise." With F. V. McNorton. *Anatomical Record* 37 (1927): 161.

"History of the Middle Piece of the Spermatozoon in the Fertilized Egg of *Echinarachnius parma*." *Anatomical Record* 37 (1927): 162.

"Methods for Experimental Embryology with Special Reference to Marine Invertebrates." *Collecting Net* 3 (1928), nos. 1-6.

"Initiation of Development in *Arbacia*. V. The Effect of Slowly Evaporating Sea-Water and Its Significance for the Theory of Auto-Parthenogenesis." *Biological Bulletin* 55 (1928): 358-68.

"Fertilization of Marine Ova in Dilute Sea-Water and its Significance for Some Problems in Cell Physiology." *Anatomical Record* 41 (1928): 29.

"Initiation of Development in *Arbacia*. IV. Some Cortical Reactions as Criteria for Optimum Fertilization Capacity and Their Significance for the Physiology of Development." *Protoplasma* 5 (1928): 97-126.

"Studies of Cell Division. II. The Period of Maximum Susceptibility to Dilute Sea-Water in the Cleavage Cycle of the Fertilized Egg of *Arbacia.*" *Physiological Zoology* 1 (1928): 26-36.

"Hydration and Dehydration in the Living Cell. I. The Effect of Extreme Hypotony on the Egg of *Nereis.*" *Physiological Zoology* 1 (1928): 122-35.

"Cortical Reactions and Attendant Physico-Chemical Changes in Ova Following Insemination." In *Colloid Chemistry*, edited by Jerome Alexander. New York: Chemical Catalog, 1928, 2: 567-74.

"Breeding Habits of *Nereis dumerilii* at Naples." *Biological Bulletin* 57 (1929): 307-10.

"The Production of Filaments by Echinoderm Ova as a Response to Insemination, with Special Reference to the Phenomenon as Exhibited by Ova of the Genus *Asterias.*" *Biological Bulletin* 57 (1929): 311-25.

"The Fertilization-Reaction in Eggs of *Paracentrotus* and *Echinus.*" *Biological Bulletin* 57 (1929): 326-31.

"Initiation of Development in *Arbacia*. VI. The Effect of Sea-Water Precipitates with Special Reference to the Nature of Lipolysin." *Biological Bulletin* 57 (1929): 422-38.

"Effects of Low Temperature on Fertilization and Development in the Egg of *Platynereis Megalops.*" *Biological Bulletin* 57 (1929): 439-42.

"Hydration and Dehydration in the Living Cell. II. Fertilization of Eggs of *Arbacia* in Dilute Sea-Water." *Biological Bulletin* 57 (1929): 443-48.

"Hydration and Dehydration in the Living Cell. III. The Fertilization Capacity of *Nereis* Eggs after Exposure to Hypotonic Sea-Water." *Protoplasma* 10 (1930): 24-32.

"Hydration and Dehydration in the Living Cell. IV. Fertilization and Development of *Nereis Eggs* in Dilute Sea-Water." *Protoplasma* 10 (1930): 33-40.

"The Present Status of the Fertilizin Theory of Fertilization." *Protoplasma* 10 (1930): 300-42.

"The Effect of Ultra-violet in Producing Fusion of Eggs of *Chaetopterus.*" *Science* n.s. 71 (1930): 72.

"A Simple Method for Experimental Parthenogenesis." *Science* n.s. 71 (1930): 72.

"The Amount of Osmic Acid in Fixing Solutions Necessary to Blacken Fat." *Science* n.s. 71 (1930): 72-73.

"The Fertilization Membrane of Echinid Ova." *Science* n.s. 71 (1930): 243.

"Die Rolle des kortikalen Cytoplasmas bei vitalen Erscheinungen." *Naturwissenschaften* 19 (1931): 953-62, 980-84, 998-1000.

"On the Origin of Mutations." *American Naturalist* 66 (1932): 61-74.

"Cortical Cytoplasm and Evolution." *American Naturalist* 67 (1933): 20-29.

"A Cytological Study of Effects of Ultra-violet Light on the Egg of *Nereis limbata.*" *Zeitschrift für Zellforschung und mikroskopische Anatomie* 17 (1933): 25-50.

"Observations on Effects of Ultra-violet Rays upon Living Eggs of *Nereis limbata* Exposed before Insemination." *Wilhelm Roux' Archiv für Entwicklungsmechanik der Organismen* 130 (1933): 495-516.

"Fertilization in *Membranipora pilosa.*" *Carnegie Institution of Washington Yearbook* 33 (1934): 268-70.

"On the Rearing of *Ciona intestinalis* under Laboratory Conditions to Sexual Maturity." *Carnegie Institution of Washington Yearbook* 33 (1934): 270.

"Effects of Ultra-violet Rays on Animal Eggs." *Archivio internazionale di radiobiologia generale* 2 (1935): 650–52.

"Nuclear Increase During Development as a Factor in Differentiation and in Heredity." *Anatomical Record* 64 (1936): 32.

"Alcoholic Solutions as Gastric Secretogogues." *Anatomical Record* 64 (1936): 69.

"A Single Theory for the Physiology of Development and Genetics." *American Naturalist* 70 (1936): 267–312.

"Protoplasmic Specificity." *Science* n.s. 84 (1936): 351–52.

"Ultra-violet Radiations as Experimental Means for the Investigation of Protoplasmic Behavior." *Anatomical Record* 67 (1937): 39.

"The Significance of Experimental Parthenogenesis for the Cell-Biology of Today." *Cytologia*, Fujii Jubilee Volume (1937): 540–50.

"Phenomena of Embryogenesis and Their Significance for a Theory of Development and Heredity." *American Naturalist* 71 (1937): 97–112.

The Biology of the Cell Surface. Philadelphia: Blakiston's, 1939.

Basic Methods for Experiments on Eggs of Marine Animals. Philadelphia: Blakiston's, 1939.

"Unsolved Problems of General Biology." *Physiological Zoology* 13 (1940): 123–42.

"On Abnormal Swimming Forms Induced by Treatment of Eggs of *Strongylocentrotus* with Lithium Salts and Other Means." *Anatomical Record* 78 (1940): 74.

"Egg-laying in *Nereis diversicolor* at Roscoff." *Anatomical Record* 78 (1940): 131.

"Fertilization-Reaction in Eggs of *Asterias rubens*." *Anatomical Record* 78 (1940): 132.

GLOSSARY OF MANUSCRIPT CITATIONS

The manuscript and records collections listed below are cited in the notes by the code letters on the left.

AHC	Austin H. Clark Papers, Archives, Smithsonian Institution, Washington, D.C.
APS	Anson Phelps Stokes Papers, Archives, Yale University, New Haven, Conn.
CC	Correspondence File of the Carnegie Corporation of New York, Carnegie Corporation, New York
CCC	Register of Mesne Conveyance, Charleston County Courthouse, Charleston, S.C.
CCHD	Charleston County Health Department, Charleston, S.C.
CCL(B)	Birth Records, Charleston County Library, Charleston, S.C.
CCL(D)	Death Records, Charleston County Library, Charleston, S.C.
CC(W)	Papers of the Carnegie Corporation of New York, Carnegie Institution, Washington, D.C.
CIW	Papers of the Carnegie Institution of Washington, Carnegie Institution, Washington, D.C.
CIW(B)	Correspondence File of the Department of Embryology, Carnegie Institution of Washington, Baltimore, Md.
CPC	Charleston Probate Court, Charleston, S.C.
CSH	Archives, Cold Spring Harbor Laboratory, Cold Spring Harbor, N.Y.
D	Dohrn Papers, Stazione Zoologica Archives, Naples
DPC	Donald P. Costello Papers, letters and manuscripts in the possession of Helen M. Costello
EEJ	Ernest Everett Just Papers, letters and manuscripts in the possession of the author
EEJ(B)	Ernest Everett Just Papers, letters and manuscripts in the possession of Margaret Just Butcher
EEJ(H)	Ernest Everett Just Papers, Moorland-Spingarn Research Center, Howard University, Washington, D.C.

EEJ(M)	Ernest Everett Just Papers, letters and manuscripts in the possession of Elisabeth Just Adèr
EGC	Edwin Grant Conklin Papers, Firestone Library, Princeton University, Princeton, N.J.
ERE	Edwin Rogers Embree Papers, Archives, Yale University, New Haven, Conn.
FRL(C)	Frank R. Lillie Papers, Archives, University of Chicago, Chicago, Ill.
FRL(W)	Frank R. Lillie Papers, Marine Biological Laboratory, Woods Hole, Mass.
GEB	Papers of the General Education Board, Rockefeller Archive Center, Pocantico Hills, N.Y.
JHG	John H. Gerould Papers, Archives, Dartmouth College, Hanover, N.H.
JL(BAS)	Jacques Loeb Papers—Book, Article, and Speech File, Manuscript Division, Library of Congress, Washington, D.C.
JL(GC)	Jacques Loeb Papers—General Correspondence, Manuscript Division, Library of Congress, Washington, D.C.
JMC	J. McKeen Cattell Papers, Manuscript Division, Library of Congress, Washington, D.C.
JR	Julius Rosenwald Papers, Archives, University of Chicago, Chicago, Ill.
JRF	E. E. Just File, Papers of the Julius Rosenwald Fund, Fisk University, Nashville, Tennessee (on microfilm from the Amistad Research Center, Dillard University, New Orleans, La.)
KWI	Correspondence Files, Bibliothek und Archiv zur Geschichte der Max-Planck-Gesellschaft, Berlin-Dahlem
LH	Leigh Hoadley Papers, Archives, Harvard University, Cambridge, Mass.
LVH	Lewis V. Heilbrunn Papers, letters and manuscripts in the possession of Constance Tolkan
MB	Margret Boveri Papers, Staatsbibliothek Preussischer Kulturbesitz, Berlin
MJH	Melville J. Herskovits Papers, Melville J. Herskovits Library of African Studies, Northwestern University, Evanston, Ill.
NA	Nancy Astor Papers, Archives, University of Reading, Reading, England
NAACP(AC)	National Association for the Advancement of Colored People—Records of the Annual Conference, Manuscript Division, Library of Congress, Washington, D.C.
NAACP(AF)	National Association for the Advancement of Colored People—Administrative Files, Manuscript Division, Library of Congress, Washington, D.C.

RF	Rosenwald Fellowship Files, National Research Council, National Academy of Sciences Archives, Washington, D.C.
RGH	Ross G. Harrison Papers, Archives, Yale University, New Haven, Conn.
RU(JL)	Rockefeller University—Jacques Loeb Correspondence, 1916–24, Rockefeller Archive Center, Pocantico Hills, N.Y.
SH	Selig Hecht Papers, letters and manuscripts in the possession of Maressa Hecht Orzack
SOM	Samuel O. Mast Papers, Ferdinand Hamburger Jr. Archives, Johns Hopkins University, Baltimore, Md.
SZ	Records of the Stazione Zoologica, Stazione Zoologica Archives, Naples
WJC	William J. Crozier Papers, Archives, Harvard University, Cambridge, Mass.
WP	William Patten Papers, Archives, Dartmouth College, Hanover, N.H.
WHW	William Henry Welch Papers, Alan Mason Chesney Medical Archives, Johns Hopkins University, Baltimore, Md.

Notes

1. THE EARLY YEARS, 1883–1907

1. For details of the celebration, see *Charleston News and Courier*, 12 Aug. 1883, pp. 1–2; 13 Aug. 1883, p. 1; 14 Aug. 1883, p. 1. See also *The Centennial of Incorporation [1783–]1883*, with a preface by W. A. C. [William A. Courtenay] (Charleston: News and Courier Book Presses, 1884).

2. *Charleston News and Courier*, 12 Aug. 1883, p. 2.

3. Ibid., 14 Aug. 1883, p. 1.

4. Ibid.

5. Birth certificate, CCHD. Notification was made on 14 Sept. 1883, mistakenly listing the date of birth as 17 Aug. 1883.

6. The earliest comprehensive discussion of the life of blacks at this time is Alrutheus A. Taylor, *The Negro in South Carolina during Reconstruction* (Washington: Association for the Study of Negro Life and History, 1924). The best recent discussions are Joel Williamson, *After Slavery: The Negro in South Carolina during Reconstruction, 1861–1877* (Chapel Hill: Univ. of North Carolina Press, 1965); George P. Tindall, *South Carolina Negroes, 1877–1900* (Columbia: Univ. of South Carolina Press, 1952); and Asa H. Gordon, *Sketches of Negro Life and History in South Carolina*, 2d ed. (Columbia: Univ. of South Carolina Press, 1971). In a book covering a larger time sequence, I. A. Newby makes some good points: *Black Carolinians: A History of Blacks in South Carolina from 1895 to 1968* (Columbia: Univ. of South Carolina Press, 1973). Many of the general histories of South Carolina examine the issue: for example, Ernest McPherson Lander, *A History of South Carolina, 1865–1960* (Chapel Hill: Univ. of North Carolina Press, 1960), pp. 10–15. Also useful is a biography by Peggy Lamson, *The Glorious Failure: Black Congressman Robert Brown Elliott and the Reconstruction in South Carolina* (New York: W. W. Norton, 1973). Reconstruction comes in for detailed attention in Ernest M. Lander, Jr., and Robert K. Ackerman, eds., *Perspectives in South Carolina History: The First 300 Years* (Columbia: Univ. of South Carolina Press, 1973), and I. A. Newby, ed., *The Civil War and Reconstruction, 1850–1877* (New York: Appleton-Century-Crofts, 1971), two collections of essays and commentaries on South Carolina history.

The most complete reference work on sources in the history of South Carolina is J. H. Easterby, *Guide to the Study and Reading of South Carolina History: A General Classified Bibliography*, 2d ed., with a supplement by Noel Polk (Spartanburg, S.C.: Reprint Co., 1975). See also Lewis P. Jones, *Books and*

Articles on South Carolina History: A List for Laymen (Columbia: Univ. of South Carolina Press, 1970).

7. Marriage certificate, CPC.

8. CCL(B), 15 March 1879, vol. 40, no. 309.

9. CCL(B), 10 June 1879, vol. 40, no. 578.

10. U.S. Dept. of Commerce, Bureau of the Census, *Census of the United States, 1880: Population.* In general, my evidence for where each of the Justs lived is taken from either the census records or various city directories: for example, *Charleston Directory* (Charleston: Walker, Evans); *Charleston City Directory* (Charleston: Jno. Orrin Lea); *Sholes' Directory of the City of Charleston* (Charleston: A. E. Sholes); *Charleston City Directory* (Charleston: Lucas, Richardson). The Charleston County Library has a good collection of old directories, as does the New England Deposit Library. Boston.

11. The *Charleston News and Courier* reported daily the names of liquor offenders and the sentences imposed on them. For a survey of the problems caused by alcohol in the community, see *Charleston News and Courier*, 21 Aug. 1883, p. 4.

12. CCL(D) 17 Jan. 1880, vol. 49, no. 120; CCL(B), 5 Dec. 1881, vol. 40, no. 1330.

13. Marion R. Hemperley, "Federal Naturalization Oaths, 1790–1860," *South Carolina Historical and Genealogical Magazine* 66 (1965): 186.

14. Henrietta P. Jervey, transc., "The Private Register of the Rev. Paul Trapier," ibid. 58 (1957): 105.

15. James B. Browning, "The Beginnings of Insurance Enterprise among Negroes," *Journal of Negro History* 22 (1937): 427–28. See also E. Horace Fitchett's articles on the activities of free blacks in South Carolina: "The Tradition of the Free Negro in Charleston, South Carolina," ibid. 25 (1940): 139–52; "The Origin and Growth of the Free Negro Population of Charleston, South Carolina," ibid. 25 (1941): 421–37; and "The Status of the Free Negro in Charleston, South Carolina, and His Descendants in Modern Society," ibid. 32 (1947): 340–51. For an interesting discussion of the relationship between slaves and free blacks, see John Hope Franklin, *From Slavery to Freedom: A History of Negro Americans* (New York: Knopf, 1967), p. 224. See also Marina Wikramanayake, *A World in Shadow: The Free Black in Antebellum South Carolina* (Columbia: Univ. of South Carolina Press, 1973).

16. Charleston Co. Deeds, S-12, pp. 283–84. These records are preserved on microfilm in the South Carolina Department of Archives and History, Columbia.

17. *Records of Wills*, vol 45, p. 136, CPC.

18. Interview with T. C. Byerly, 19 March 1978.

19. Secretary of State Miscellaneous Records, vol 6G, p. 82, South Carolina Department of Archives and History, Columbia.

20. See Wikramanayake, *World in Shadow*, pp. 36–37, 43–44.

21. *Record of Wills*, vol. 45, pp. 299–300, CPC.

22. Book A-18, p. 156, CCC; also, *Record of Wills*, vol. 45, pp. 299–300, CPC.

23. See Will No. 319-0024, CPC; also Book L-15, pp. 761 and 779, CCC.

24. Tindall, *South Carolina Negroes*, p. 55.

25. Will No. 319-0024, CPC.

26. *Charleston News and Courier*, 18 Jan. 1884, p. 4.

27. See *Charleston News and Courier*, 19 Aug. 1883, p. 4; 31 Aug. 1883, p. 3; 22 May 1884, p. 3.

28. CCL(D), vol. 48, no. 1941; vol. 49, no. 120.

29. *Charleston News and Courier*, 28 Dec. 1883, p. 4; 31 January 1884, p. 4.

30. Robert Molloy, *Charleston: A Gracious Heritage* (New York: Appleton-Century, 1947), p. 219; also *Charleston News and Courier*, 12 Dec. 1883, p. 4.

31. Molloy, *Charleston*, pp. 218–19.

32. A weekly record of deaths in the city was published in *Charleston News and Courier*, e.g., 6 Oct. 1884, p. 8.

33. CCL(D), vol. 49, no. 399.

34. See "Letters of Administration," Will No. 319-0024, CPC. These documents contain details on many of Charles's financial transactions.

35. CCL(D), vol. 52, no. 414.

36. See "Letters of Administration," Will No. 319-0024, CPC.

37. *Charleston News and Courier*, 3 Sept. 1886, p. 1; also, 4 Sept. 1886, pp. 1–3; 5 Sept. 1886, pp. 1–2; 7 Sept. 1886; pp. 1, 8; 8 Sept. 1886, pp. 2, 4; 9 Sept. 1886, p. 1; 10 Sept. 1886, pp. 1, 8. See also "A Descriptive Narrative of the Earthquake of August 31, 1886," *Charleston Year Book*, 1886, pp. 343–441.

38. *Charleston News and Courier*, 4 Sept. 1886, p. 2.

39. Ibid., 30 May 1887, p. 8.

40. Ibid., 21 June 1887, p. 8; 27 July 1887, p. 8.

41. Just to Julius Rosenwald, 24 May 1926, JR, box 19, folder 4; Just to Raymond B. Fosdick, 22 July 1936, GEB, box 695, folder 6969. See also Mary White Ovington, "Ernest Everett Just," in *Portraits in Color* (New York: Viking, 1927), p. 167.

42. Tindall, *South Carolina Negroes*, p. 127. For further information on the phosphate industry, see *Charleston News and Courier*, 16 Feb. 1883, p. 2; 2 April 1883, p. 1; 3 April 1883, p. 2. Also useful is A. R. Guerard, *A Sketch of the History, Origin, and Development of the South Carolina Phosphates* (Charleston: Walker, Evans, and Cogswell, 1884).

43. See *Charleston News and Courier*, 4 May 1888, p. 8; Henry A. M. Smith, "Old Charlestown and Its Vicinity, Accabee and Wappoo, Where Indigo Was First Cultivated with Some Adjoining Places in Old St. Andrew's Parish," *South Carolina Historical and Genealogical Magazine* 16 (1915): 49–67; Burnet R. Maybank et al., *South Carolina: A Guide to the Palmetto State* (New York: Oxford Univ. Press, 1941), p. 285.

44. See Just to Julius Rosenwald, 10 March 1927, JR, box 19, folder 4.

45. See *Charleston News and Courier*, 1 May 1888, p. 8.

46. See "A Negro Biologist Has Won Distinction in His Field," *New York Times*, 3 Feb. 1929, sec. 9, p. 2; reprinted in *Collecting Net* 6 (1930): 147–48; summarized by Rosenwald administrators, JRF, folder 1. For further information on life among the islanders, see Mason Crum, *Gullah: Negro Life in the Carolina Sea Islands* (Durham, N.C.: Duke Univ. Press, 1940); Albert H. Stoddard, "Origin, dialect, beliefs, and characteristics of the Negroes of the South Carolina and Georgia coasts," *Georgia Historical Quarterly* 28 (1944): 186–95; and J. K. Blackman, *The Sea Islands of South Carolina. Their Peaceful and Prosperous Condition. A Revolution in the System of Planting. The Condition and Progress of the Whites and Freedmen* (Charleston: News and Courier Book Presses, 1880). The *Charleston News and Courier* often addressed the issue of education on the islands, e.g., 27 May 1886, p. 8.

47. *Charleston News and Courier*, 4 May 1888, p. 8.

48. Just to Hedwig Schnetzler, 23 Feb. 1933, EEJ(M).

49. Ibid., and 13 Feb. 1937.

50. E. E. Just, "On Going Home," 1907, p. 289.

51. Just to Hedwig Schnetzler, 23 Feb. 1933, EEJ(M).

52. Just to Julius Rosenwald, 10 March 1927, JR, box 19, folder 4; also Just to Anson Phelps Stokes, 7 April 1936, EEJ(H), box 125-8, folder 137.

53. Just to Julius Rosenwald, 10 March 1927, JR, box 19, folder 4.

54. Just to Hedwig Schnetzler, 5 Jan. 1934, EEJ(M).

55. Just to Schnetzler, 19 March 1936, ibid.

56. Just to Oswald Garrison Villard, 10 March 1916, GEB, box 702, folder 7020.

57. Just to Julius Rosenwald, 10 March 1927, JR, box 19, folder 4.

58. Just to Rosenwald, 24 May 1926 and 10 March 1927, ibid.

59. Lewis K. McMillan, *Negro Higher Education in the State of South Carolina* (n.p., 1952), pp. 4–5; also, *Charleston News and Courier*, 8 Oct. 1884, p. 8; 23 Jan. 1884, p. 1. Cf. Fred L. Brownlee, *New Day Ascending* (Boston: Pilgrim Press, 1946), pp. 134–38.

60. For further historical information, see McMillan, *Negro Higher Education*, pp. 167–200; also the analysis of the work done at Claflin University in *Charleston News and Courier*, 30 May 1887, p. 4.

61. See *Plessy v. Ferguson*, 163 U.S. 537 *United States Reports: Cases Adjudged in the Supreme Court* (New York: William Morrow, 1970), pp. 339–42.

62. Details on the curriculum were provided by M. Maceo Nance, Jr., president of State College, from catalogs in his office. Unfortunately, files on individual students for the period in question are no longer available; a major fire at the institution destroyed all academic records shortly after the turn of the century (Nance to the author, 19 Oct. 1976).

63. *Charleston News and Courier*, 4 Jan. 1899, p. 6.

64. Ibid., 3 May 1899, p. 6.

65. See Ovington, "Just," p. 161; also *Charleston News and Courier*, 5 Feb. 1899, p. 8.

66. Just to Oswald Garrison Villard, 10 March 1916, GEB, box 702, folder 7020. The schoolhouse incident was related to me by Mary Just, 2 April 1979; see also Ovington, "Just," pp. 167–68.

67. Ovington, "Just," p. 161.

68. Just, "On Going Home," p. 290.

69. *Catalog of Kimball Union Academy*, 1901, p. 6. Unless otherwise specified, details on Kimball history, curriculum, faculty, etc., have been taken from these catalogs (1900–1903); also, letters to the author from Mrs. Charles Hill, 19 Nov. 1976, and Charles Laughton, 18 Dec. 1976—both Kimball class of 1903.

70. See Just to Converse Chellis, 7 June 1941, EEJ(H), box 125-2, folder 47.

71. Adolf von Harnack, *What Is Christianity? Sixteen Lectures Delivered in the University of Berlin During the Winter Term, 1899–1900*, trans. Thomas Bailey (New York: G. P. Putnam's Sons, 1901). First published Feb. 1901, reprint ed. Sept. 1901.

72. See Just to Hedwig Schnetzler, 23 Feb. 1933, EEJ(M). Also Just to Edwin R. Embree, 28 Jan. and 16 July 1930, JRF, folder 3; Just to W. C. Curtis, 25 March 1931, RF. On the life and views of Harnack, see Agnes von Zahn-Harnack, *Adolf von Harnack* (Berlin: W. de Gruyter, 1951).

73. Interview with M. Wharton Young, 17 Jan. 1977. Also Philip St. Laurent, "The Negro in World History: Dr. Ernest E. Just," *Tuesday Magazine* (monthly

Supp. to *Washington Sunday Star*), June 1970, p. 11. Much of St. Laurent's information on Just's childhood came from an interview with his daughter, Margaret Just Butcher.

74. Henry E. Burnham to Just, 18 Feb. 1903, EEJ(H), box 125-2, folder 38.

75. The details of commencement week have been taken from *Kimball Union* 11 (1903), no. 6; those on prizes and awards from *Catalog of Kimball Union Academy*, 1903, p. 18.

76. *Kimball Union* 11 (1903): 87–90.

77. Ovington, "Just," p. 162.

78. For Dartmouth's emphasis on the fundamental importance of sports to the college experience, see *Dartmouth* 25 (1903): 2–3, 19, 92, etc. Also Robert French Leavens and Arthur Hardy Lord, *Dr. Tucker's Dartmouth* (Hanover, N.H.: Dartmouth, 1965), pp. 228–36.

79. Leon Burr Richardson, *History of Dartmouth College* (Hanover, N.H.: Dartmouth, 1932) 1: 37. For further historical information on the college, see "A historical note," *Catalog of Dartmouth College*, 1903–4, pp. 20–22; John King Lord, *A History of Dartmouth College, 1815–1909* (Concord, N.H.: Rumford Press, 1913), which is vol. 2 of a comprehensive work begun by Frederick Chase, *A History of Dartmouth College and the Town of Hanover, New Hampshire* (Cambridge, Mass.: John Wilson, 1891); William Jewett Tucker, *My Generation: An Autobiographical Interpretation* (Boston: Houghton Mifflin, 1919); and Ralph Nading Hill, *The College on the Hill: A Dartmouth Chronicle* (Hanover, N.H.: Dartmouth, 1964). A good general essay is "A Description of Dartmouth College," *Dartmouth College Bulletin* 7 (1941): 1–83.

80. Most of the specific information on Just's college life—housing, courses, requirements, prizes, scholarships, honors, etc.—has been taken from *Catalog of Dartmouth College*, 1903–7. Unless otherwise specified, details can be located in these catalogs. Also, letters to the author from Arthur H. Leavitt, 27 March 1979; and T. T. Redington, 16 April 1979—both Dartmouth class of 1907.

81. Just to Hedwig Schnetzler, 11 Jan. 1936, EEJ(M).

82. *Dartmouth* 25 (1903): 139.

83. Just's course and grade record, Dartmouth College Archives.

84. A survey of surviving members of the Dartmouth classes of 1904 to 1910 drew comments on black football stars, but nothing on black scholars: letters to the author. from Arthur H. Leavitt, '07 (27 March 1979); Guy C. Blodgett, '08 (13 May 1979); John H. Hinman, '08 (16 May 1979); Herbert G. Coas, '10 (16 May 1979); Whitney Eastman, '10 (17 May 1979); Arthur H. Lord, '10 (18 May 1979); Wilbur L. Taylor, '10 (16 May 1979).

85. *Dartmouth* 25 (1904): 353.

86. Ibid. 25 (1903): 119; 25 (1904): 409.

87. Albert W. Bates to the author, 7 July 1979.

88. Lord, *History of Dartmouth*, p. 360.

89. Ray Spencer to Just, n.d., EEJ(H), box 125-9, folder 161.

90. See *Dartmouth* 25 (1905): 225. For further information on the elaborate celebration surrounding Lord Dartmouth's visit, see Ernest Martin Hopkins, ed., *Exercises and Addresses Attending the Laying of the Corner-Stone of the New Dartmouth Hall and the Visit of the Earl of Dartmouth to the College, October 25 and 26, 1904* (Hanover, N.H.: Dartmouth, 1905).

91. See Just, "Dr. Patten Lectures on his Spitzbergen Trip," *Collecting Net* 1

(1926): 4. For college advertisements of Patten's popular lectures, see "Heredity and Evolution," *Dartmouth* 25 (1905): 338; "The Mosaic Theory and Germinal Localization," ibid. 26 (1905): 387.

92. Ovington, "Just," p. 163; also interview with Margaret Just Butcher, 3 Dec. 1976.

93. Just to Gerould, 5 Jan. 1941, JHG.

94. Gerould to F. R. Lillie, 2 April 1943, FRL(W).

95. For further biographical information on Patten, see E. M. Hopkins, "Statement," WP; also "William Patten," *National Cyclopedia of American Biography* 24: 121–22.

96. Patten to Just, 4 Oct. 1915, FRL(W).

97. Patten, *Evolution of the Vertebrates and Their Kin* (Philadelphia: Blakiston's Sons, 1912), pp. 257–61.

98. Just, "The Crucifixion" (1907); "On Going Home" (1907); "When the Light Went Out" (1907). "On Going Home" was favorably reviewed in *Dartmouth* 28 (1907): 522.

99. "On Going Home," p. 292.

100. Ibid., p. 248.

101. See the commencement program, *Dartmouth College—The One Hundred and Thirty-eighth Commencement*, Wed., 26 June 1907, Dartmouth College Archives.

102. For regulations concerning speaker appointments, see *Dartmouth* 28 (1907): 377; also faculty minutes, 25 Feb., 11 March, 22 April, 22 June 1907, in the Office of the Registrar.

2. THE BEGINNINGS OF A PROFESSIONAL CAREER, 1907–16

1. The standard histories of Howard University are Walter Dyson, *Howard University, The Capstone of Negro Education: A History, 1867–1940* (Washington, D.C.: Graduate School, Howard Univ., 1941); and Rayford W. Logan, *Howard University: The First Hundred Years, 1867–1967* (New York: New York Univ. Press, 1969). For information on Morehouse College, see Benjamin Brawley, *History of Morehouse College* (1917, reprint ed., College Park, Md.: McGrath, 1970).

2. Wilbur P. Thirkield, "Doctor Just," *Washington Afro-American*, 16 June 193?, in the E. E. Just file, Moorland-Spingarn Research Center, Howard University.

3. See Moorland-Spingarn appointment card (19 March 1915) and Moorland-Spingarn personnel card (27 March 1915), in the E. E. Just file, Moorland-Spingarn Research Center, Howard University, for changes in Just's position and salary during the early years.

4. For detailed information on the courses taught by Just, see *Howard University Catalog* (published annually).

5. Thirkield, "Doctor Just."

6. Ibid.; also Thirkield to Just, 28 March 1932, EEJ(H), box 125-8, folder 140.

7. The ceremony is reproduced in *Howard University Record* 5 (1911): 3–18.

8. Dyson, *Howard University*, pp. 146–47.

9. *The Crisis,* official organ of the NAACP, often publicized work in drama at Howard: e.g., 10 (1915): 61; 16 (1918): 31, 189; 25 (1922): 66–68.

10. For Just's part in the fraternity movement, see Herman Dreer, *The History of the Omega Psi Phi Fraternity, 1911–1939* (n.p.: Omega Psi Phi Fraternity, 1940), pp. 7–25; and Robert L. Gill, *The Omega Psi Phi Fraternity and the Men Who Made Its History* (n.p.: Omega Psi Phi Fraternity, 1963), pp. 1, 3, 50; also *Oracle* 61 (1976): 1, 65.

11. Dreer, *History of Omega Psi Phi,* p. 14.

12. See William Patten to Just, 7 June 1916, EEJ(H), box 125-7, folder 117.

13. Information on the courses can be found in two of Just's notebooks (93 pp. and 37 pp.), EEJ. The atmosphere at Woods Hole was recreated for me in letters from former students Rosamond Clark, 2 Sept. 1977; Seymour M. Farber, 15 April 1977; William E. MacFarland, 27 May 1977; Bernice F. Pierson, 3 Aug. 1977; C. S. Shoup, 8 July 1977; and Grace D. Sterne, 24 Aug. 1977.

14. A fair amount of biographical information is available on members of the "old guard" at the MBL, especially Conklin, Morgan, and Wilson. For information on Conklin, see F. Bard, "Edwin Grant Conklin," *Proceedings of the American Philosophical Society* 208 (1964): 55–56; Elmer G. Butler, "Edwin Grant Conklin (1863–1952)," *Yearbook: American Philosophical Society, 1952,* pp. 5–12; and E. N. Harvey, "Edwin Grant Conklin," *Biographical Memoirs: National Academy of Sciences* 31 (1958): 54–91. For Morgan, see A. H. Sturtevant, "Thomas Hunt Morgan," *Biographical Memoirs: National Academy of Sciences* 33 (1959): 283–325; Garland E. Allen, *Thomas Hunt Morgan: The Man and His Science* (Princeton: Princeton Univ. Press, 1978). For Wilson, see T. H. Morgan, "Edmund Beecher Wilson, 1856–1939," *Biographical Memoirs: National Academy of Sciences* 21 (1941): 315–42; and H. J. Muller, "Edmund B. Wilson—An Appreciation," *American Naturalist* 77 (1943): 5–37, 142–72. Also useful are the articles on all three of these men by Garland E. Allen in *Dictionary of Scientific Biography.*

15. Report by W. C. Curtis [October 1915], FRL(W).

16. Gilman A. Drew to F. R. Lillie, 2 Oct. 1915, FRL(W); Lillie to W. C. Graves, 31 Oct. 1921, FRL(C), box 4, folder 20.

17. See *Annual Register,* University of Chicago. Just was first listed as a graduate student in 1911–12 under the following category: "*Students not yet admitted to candidacy* (students are admitted to candidacy for higher degrees by vote of the faculty on approval of the thesis subject and fulfillment of other conditions under the regulations)." Just kept that listing, except in 1912–13 (when he was not listed at all), until 1915–16, the year he was admitted to candidacy for the Ph.D.

18. See Neil S. Dungay, "A Study of the Effects of Injury upon the Fertilizing Power of the Sperm," *Biological Bulletin* 25 (1913): 219; Marie Goldsmith, "Revue," *L'Année Biologique* 17 (1912): 93.

19. T. H. Morgan and Albert Tyler, "The Point of Entrance of the Spermatozoön in Relation to the Orientation of the Embryo in Eggs with Spiral Cleavage," *Biological Bulletin* 58 (1930): 71–72. The author found other references to Just's first article by going through the major domestic and foreign journals. Among the scientists who cited this article (and the year of citation) are Marie Goldsmith (1912); F. R. Lillie (1912); Neil S. Dungay (1913);

Charles Packard (1914); W. Schleip (1925); Sven Hörstadius (1928); Albert Tyler (1930, 1931); Daniel C. Pease (1938, 1940); Albert Tyler (1941); Pierre Guerrier (1970); Christopher A. Gabel, E. M. Eddy, and Bennett M. Shapiro (1979).

20. For details on the Montgomery episode, see Just to F. R. Lillie, 19 Feb. and 15 April 1912, FRL(W); Lillie to Just, 24 Feb. 1912, ibid.

21. Just to Lillie, 15 April 1912, ibid.

22. Lillie to Just, 18 April 1912, ibid.

23. The status of each member in the laboratory is recorded in the *Annual Report of the Marine Biological Laboratory*. See also *Science* 36, no. 921 (23 Sept. 1912).

24. This author found references to the *Nereis limbata* article by going through the major domestic and foreign biological journals. Among the scientists who cited this article (and the year of the citation) are the following: Marie Goldsmith (1912); Neil S. Dungay (1913); Wilhelm Roux (1913); W. C. Allee (1919); B. H. Grave (1922); Charles Gravier (1923); W. C. Allee (1923); H. Munro Fox (1924); W. C. Allee (1927); Louis Fage and René Legendre (1927); Benjamin H. Grave (1927); C. Amirthalingam (1928); Silvio Ranzi (1931); Hans-Joachim Elster (1935); Louise Palmer (1937); Grace Townsend (1938, 1939); Paul S. Galtsoff (1940); Martin W. Johnson (1943); Paul G. Lefevre (1945); Maurice Durchon (1948); Gunnar Thorson (1950); M. Jean Allen (1957); Maurice Durchon (1957); J. B. Gilpin-Brown (1959); R. B. Clark (1961); Yolande Boilly-Marer (1969); S. M. Evans (1971); Y. Boilly-Marer (1974); A. Cram and S. M. Evans (1980).

Among the scientists who cited the *Platynereis megalops* article are: Charles Gravier (1923); Sally Hughes-Schrader (1924); Louis Fage and René Legendre (1927); W. C. Allee (1927); C. Amirthalingam (1928); Silvio Ranzi (1931); Hans-Joachim Elster (1935); Paul S. Galtsoff (1938, 1940); Ralph I. Smith (1950); François Rullier (1954); Ralph I. Smith (1958); J. B. Gilpin-Brown (1959); R. B. Clark (1961); John F. Fallon and C. R. Austin (1967); S. M. Evans (1971); A. Cram and S. M. Evans (1980).

25. Monroe Alphus Majors, *Noted Negro Women: Their Triumphs and Activities* (Chicago: Donohue and Henneberry, 1893), p. 148.

26. On blacks in Ohio during the nineteenth century, see Frank U. Quillin, *The Color Line in Ohio: A History of Race Prejudice in a Typical Northern State* (Ann Arbor, Mich.: Wahr, 1913); W. A. Joiner, comp., *A Half Century of Freedom of the Negro in Ohio* (Xenia, Ohio: Smith Advertising, 1915); Wendell P. Dabney, *Cincinnati's Colored Citizens* (Cincinnati: Dabney, 1926); and Charles T. Hickok, *The Negro in Ohio, 1802-1870* (New York: AMS, 1975).

27. See Felix J. Koch, "Where Eliza Crossed the Ice," *Crisis* 19 (1920): 118-20.

28. See *Howard University Catalog, 1907-8,* p. 41.

29. For detailed accounts of Loeb's life and work, see W. J. V. Osterhout, "Jacques Loeb," *Journal of General Physiology* 8 (1928): ix-xcii; Nathan Reingold, "Jacques Loeb, the Scientist: His Papers and His Era," *Quarterly Journal of Current Acquisitions in the Library of Congress* 10 (1962): 119-30; Donald Fleming, "Introduction," in Jacques Loeb, *The Mechanistic Conception of Life* (Cambridge: Harvard Univ. Press, 1964), pp. vii-xli; Philip J. Pauly, "Jacques Loeb and the Control of Life: An Experimental Biologist in Germany and America" (Ph.D. diss., Johns Hopkins, 1980).

30. Loeb to Alice and Justus Gaule, 4 Dec. 1914, JL(GC), box 5; Loeb to Richard Goldschmidt, 16 Feb. 1920, ibid.

31. Loeb to Erving Winslow, 24 Oct. 1913, JL(GC), box 16; Loeb to Joel E. Spingarn, 14 May 1914, box 3; Loeb to Charles W. Wendte, 2 Dec. 1914, box 16; Loeb to Svante Arrhenius, 14 Dec. 1914, box 1; Loeb to Sakyo Kanda, 23 July 1915, box 7; Loeb to Richard Pearce, 9 April 1918, box 12. See also ms. of speech "Can Socialism Obliterate Race Antagonism?" delivered at the Socialist Press Club, 22 March 1915, JL(BAS), box 44.

32. Just to Loeb, 16 Jan. and 8 April 1913, 1 May, 12 and 18 Oct. 1914, 15 April 1915, JL(GC), box 7.

33. Loeb to Just, 20 May 1914; 16, 20, 23 Oct. 1914; 29 Dec. 1914, ibid.

34. See Loeb to Just, 16 Oct. 1915, ibid.

35. Loeb to Jerome D. Greene, 17 Oct. 1914, GEB, box 28, folder 256.

36. Ibid.

37. Ibid.

38. Loeb to Greene, 20 Oct. 1914, ibid.

39. Just to Loeb, 12 Oct. 1914, JL(GC), box 7.

40. See remarks by Du Bois at the seventh session of the conference held in May 1914, NAACP(AC), box B-1. Some of the comprehensive studies of the NAACP include Langston Hughes, *Fight for Freedom: The Story of the NAACP* (New York: W. W. Norton, 1962); Robert L. Jack, *History of the National Association for the Advancement of Colored People* (Boston: Meader, 1943); Mary White Ovington, *The Walls Came Tumbling Down* (reprint ed., New York: Schocken, 1970); and Barbara Joyce Ross, *J. E. Spingarn and the Rise of the NAACP, 1911–39* (New York: Atheneum, 1972). The first part of a mammoth study has been completed: Charles Flint Kellogg, *NAACP: A History of the National Association for the Advancement of Colored People*, vol. 1, 1909–20 (Baltimore: Johns Hopkins Univ. Press, 1967). For a brief overview, see "Profile of the NAACP, America's Oldest and Largest Civil Rights Organization," *Negro History Bulletin* 27 (1964): 74–76.

41. The life of Du Bois has been well documented. The full-length studies include Francis Lyons Broderick, *W. E. B. Du Bois: Negro Leader in a Time of Crisis* (Stanford: Stanford Univ. Press, 1959); Shirley Graham, *His Day Is Marching On: A Memoir of W. E. B. Du Bois* (Philadelphia: Lippincott, 1971); Emma Sterne, *His Was the Voice: The Life of W. E. B. Du Bois* (New York: Crowell-Collier, 1971); and Arnold Rampersad, *The Art and Imagination of W. E. B. Du Bois* (Cambridge: Harvard Univ. Press, 1976). A good anthology of shorter articles is Rayford W. Logan, ed., *W. E. B. Du Bois: A Profile* (New York: Hill and Wang, 1971). See also R. McGill, "W. E. B. Du Bois," *Atlantic* 216 (1965): 78–81; Irving Howe, "Remarkable Man, Ambiguous Legacy," *Harper's* 236 (1968): 143–49; Martin Duberman, "Du Bois as Prophet," *New Republic* 158 (1968): 36–39.

42. See Du Bois to Loeb, 28 April 1914, JL(GC), box 3.

43. Loeb to Oswald Garrison Villard, 1 May 1914, JL(GC), box 15; also Loeb to Villard, n.d., NAACP(AC), box B-1.

44. *Crisis* 8 (1914): 84–85.

45. The longest of these articles, "Science and Race," was published in *Crisis* 9 (1914): 92–93. See also Loeb MSS, JL(BAS), box 44.

46. Loeb to Joel E. Spingarn, 17 March 1915, JL(GC), box 14.

47. Spingarn to Loeb, 18 March 1915, JL(GC), box 14.

48. Loeb to Spingarn, 23 March 1915, JL(GC), box 14.

49. *Biological Bulletin* 30 (1916): 13.

50. Loeb to F. R. Lillie, 2 Feb. 1916, FRL(W); Loeb to T. H. Morgan, 3 Feb. 1916, JL(GC), box 9.

51. Lillie to Loeb, 6 Feb. 1916, FRL(W).

52. Morgan to Loeb, 4 Feb. 1916, JL(GC), box 9.

53. Loeb to Lillie, 6 March 1916, JL(GC), box 8.

54. See "The Number of Chromosomes and the Biological Status of the Negro," JL(BAS), box 44.

55. Loeb to Lillie, 8 Feb. 1916, FRL(W).

56. See circular (April 1913) and public announcement (May 1913), NAACP (AF), box C-209; also official program, box C-212.

57. *Crisis* 9 (1915): 284.

58. See *Science* 39 (1914): 158–59; *Crisis* 9 (1914): 81.

59. Loeb to May Childs Nerney, 24 Sept. 1914, JL(GC), box 10.

60. Loeb to Oswald Garrison Villard, 14 Jan. 1915, ibid., box 15.

61. Just to F. R. Lillie, 11 Jan. 1915, FRL(W).

62. Just to Loeb, 12 Oct. 1914, JL(GC), box 7.

63. See Mabel M. Smythe, ed., *The Black American Reference Book* (Englewood Cliffs, N.J.: Prentice-Hall, 1976), pp. 554–55, 851–52. For further information on Trotter see *Crisis* 9 (1915): 119–20, 129; 10 (1915): 10; 13 (1916): 32; 22 (1921): 81.

64. The telegrams are dated 14–15 Feb. 1915, NAACP(AF), box C-209. For other expressions of surprise at the committee's selection, see *Crisis* 9 (1915): 281.

65. *Crisis* 9 (1915): 284.

66. *Crisis* 4 (1912): 171.

67. See Ovington, "Just," p. 160; also *Crisis* 9 (1915): 284; 14 (1917): 131.

68. *New York Evening Post*, 13 Feb. 1915, p. 4. See also *New York Evening Post*, 12 Feb. 1915, p. 9.

69. News release by *Crisis*, Feb. 1916, NAACP(AF), box C-1. For a detailed summary of articles in various newspapers, including *New York Evening Post, Jamestown* (N.Y.) *Post, Pittsburgh Chronicle Telegram*, and *San Francisco Bulletin*, see *Crisis* 9 (1915): 281–82.

70. Quoted in *Crisis* 11 (1916): 301.

71. Lillie to Just, 19 Feb. 1915, FRL(W).

72. Just to Loeb, 22 Feb. 1915, JL(GC), box 7.

73. Just to Lillie, 22 Feb. 1915, FRL(W).

74. Just to Spingarn, 22 Feb. 1915, quoted in W. Montague Cobb, "Ernest Everett Just, 1883–1941," *Journal of the National Medical Association* 49 (1957): 349.

75. Lillie to Just, 3 June 1915, FRL(W); Lillie to S. M. Newman, 28 July 1915, ibid.

76. Lillie to W. C. McNeill, 26 Aug. 1915, FRL(W).

77. McNeill to Lillie, 30 Aug. 1915, FRL(W); also Edward A. Ballock to Lillie, 21 Sept. 1915, ibid.

78. See Lillie to McNeill, 26 Aug. 1915, ibid.

79. See Just to Loeb, 8 Dec. 1915, JL(GC), box 7.

80. See *Crisis* 24 (1922): 108, 110.

81. *The Crisis* published several articles on the professional situation of blacks in Chicago. The most comprehensive of these is "Colored Chicago," *Crisis* 10 (1915): 234-36; but see also "Employment of Colored Women in Chicago," *Crisis* 1 (1911): 24-25; and "Chicago," *Crisis* 7 (1914): 130-31.

82. For further information on Turner, Overton, and Williams, see "Some Chicagoans of Note," *Crisis* 10 (1915): 237-42. References to the others may also be found in *The Crisis*: Chandler—15 (1918): 297; Quinn—17 (1919): 142; Binga—23 (1922): 253; Wright—22 (1921): 84; Landry—21 (1921): 226; Abbott— 24 (1922): 74-75; James—29 (1925): 268; and Duke—25 (1922): 75.

83. W.E.B. Du Bois, "Chicago," *Crisis* 35 (1928): 346-48.

84. See *Crisis* 12 (1916): 193-94.

85. *Crisis* 22 (1921): 72.

86. For a summary of Bentley's life, see Du Bois's obituary, "Bentley," *Crisis* 36 (1929): 423. See also *Crisis* 10 (1915): 238-39; 11 (1916): 111; 21 (1921): 174, 276.

87. *Crisis* 10 (1915): 242; 14 (1917): 259.

88. See note 65 for articles on the employment situation. An analysis of the housing situation can be found in "Housing the Negro in Chicago," *Crisis* 5 (1912): 25-26.

89. *Chicago Tribune*, 5 Oct. 1915, p. 5; 16 Oct. 1915, p. 7; 17 Oct. 1915, p. 1; 20 Oct. 1915, p. 13; 24 Oct. 1915, p. 1. The lynching incidents took place not in Chicago proper but in outlying areas such as Murphysboro, Mt. Vernon, and Centralia.

90. For a detailed summary of the case see *Chicago Tribune*, 28 Nov. 1915, p. 1.

91. See F. A. Moore to Rosenwald, 28 Feb. 1916, JR, box 14, folder 5.

92. See articles in *The Crisis*: "The Riots," 18 (1919): 241-44; "Race War," 18 (1919): 247-49; "Chicago and its Eight Reasons," 18 (1919): 293-97; "Aftermath at Washington and Chicago," 19 (1920): 129-30. See also T. Arnold Hill to L. Hollingsworth Wood, 29 July 1919, JR, box 6, folder 4; Stanley B. Norvell to Victor F. Lawson, 22 Aug. 1919, ibid., folder 3.

93. Just to L. V. Heilbrunn, 27 June 1916, LVH.

94. *Crisis* 36 (1929): 423.

95. Ibid., 2 (1911): 77; 8 (1914): 141-42.

96. Ibid., 6 (1913): 38-39.

97. Ibid., 5 (1912): 18, 41; 6 (1913): 41.

98. Ibid., 10 (1915): 18, 41; 10 (1915): 61, 85-86; 11 (1916): 257.

99. Ibid., 31 (1926): 270.

100. Just to Loeb, 8 Dec. 1915, JL(GC), box 7.

101. *Chicago Tribune*, 22 Nov. 1915, p. 1.

102. Ibid., 14 Oct. 1915, p. 13.

103. Ibid., 4 Dec. 1915, p. 2; see also 10 Nov. 1915, p. 1; 17 Nov. 1915, p. 15.

104. Ibid., 7 Dec. 1915, pp. 1-2.

105. Ibid., 8 Dec. 1915, p. 17.

106. See graduate prospectus, FRL(C), box 8, folder 13.

107. For the early history of the university, see Thomas Wakefield Goodspeed, *A History of the University of Chicago* (Chicago: Univ. of Chicago Press, 1916); see also David Allan Robertson, *The University of Chicago: An Official Guide* (Chicago: Univ. of Chicago Press, 1916).

108. *The President's Report Covering the Academic Year Ending June 30, 1916* (Chicago: Univ. of Chicago Press, 1917).

109. Lillie to University Recorder, 7 Oct. 1915, FRL(C), box 8, folder 13; also "Examiners' and Instructors' Grade Reports," summer 1909 and summer 1910 (updated 7 Oct. 1915). Much of the information on the Chicago courses and Just's performance in them is taken from these reports and from *Annual Register of the University of Chicago* and *Official Publications of the University of Chicago* (published annually).

110. Course information can be found in Just's notebooks: Zoology 45 (37 pp.), Physiology 13b (66 pp.), Chemistry 31 (171 pp.), Chemistry 32 (104 pp.), Physiology 31 (68 pp.), Physiology 12 (65 pp.), Anatomy 31 (99 pp.), EEJ.

111. Just to Lillie, 30 Aug. 1916, FRL(W).

112. Just to L. V. Heilbrunn, 27 June 1916, LVH.

113. Mathews to Just, 13 Dec. 1937, EEJ(M); Mathews to F. P. Keppel, 20 Nov. 1936, CC(N).

114. Just, "Studies of Fertilization in *Platynereis megalops*," (1915). The two articles in question were "Breeding Habits of the Heteronereis Form of *Platynereis megalops* at Woods Hole, Mass." (1914), and "An Experimental Analysis of Fertilization in *Platynereis megalops*," (1915). A third article, "The Morphology of the Normal Fertilization in *Platynereis megalops*" (1915), was not incorporated in the thesis but was an integral part of the series.

115. James M. Jay, *Negroes in Science: Natural Science Doctorates, 1876-1969* (Detroit: Balamp, 1971).

116. *General Program for the Celebration of the Quarter Centennial of the University of Chicago, June 2 to 6, 1916.*

3. THE EXPANSION OF JUST'S SCIENTIFIC WORLD: WOODS HOLE, 1909–29

1. From *Songs and Poems of Woods Hole*, ed. Donnell B. Young (Woods Hole, Mass.: Book Shop, 1921), p. 8. A copy of this rare little book was loaned to me by Bernice F. Pierson.

2. Frank R. Lillie, *The Woods Hole Marine Biological Laboratory* (Chicago: Univ. of Chicago Press, 1944), p. 35. This book is the most comprehensive treatment of the origin and development of the MBL. For further information, see J. S. Kingsley, "The Marine Biological Laboratory," *Popular Science Monthly* 41 (1892): 604–15; C. O. Whitman, "A Marine Biological Observatory," *Popular Science Monthly* 42 (1893): 459–71; C. O. Whitman, "The Work and the Aims of the Marine Biological Laboratory," in *Biological Lectures Delivered at the Marine Biological Laboratory of Woods Hole in the Summer Session of 1893* (Boston: Ginn, 1894), pp. 235–42; E. G. Conklin, "The Marine Biological Laboratory," *Science* n.s. 11 (1900): 333–43; H. F. Osborn, "A Sea-shore Laboratory," *Harper's* 104 (1902): 552–58; F. R. Lillie, "The Marine Biological Laboratory at Woods Hole," *Internationale Revue der gesamten Hydrobiologie und Hydrographie* 5 (1913): 583–89; E. G. Conklin, "The Story of Woods Hole and the Marine Biological Laboratory," *Collecting Net* 2 (1927) no. 1: 1–3; no. 2: 1, 3, 5; no. 4: 3, 10; also, "Supplement" (1929), *13th International Physiological Congress*, pp. xi–xx; F. Baltzer, "Wissenschaft und Sommerfrische bei den amerikanische Biologen: Das marine-biologische

Laboratorium im Woods Hole," *Der kleine Bund: Literarische Beilage des 'Bund'* 17 (1930): 158-60. Two more recent articles are Luther J. Carter, "Woods Hole: Summer Mecca for Marine Biology," *Science* n.s. 157 (1967): 1288-92; and Detlev W. Bronk, "Marine Biological Laboratory: Origins and Patrons," *Science* n.s. 189 (1975): 613-17.

3. Lillie, *Woods Hole*, pp. 16-17. For further biographical information on Agassiz, see Elizabeth Cary Agassiz, ed., *Louis Agassiz: His Life and Correspondence* (Boston: Houghton, Mifflin, 1885); Edward Lurie, *Louis Agassiz: A Life in Science* (Chicago: Univ. of Chicago Press, 1960); and Jules Marcou, *Life, Letters, and Works of Louis Agassiz* (New York: Macmillan, 1896; reprint ed., Farnborough, Eng.: Gregg, 1972). A good brief article is Edward Lurie, "Louis Agassiz," *Dictionary of Scientific Biography* 1: 72-74. The Penikese experiment is described in David Starr Jordan, "Agassiz at Penikese," *Popular Science Monthly* 40 (1892): 721-29; B. G. Wilder, "Agassiz at Penikese," *American Naturalist* 32 (1898): 189-96; and Edward S. Morse, "Agassiz and the School at Penikese," *Science* n.s. 58 (1923): 273-75.

4. Lillie, *Woods Hole*, p. 36. For further biographical information on Whitman, see F. R. Lillie, "Charles Otis Whitman," *Journal of Morphology* 22 (1911): xv-lxxvii; Edward S. Morse, "Charles Otis Whitman," *Biographical Memoirs: National Academy of Sciences* 7 (1912): 269-88; and C. B. Davenport, "The Personality, Heredity, and Work of Charles Otis Whitman," *American Naturalist* 51 (1917): 5-30.

5. Lillie, *Woods Hole*, pp. 13-14, 37. For further biographical information on Dohrn, see Stewart Paton, "Anton Dohrn, Founder and Director of the Naples Aquarium," *Science* n.s. 30 (1909): 833-35; E. B. Wilson, "The Memorial to Anton Dohrn," *Science* n.s. 34 (1911): 632-33; Theodor Boveri, "Anton Dohrn," *Science* n.s. 36 (1912): 453-68; and Theodor Heuss, *Anton Dohrn* (n.p.: Atlantis Verlag, 1945; reprint ed., Tübingen: Rainer, Wunderlich Verlag, 1962). The most complete studies of the origin and development of the European stations are C. A. Kofoid, "Biological Stations of Europe," *United States Bureau of Education Bulletin* no. 440 (1910): 1-360; and T. W. Vaughan et al., *International Aspects of Oceanography: Oceanographic Data and Provisions for Oceanographic Research* (Washington, D.C.: National Academy of Sciences, 1937). For discussions of the Naples station in particular, see Anton Dohrn, "The Zoological Station at Naples," *Nature* 6 (1872): 535-36; 8 (1873): 81; Emily Nunn Whitman, "The Zoological Station at Naples," *Century Illustrated Monthly Magazine*, Sept. 1886, pp. 791-99; T. H. Morgan, "Impressions of the Naples Zoological Station," *Science* n.s. 3 (1896): 16-18; Charles Lincoln Edwards, "The Zoological Station at Naples," *Popular Science Montly* 77 (1910): 209-25. See also M. Leiner, "Die Bedeutung der Zoologischen Station in Neapel für die deutschen Biologen," *Bremerbeiträge zur Naturwissenschaft* 4 (1937): 3-35; Homer A. Jack, "The Biological Field Stations of Italy and Monaco," *Collecting Net* 15 (1940): 184-86; Reinhard Dohrn, "Stazione Zoologica di Napoli," *Notes and Records of the Royal Society of London* 8 (1951): 277-82; Reinhard Dohrn, "The Zoological Station at Naples," *Endeavour* 13 (1954): 22-26; C. M. Yonge, "The Zoological Station at Naples," *Discovery* 17 (1956): 187-88; and Andrew Packard, "Marine Biological and Oceanographic Institutions of the World. IV. The Naples Zoological Station," *Journal of the Marine Biological Association of India* 2 (1960): 259-62. An

excellent summary is contained in *The Naples Zoological Station at the Time of Anton Dohrn*, Exhibition and Catalogue by Christiane Groeben, Naples Zoological Station, in collaboration with Irmgard Müller, Düsseldorf, trans. Richard and Christl Ivell (edition for the centenary of the Naples Zoological Station, 1975).

6. Lillie, *Woods Hole*, pp. 31, 34-35.

7. Ibid., pp. 48-49, 51-52.

8. Seymour M. Farber to the author, 15 April 1977. The unique Woods Hole atmosphere was recreated for me by nearly 250 former MBL investigators and students who took the time to detail their reminiscences in letters and in interviews.

9. Whitman folder, FRL(W); also Lillie to E. G. Conklin, 20 and 25 Oct. 1910, EGC, box 32.

10. Interview with Donald P. Costello, 3 April 1977.

11. Just to Hedwig Schnetzler, 11 Jan. 1936, EEJ(M).

12. Ibid.

13. Just to L. V. Heilbrunn, 27 June 1916, LVH.

14. Fragment left inside MBL notebook, EEJ.

15. Just to L. V. Heilbrunn, 27 June 1916, LVH.

16. Ibid., 12 Sept. 1916.

17. Ibid., 27 June 1916.

18. Ibid., 10 June 1917.

19. Ibid., 12 Aug. 1917.

20. Ibid.; see also F. R. Lillie to E. G. Conklin, 23 April 1917, EGC, box 32.

21. Interview with Sally Hughes-Schrader, 8 July 1977.

22. For Just's interest in Schrader's professional prospects, see Just to L. V. Heilbrunn, 13 Sept. 1920, 11 Dec. 1920, and 14 Jan. 1921, LVH.

23. Ibid., 20 May 1936.

24. Interview with Sally Hughes-Schrader, 8 July 1977.

25. Interview with Donald and Rebecca Lancefield, 8 July 1977.

26. On Just's loneliness, see Just to Hedwig Schnetzler, 11 Jan. 1936, EEJ(M). On his "two residences" argument for support, see for example Just to Abraham Flexner, 28 Dec. 1922 and 18 April 1925, GEB, box 695, folder 6967.

27. Just to Hedwig Schnetzler, 11 Jan. 1936, EEJ(M).

28. Just to F. R. Lillie, 18 Sept. 1920, FRL(C), box 4, folder 20.

29. See Just to L. V. Heilbrunn, 3 March 1915, LVH.

30. Just to F. R. Lillie, 14 Oct. 1921, FRL(C), box 4, folder 20.

31. Lillie to W. C. Graves, 31 Oct. 1921, ibid.

32. Ibid.

33. Just to Lillie, 14 Oct. 1921, ibid.

34. Just to W. C. Graves, 14 Jan. 1921, JR, box 19, folder 3.

35. From an article Just published in *Oracle* (spring 1919), official organ of Omega Psi Phi: quoted in Dreer, *History of Omega Psi Phi*, p. 94.

36. Just to W. C. Graves, 14 Jan. 1921, JR, box 19, folder 3.

37. For example, Marie Goldsmith, "Revue," *L'Année Biologique* 17 (1912): 93; Wilhelm Roux, "Literaturverzeichnis 1912 zu 1913," *Archiv für Entwicklungsmechanik der Organismen* 38 (1913): 170.

38. Just to Loeb, 22 Dec. 1916, JL(GC), box 7.

39. Loeb to Just, 5 Jan. 1917, ibid.

40. Just to Loeb, 15 Dec. 1918, ibid.

41. Lillie to Just, 19 Dec. 1918, FRL(W).

42. Just to Lillie, 22 Dec. 1918, FRL(C), box 4, folder 20.

43. See Lillie to Just, 14 Jan. 1919, FRL(C), box 4, folder 20.

44. Francisco J. Ayala (hon. secy., Amer. Soc. of Naturalists) to the author, 7 Jan. 1977; John L. Roberts (hon. secy., Amer. Soc. of Zoologists) to the author, 26 Dec. 1976; Hans Nussbaum (bus. man., Amer. Assoc. for the Advancement of Science) to the author, 1 Dec. 1976; J. Ancellin (assoc. dir., *Société nationale des sciences naturelles et mathématiques de Cherbourg*) to the author, 8 March 1979. For Just's membership in the American Ecological Society, see Just to Raymond B. Fosdick, 22 July 1936, GEB, box 695, folder 6969.

45. "The Fertilization-Reaction in *Echinarachnius parma*. I–IV" (1919–20). For further information on what Just considered to be the major "inaccuracies" of Loeb's work, see Just to F. R. Lillie, 3 Feb. 1920, FRL(C), box 4, folder 20; Lillie to Just, 18 Feb. 1920, ibid. The other four papers in the eight-part series, published in 1922 and 1923, are less devastating to Loeb. This author found references to Just's eight *Echinarachnius parma* articles by going through the major domestic and foreign biological journals. Among the scientists who cited these articles (and the year of the citation) are: Marie Goldsmith (1919); Charles Pérez (1920–21); Otto Glaser (1921); Frank R. Lillie (1921); W. J. Crozier (1922); Hope Hibbard (1922); Myra M. Sampson (1922); Paul Boyer (1922–23); Leigh Hoadley (1923); Libbie H. Hyman (1923); Charles Pérez (1923–24); G. S. Carter (1924); Otto Glaser (1924); L. V. Heilbrunn (1924); Ralph S. Lillie (1924); J. Runnström (1924); Henry J. Fry (1925); Emile Godlewski (1926); Marie A. Hinrichs (1926); C. E. Tharaldsen (1926); Henry J. Fry (1927); Benjamin H. Grave (1927); Marie A. Hinrichs (1927); Reuben Blumenthal (1928); Henry J. Fry (1928); John Runnström (1928–29); A. J. Goldforb (1929); Irvine H. Page (1929); Leigh Hoadley (1930); A. R. Moore (1930); John Runnström (1930); W. J. Baumgartner and M. Anthony Payne (1931); Albert Tyler (1931); D. M. Whitaker (1931); G. S. Carter (1932); W. A. Dorfman (1932); W. A. Dorfman and W. Saranow (1932); A. D. Hobson (1932); Per Eric Lindahl (1932); Albert Tyler and Jack Schultz (1932); Robert Chambers (1933); Ernst Wertheimer (1933); Émile Godlewski (1934); H. Y. Chase (1935); Donald P. Costello (1935); Hans-Joachim Elster (1935); G. E. MacGinitie (1935); J. Porte (1935, 1938); Walter E. Southwick (1939); Floyd Moser (1939, 1940); J. A. Kitching and Floyd Moser (1940); Albert Tyler (1940, 1941); Ralph S. Lillie (1941); Ethel Browne Harvey (1942); Johannes Holtfreter (1943); Paul G. Lefevre (1945); Albert Tyler and Charles B. Metz (1945); Alberto Monroy and Anna Monroy Oddo (1946); Albert Tyler (1946); Alberto Monroy and Giuseppe Montalenti (1947); Willis E. Pequegnat (1948); Iben Browning (1949); Lord Rothschild and M. Swann (1949); Katsuma Dan and Kayo Okazaki (1951); Masao Sugiyama (1953); Eizo Nakano (1956); Y. Endo (1961).

46. "On the Nature of the Process of Fertilization and Artificial Production of Normal Larvae (Plutei) from the Unfertilized Eggs of the Sea Urchin," *American Journal of Physiology* 3 (1899): 135–38. A complete bibliography of Loeb's scientific work was compiled by Nina Kobelt and published in *Journal of General Physiology* 8 (1928): lxiii–xcii.

47. *Biological Bulletin* 36 (1919): 51–52.

48. "Observations and Experiments Concerning the Elementary Phenom-

ena of Embryonic Development in *Chaetopterus*," *Journal of Experimental Zoology* 3 (1906): 153–268. The most thorough analysis of Lillie's theory is still Just, "The Present Status of the Fertilizin Theory of Fertilization" (1930). See also "Specific Egg and Sperm Substances and Activation of the Egg," in A. Tyler, R. C. von Borstel, and C. B. Metz, eds., *The Beginnings of Embryonic Development* (Washington, D.C.: American Association for the Advancement of Science, 1957), pp. 22–69; and "Gamete Surface Components and Their Role in Fertilization," in C. B. Metz and A. Monroy, eds., *Fertilization: Comparative Morphology, Biochemistry, and Immunology* (New York: Academic Press, 1967–69) 1: 163–236.

A number of studies, contemporary and historical, are useful aids in understanding the fertilizin theory and some of the other biological controversies in which Just had a part. See, in particular, Frank R. Lillie, *Problems of Fertilization* (Chicago: Univ. of Chicago Press, 1919); T. H. Morgan, *Experimental Embryology* (New York: Columbia Univ. Press, 1927); E. B. Wilson, *The Cell in Development and Heredity* (New York: Macmillan, 1925); J. Gray, *A Textbook of Experimental Embryology* (New York: Macmillan, 1931); Ethel Browne Harvey, *The American Arbacia and Other Sea Urchins* (Princeton: Princeton Univ. Press, 1956); Lord Rothschild, *Fertilization* (New York: Wiley, 1956); J. Runnström, B. E. Hagström, and P. Perlmann, "Fertilization," in Jean Brachet and Alfred E. Mirsky, eds., *The Cell: Biochemistry, Physiology, Morphology* (New York: Academic Press, 1961), vol. 1; Jane M. Oppenheimer, *Essays in the History of Embryology and Biology* (Cambridge: M.I.T. Press, 1967); Paul C. Schroeder and Colin O. Hermans, "Annelida: Polychaeta," in Arthur C. Giese and John S. Pearse, eds., *Reproduction of Marine Invertebrates* (New York: Academic Press, 1975), vol. 3; and Garland E. Allen, *Life Science in the Twentieth Century* (New York: Wiley, 1975).

49. *Biological Bulletin* 22 (1912): 239–52.

50. Ibid. 24 (1913): 147–67; 27 (1914): 201–12.

51. "Initiation of Development in *Nereis*" (1915); "An Experimental Analysis of Fertilization in *Platynereis megalops*" (1915); "The Morphology of Normal Fertilization in *Platynereis megalops*" (1915).

This author found references to the three articles by going through the major domestic and foreign biological journals. The following cited the first article (with the year of citation): A. J. Goldforb (1918); Alvalyn E. Woodward (1918); H. B. Goodrich (1920); Alvalyn E. Woodward (1921); L. V. Heilbrunn (1925); Myra M. Sampson (1926); S. Prát and K. M. Malkovský (1927); John Runnström (1928–29); Josef Spek (1930); Vera Koehring (1931); Albert Tyler (1931); Yoshi Kuni Hiraiwa and Toshijiro Kawamura (1936); L. V. Heilbrunn and Karl M. Wilbur (1937); Albert Tyler (1941); Paul G. Lefevre (1945); Paul G. Lefevre (1948); R. Phillips Dales (1950); R. B. Clark (1961). References to the second article were made by: Carl R. Moore (1916, 1917); Alvalyn E. Woodward (1918); Myra M. Sampson (1922); Myra M. Sampson (1926); John Runnström (1928–29); Wilhelm Einsele (1930); Josef Spek (1930); Hans-Joachim Elster (1935); Walter E. Southwick (1939); Albert Tyler (1940); Ralph I. Smith (1958); and R. B. Clark (1961). Among the scientists who cited the third article are: Vishna Nath (1926); Edward Drane Crabb (1927); Gerhard Beissler (1944); Gerald P. Pesch and Carol E. Pesch (1980).

52. *Biological Bulletin* 28 (1915): 112.

53. Ibid., p. 16.

54. "The Fertilization-Reaction in *Echinarachnius parma.* III" (1919); see also "The Present Status of the Fertilizin Theory of Fertilization" (1930).

55. *Biological Bulletin* 36 (1919): 52.

56. Interview with Donald Lancefield, 8 July 1977; Herbert Pollack to the author, 27 July 1977.

57. *Biological Bulletin* 39 (1920): 298.

58. Ibid., p. 303.

59. Lillie and Just, "Fertilization," in E. V. Cowdry, ed., *General Cytology* (Chicago: Univ. of Chicago Press, 1924), pp. 449–536. Among the scientists who have cited Lillie and Just's article (and the year of the citation) are the following: Hugo Miehe (1925); Marie A. Hinrichs (1926); Myra M. Sampson (1926); Edward Drane Crabb (1927); Marie A. Hinrichs (1927); Fernandus Payne (1927); Reuben Blumenthal (1928); E. Bataillon and Tchou-Su (1929); Henry J. Fry (1929); A. J. Goldforb (1929); Robert Chambers (1930); Jean Pasteels (1930); E. S. Guzman Barron (1932); G. S. Carter (1932); Giuseppe Montalenti (1933); Paul Wintrebert (1933); Emile Godlewski (1934); H. Y. Chase (1935); Giuseppe Reverberi (1936); D. M. Whitaker (1937); H. Y. Chase (1938); Sven Hörstadius (1939); Albert Tyler (1940); Albert Tyler and Sidney W. Fox (1940); Ivor Cornman (1941); Roberta Lovelace (1949); Wilhelm Goetsch (1950); Pierre Couillard (1952); A. Bohus Jensen (1953); C. V. Harding, Drusilla Harding, and P. Perlmann (1954); Ernst Caspari and Ingbritt Blomstrand (1956); Ray Leighton Watterson (1979).

60. Just to Julius Rosenwald, 27 Aug. 1920, JR, box 19, folder 3.

61. Just to Lillie, 16 Nov. 1920, FRL(C), box 4, folder 20.

62. Just to L. V. Heilbrunn, 11 Dec. 1920, LVH.

63. See St. Laurent, "The Negro in World History: Dr. Ernest E. Just," p. 12.

64. See "Proceedings of the American Society of Zoologists," *Anatomical Record* 20 (1921): 175–231.

65. Just to L. V. Heilbrunn, 14 Jan. 1921, LVH.

66. See Selig Hecht to W. J. Crozier, 27 Jan. 1921, WJC; Libbie Hyman to W. J. Crozier, 11 Jan. 1921, WJC.

67. See Hyman to Crozier, ibid.; Just to Heilbrunn, 14 Jan. 1921, LVH.

68. Just to W. C. Graves, 14 Jan. 1921, JR, box 19, folder 3.

69. Hyman to W. J. Crozier, 11 Jan. 1921, WJC.

70. Ibid.

71. Ibid., [January 1921].

72. Ibid., 26 Feb. 1921.

73. Hecht to Crozier, 17 Jan. 1921, WJC.

74. A. R. Moore to Hecht, 31 Jan. [1921], SH.

75. See Just to Heilbrunn, 22 May 1921, LVH.

76. Loeb to Osterhout, 21 Feb. 1921, JL(GC), box 11.

77. Flexner to Loeb, 9 Nov. 1920, GEB, box 695, folder 6967.

78. Loeb to Flexner, 11 Nov. 1920, JL(GC), box 4.

79. Lillie to Flexner, 29 Oct. 1920, FRL(C), box 4, folder 20.

80. Flexner to Lillie, 15 Nov. 1920, ibid.

81. See Just to Flexner, 10 Jan. 1923, GEB, box 695, folder 6967.

82. Gay to Flexner, 10 Jan. 1923, ibid.

83. Abraham Flexner to Loeb, 15 Jan. 1923, JL(GC), box 4.

84. Loeb to Abraham Flexner, 16 Jan. 1923, ibid.

85. The controversy with Jennings has been dealt with at length: see Donald D. Jensen, Foreword, in Jennings, *Behavior of the Lower Organisms* (Bloomington: Indiana Univ. Press, 1962), pp. ix–xvii; for a different view of Loeb's work, see Philip J. Pauly, "The Loeb-Jennings Debate and the Science of Animal Behavior," *Journal of the History of the Behavioral Sciences* 17 (1981): 504–15.

86. W. C. Curtis to Just, 10 Aug. 1918, EEJ(H), box 125-3, folder 54.

87. Loeb to Osterhout, 19 Dec. 1917, JL(GC), box 11; Loeb to T. H. Morgan, 21 Dec. 1917, ibid., box 9.

88. Loeb to Ross G. Harrison, 17 and 18 May 1916, ibid., box 6.

89. Loeb to Osterhout, 26 Sept. 1917, ibid., box 11; also Loeb to A. R. Moore, 26 Sept. 1917, box 9.

90. Loeb to J. Pickering, 16 April 1921, ibid., box 11.

91. Ibid.

92. Loeb to M. G. Banus, 7 Nov. 1919, ibid., box 2.

93. Loeb to Harry Friedenwald, 5 May 1915, ibid., box 5; Loeb to Richard Gottheil, 5 and 8 May 1915, ibid.

94. Loeb to Just, 3 April 1916, ibid., box 7.

95. See Loeb to Michelson, 16 May 1916, ibid., box 10; Loeb to Rutherford, 11 March 1918, box 13. For another effort at job placement for Leonard, see Loeb to E. G. Conklin, 23 May 1916, EGC, box 32.

96. See Loeb to Ernest Rutherford, 31 Aug. 1920, JL(GC), box 13; Loeb to T. H. Morgan, 18 May 1921, box 9.

97. Morgulis to Loeb, 19 Oct. 1918, ibid., box 10.

98. Ibid.

99. Hecht to W. J. Crozier, 27 Jan. 1921, WJC.

100. Loeb to Simon Flexner, 7 Feb. 1921, JL(GC), box 4.

101. Loeb to Hecht, 19 April 1920, ibid., box 6.

102. Loeb to Franklin C. McLean, 8 May 1919, ibid., box 10; Loeb to G. H. Parker, 17 Feb. 1920, box 11.

103. Loeb to Franklin C. McLean, 8 May 1919, ibid., box 10. The request was in a letter from McLean, 6 May 1919, ibid.

104. Loeb to G. H. Parker, 5 and 17 Feb. 1920, ibid., box 11.

105. Loeb to Frank A. Hartman, 19 April 1920, ibid., box 6.

106. Norbert Wiener, *I Am a Mathematician* (Cambridge: MIT Press, 1956), pp. 27–29.

107. Hecht to Crozier, 3 July 1922, WJC.

108. Loeb to F. G. Cottrell, 3 Feb. 1921, JL(GC), box 3.

109. See correspondence between Loeb and J. McKeen Cattell, 30 April to 24 July 1913, ibid., box 2.

110. Loeb to Hecht, 21 Jan. 1921, ibid., box 6.

111. Loeb to Leonor Michaelis, 19 April 1923, ibid., box 9.

112. Hecht to Crozier, 2 Aug. 1922, WJC.

113. Hecht to Crozier, 5 Feb. 1923, ibid.

114. See Loeb to Simon Flexner, 1 Feb. 1924, JL(GC), box 4.

115. Ibid.

116. B. A. Younker to Loeb, 29 Jan. 1924, RU(JL).

117. Donald S. Fleming, "Jacques Loeb," *Dictionary of Scientific Biography* 8:446.

118. *Journal of General Physiology* 8 (1928): lix.

119. See Loeb to Just, 20 May, 23 Oct., and 29 Dec. 1914, JL(GC), box 7.

120. See Just to Lillie, 12 Aug. 1936, FRL(W).

121. See Loeb to Maurice Caullery, 4 Feb. 1918, JL(GC), box 3; also, J. McKeen Cattell to Loeb, 30 April 1913, box 2.

122. At the NAACP Sagamore conference in 1914, Loeb based his argument for intermarriage on the concept of hybrid vigor: see Loeb to George W. Coleman, with enclosure, 13 Oct. 1914, JL(GC), box 3. See also his correspondence with Charles W. Eliot, 6 to 10 Feb. 1915, box 4. Attacks on "racial biology" appear frequently in his correspondence with European scientists, especially Svante Arrhenius (box 1) and Wilhelm Ostwald (box 11). For his speeches, see MSS in JL(BAS), box 44.

123. See Just to Julius Rosenwald, 16 April 1922, JR, box 19, folder 3?

124. Letters to the author from W. B. Baker, 6 May 1977, and Alfred M. Lucas, 22 June 1977.

125. Letters to the author from E. G. Stanley Baker, 11 July 1977; George E. Hutchinson, 17 May 1977; Robert L. Kroc, 26 May 1977; Bernice F. Pierson, 3 Aug. 1977; Mary Warters, 17 Aug. 1977; Allyn J. Waterman, 9 Aug. 1977. Just had written Lillie (18 May 1922) about using the *Arbacia* material for these lectures: FRL(C), box 4, folder 20.

Just prepared a series of articles on the subject, "Initiation of Development in the Egg of *Arbacia*, I–VI" (1922–29). This author found references to these articles in the major domestic and foreign biological journals by the following scientists (with year of citation): Charles Pérez (1923–24); G. S. Carter (1924); L. V. Heilbrunn (1924); T. Péterfi (1924); J. Runnström (1924); Henry J. Fry (1925); Myra M. Sampson (1926); John Runnström (1928); L. Genevois (1928–29); John Runnström (1928–29); E. Bataillon (1929); Reuben Blumenthal (1930); Charles Pérez (1930); Ferdinand Reith (1930); John Runnström (1930); Charles Pérez (1931); Albert Tyler (1931); D. M. Whitaker (1931); G. S. Carter (1932); E. Newton Harvey (1932); Ethel Browne Harvey (1932); A. D. Hobson (1932); Dorothy R. Stewart and M. H. Jacobs (1932); Albert Tyler (1932); Albert Tyler and Jack Schultz (1932); W. A. Dorfman (1933); E. Bataillon and Tchou -Su (1934); Emile Godlewski (1934); H. Y. Chase (1935); Hans-Joachim Elster (1935); John Runnström (1935); Jean M. Clark (1936); A. J. Waterman (1936); John Runnström (1937); Albert Tyler and Bradley T. Scheer (1937); Floyd Moser (1939); John Runnström (1939); Donald P. Costello (1940); M. J. Kopac (1940); Floyd Moser (1940); Ralph S. Lillie (1941); Albert Tyler (1941); Robert Chambers (1942); Teru Hayashi (1945); Willis E. Pequegnat (1948); Charles Thibault (1949); R. Phillips Dale (1950); John Runnström (1950); R. D. Allen and J. L. Griffin (1958); J. M. Butros (1959); Y. Hiramoto (1962); Masaru Ishikawa (1962); Takashi Iwamatsu (1966); Martin I. Sachs and Everett Anderson (1970); Frank J. Longo and William Plunkett (1973); A.F.W. Hughes (1976); R. Rappaport (1976).

126. Mary L. Austin to the author, 24 Aug. 1977. MacDougall acknowledged Just's help in her article "Modifications in *Chilodon uncinatus* Produced by Ultraviolet Radiation," *Journal of Experimental Zoology* 54 (1929): 95.

127. Interview with Sally Hughes-Schrader, 8 July 1977.

128. Beerman to the author, 28 March 1977.

129. Letters to the author from K. S. Cole, 13 April 1977, and W. B. Baker, 6 May 1977.

130. For details on Just's income at this time, see Just to Abraham Flexner, 18 April 1925, GEB, box 695, folder 6967. For information on MBL salaries, see Lillie to E. G. Conklin, 8 Dec. 1925, EGC, box 32.

131. Lillie to Heilbrunn, 6 March 1925, LVH.

132. Letters to the author from Wilfred M. Coperhaver, 30 April 1977; Harriet W. Hoadley, 17 Oct. 1977; Janet K. Nelson, 19 April 1977; Sally Hughes-Schrader, 25 May 1977; Don C. Warren, 23 Aug. 1977.

133. Letters to the author from Ethel S. Carpenter, 15 July 1977; Aurin M. Chase, 20 April 1977; Francis F. Dunbar, 15 Aug. 1977; John M. Fogg, Jr., 18 April 1977; Mark Graubard, 12 May 1977; Jessie S. Hendry, 14 Aug. 1977; S. A. Matthews, 9 June 1977; Irvine H. Page, 11 Aug. 1977; Madelene E. Pierce, 28 July 1977; Herbert Pollack, 27 July 1977; Maurice N. Richter, 26 Sept. 1977; Lyman S. Rowell, 27 July 1977; Emily E. Trueblood, 29 April 1977; Shields Warren, 2 Aug. 1977.

134. Ruth H. Romer to the author, 19 March 1978.

135. Mary L. Austin to the author, 24 Aug. 1977.

136. W. C. Curtis to Just, 1 Jan. 1925, EEJ(H), box 125-3, folder 54.

137. As told by his wife, Evelyn Stern, to the author, 23 July 1977.

138. Sally Hughes-Schrader to the author, 25 May 1977; also interview, 8 July 1977.

139. Interview with Sally Hughes-Schrader, 8 July 1977.

140. T. H. Morgan, *Experimental Embryology* (New York: Columbia Univ. Press, 1927), pp. 18, 26, 38, 44–45, 48–49, 52, 95, 99–101, 180, 186, 547–48, 550, 579–80; E. B. Wilson, *The Cell in Development and Heredity*, 3d ed. (New York: Macmillan, 1925), pp. 421–22, 476, 484, 1104, 1172–73.

141. Interview with Paul Reznikoff, 11 May 1977.

142. Lillie to L. V. Heilbrunn, 6 March 1925, LVH.

143. Heilbrunn to F. R. Lillie, 8 March 1925, ibid.

144. Interview with Janet K. Nelson, 19 April 1977.

145. Interview with Sally Hughes-Schrader, 8 July 1977.

146. Letters to the author from W. B. Baker, 6 May 1977; William F. Diller, 25 April 1977; Henry Guerlac, 12 May 1977; Maurice N. Richter, 26 Sept. 1977; Roland Walker, 10 Aug. 1977. Also, interview with Lester G. Barth, 25 Jan. 1977.

147. Interview with S. M. Nabrit, 5 July 1977. The German biologist may have been Felix Bernstein, a visiting biostatistician from the University of Göttingen. Bernstein delivered a lecture on heredity and human races, using "the Negritos" of the Philippines as a case study: see *Collecting Net* 3 (1928): 18–20.

148. Interview with Sally Hughes-Schrader, 8 July 1977; Paul Reznikoff, 11 May 1977; Donald P. Costello, 3 April 1977.

149. Interview with Sally Hughes-Schrader, 8 July 1977.

150. For one Woods Hole scientist's deference to Just on these matters, see W. C. Allee, "Note on Animal Distribution Following a Hard Winter," *Biological Bulletin* 36 (1919): 103.

151. Interview with Sally Hughes-Schrader, 8 July 1977.

152. Interview with Costello, 3 April 1977.

153. Interview with Nabrit, 5 July 1977; also, Nabrit, "Phylon Profile VIII: Ernest E. Just," *Phylon* 7 (1946): 124. Plough's lecture, "New Facts on Differentiation in the Egg of the Sea-Urchins *Strongylocentrotus* and *Echninus*," was summarized and reviewed by Conklin in *Collecting Net* 3 (1928): 4-5.

154. Interview with Paul Reznikoff, 11 May 1977.

155. *Collecting Net* 3 (1928) no. 1: 4. The articles appeared under the title "Methods for Experimental Embryology with Special Reference to Marine Invertebrates," in *Collecting Net* 3 (1928).

156. Just to Benjamin M. Duggar, 8 April 1926, RF.

157. Ware Cattell to Just, 29 February 1932, EEJ(H), box 125-1, folder 45; also *Collecting Net* 5 (1930): 4.

158. *Basic Methods for Experiments on Eggs of Marine Animals* (Philadelphia: Blakiston's, 1939). It was quoted and followed up by Donald P. Costello and Catherine Henley, *Methods for Obtaining and Handling Marine Eggs and Embryos*, 2d ed. (Woods Hole: Marine Biological Laboratory, 1971). For Just's early correspondence with Blakiston's and Darwin, see Robert Bowman to Just, 28 July 1928, EEJ(H), box 125-2, folder 36; Ware Cattell to Just, 29 Feb. 1932, and Just to Ware Cattell, 9 March 1932, folder 45.

159. Just to F. R. Lillie, 9 Sept. 1929, FRL(W).

160. Just to Lillie, 9 Nov. 1930, ibid.; see also Gregory Pincus to Just, 26 Oct. 1930, EEJ(H), box 125-7, folder 119.

161. Alfred C. Redfield to Just, 12 Dec. 1929, DPC.

162. Just to Lillie, 9 Nov. 1930, FRL(W).

163. Lillie to Just, 11 Nov. 1930, ibid.

164. Just to Lillie, 6 Dec. 1928, ibid.

4. THE ROLE OF FOUNDATION SUPPORT, 1920-29

1. The general history of American philanthropy, particularly as it relates to the development of universities and research institutions, has been well documented in Merle Curti and Roderick Nash, *Philanthropy in the Shaping of American Higher Education* (New Brunswick, N.J.: Rutgers Univ. Press, 1965); Warren Weaver, *U.S. Philanthropic Foundations: Their History, Structure, Management, and Reward* (New York: Harper and Row, 1967); Waldemar A. Nielson, *The Big Foundations* (New York: Columbia Univ. Press, 1972); Abraham Flexner, *Funds and Foundations: Their Policies Past and Present* (reprint ed., New York: Arno Press, 1976); Frederick P. Keppel, *The Foundation: Its Place in American Life* (New York: Macmillan, 1930). See also Stanley Coben, "Foundation Officials and Fellowships: Innovation in the Patronage of Science," *Minerva* 14 (1976): 225-40.

2. Walter Dyson, *Howard University, The Capstone of Negro Education: A History, 1867-1940* (Washington: Howard Univ., 1941), p. 37. This study is useful, though less comprehensive than Rayford W. Logan, *Howard University: The First Hundred Years, 1867-1967* (New York: New York Univ. Press, 1969). Logan's notes, pp. 604-16, provide useful leads to many of the important primary and secondary sources on the history of Howard University.

3. Dyson, *Howard University*, p. 305. For a summary of the early history of Howard's congressional appropriations, see Babalola Cole, "Appropriation Politics and Black Schools: Howard University and the U.S. Congress, 1879–1928," *Journal of Negro Education* 46 (1977): 7–23.

4. Abraham Flexner, *Medical Education in the United States and Canada: A Report to the Carnegie Foundation for the Advancement of Teaching* (New York: Carnegie Foundation for the Advancement of Teaching, Bulletin no. 4, 1910), pp. 88, 237. For an extensive analysis of Flexner's educational ideas, see Michael R. Harris, *Five Counterrevolutionists in Higher Education* (Corvallis: Oregon State Univ. Press, 1970). See also Carleton B. Chapman, "The Flexner Report," *Daedalus* 103 (1974): 105–17. Flexner's particular influence on black medical education is the subject of a more recent study: William Harvey Carson, Jr., "Medical Education Reform in America, 1868–1928: A Study of Abraham Flexner and his Direct Effect on the Reform of Medical Education for Blacks" (honors thesis, Harvard Univ., 1980). See also Marlene Yvonne MacLeish, "Medical Education in Black Colleges and Universities in the United States of America: An Analysis of Black Medical Schools Between 1867 and 1967" (Ed.D. diss., Harvard Univ., 1978).

5. Flexner, *Medical Education*, p. 180.

6. For a detailed transcript of the speeches given by Pritchett and Washington, see *Howard University Record* 5 (1911): 3–18. Excerpts from the other speeches ·can be found in *Howard University Catalog*, 1911–12, pp. 250–53.

7. Dyson, *Howard University*, p. 394.

8. S. M. Newman to Just, 28 Jan. 1916, EEJ(H), box 125-7, folder 114.

9. Just to Lillie, May 1919, FRL(C), box 4, folder 20. Just mentioned the range of his medical school salary in a letter to William H. Welch, 23 Nov. 1919, GEB, box 695, folder 6967. See also manuscript appointment card (19 March 1915) in the E. E. Just file, Moorland-Spingarn Research Center, Howard University.

10. Just to W. C. Graves, 15 Jan. 1921, JR, box 19, folder 3. For a reference to the war program in universities, see Just to Lillie, 5 March 1919, FRL(W). The role of blacks in the war effort is documented in Emmett J. Scott, "The Negro and the War Department," *Crisis* 15 (1918): 76. See also *Crisis* 17 (1919): 126, 137, 215, 221–23; 18 (1919): 87, 95–96. One of Just's students, the late Roscoe C. McKinney, provided interesting insights into the issue of segregation in the U.S. Army: interview, 5 December 1976.

11. Just to Lillie, [May 1919], FRL(C), box 4, folder 20.

12. See Just to Lillie, 5 March 1919, FRL(W); Just to Lillie, [May 1919], FRL(C), box 4, folder 20; Just to Maynard M. Metcalf, 13 March 1925, RF.

13. See Durkee to Just, 13 Dec. 1918, EEJ(H), box 125-4, folder 65.

14. Just to Lillie, 4 Sept. 1919, FRL(C), box 4, folder 20; also, Just to F. R. Lillie, [May 1919], ibid.

15. *Crisis* 36 (1929): 203.

16. See Just to F. R. Lillie, 24 Sept. and 23 Nov. 1919, FRL(C), box 4, folder 20; Lillie to Just, 29 Nov. 1919, ibid.

17. There is no full-length study of the National Research Council. The *Bulletin of the National Research Council* and the *Reprint and Circular Series of the National Research Council* contain several commentaries on the history and early development of the organization. The most detailed of these is

W. H. Howell, ed., "A History of the National Research Council, 1919–1933," *Reprint and Circular Series* no. 106 (1933), reprinted from *Science* n.s. 77 (1933): 355–60, 500–503, 552–54, 618–20; 78 (1933): 26–29, 93–95, 158–61, 203–6, 254–56. See also George Ellery Hale, "A National Focus of Science and Research," *Reprint and Circular Series* no. 39 (1922), reprinted from *Scribner's Magazine*, November 1922; "The National Importance of Scientific and Industrial Research: The Purpose of the National Research Council," *Bulletin* 1 (1919): 1–7. There are some historical summaries in the organization's official published records, e.g., *National Research Council: Organization and Members* (Washington, D.C.: National Research Council) and *Report of the National Academy of Sciences* (Washington, D.C.: Government Printing Office). Philip Miller Boffey, *The Brain Bank of America: An Inquiry into the Politics of Science* (New York: McGraw-Hill, 1975) focuses on the later development of the council, beginning in the 1940s. Daniel J. Kevles, *The Physicists: The History of a Scientific Community in Modern America* (New York: Knopf, 1978) provides a discussion on its history and development.

18. The only full-length biography of Rosenwald is M. R. Werner, *Julius Rosenwald: The Life of a Practical Humanitarian* (New York: Harper and Brothers, 1939). The best summary of Rosenwald's philanthropic activities is Edwin R. Embree, "Julius Rosenwald," in *Julius Rosenwald Fund: Review for the Two-Year Period, 1931–1933* (Chicago, 1933), pp. 1–14. See also Pauline K. Angell, "Julius Rosenwald," *American Jewish Yearbook* 34 (1932–33): 141–76; W. E. B. Du Bois, "Rosenwald," *Crisis* 39 (1932): 58; "Julius Rosenwald: In Memoriam" (record of a memorial held 27 March 1932 and broadcast by the National Broadcasting Company under the auspices of the American Jewish Joint Distribution Committee) (n.p., 1932); "The Julius Rosenwald Centennial Observance at the University of Chicago, Oct. 15, 1962" (Chicago: Univ. of Chicago Press, 1963); Alfred Q. Jarrette, *Julius Rosenwald, Son of a Jewish Immigrant, a Builder of Sears, Roebuck and Company, Benefactor of Mankind: A Biography Documented* (Greenville, S.C.: Southeastern Univ. Press, 1975).

19. Edwin R. Embree, *Julius Rosenwald Fund: A Review to June 30, 1928* (Chicago, 1928), p. 6. For a full history of the fund, see Edwin R. Embree and Julia Waxman, *Investment in People: The Story of the Julius Rosenwald Fund* (New York: Harper and Brothers, 1949).

20. Edsall to Abraham Flexner, 29 Dec. 1919, GEB, box 702, folder 7221.

21. See *Crisis* 20 (1920): 193.

22. Lillie to McClung, 29 Nov. and 8 Dec. 1919, FRL(C), box 4, folder 20.

23. Just to Lillie, 23 Nov. 1919, ibid.

24. There are two full-length biographies of Welch: Simon Flexner and James Thomas Flexner, *William Henry Welch and the Heroic Age of American Medicine* (New York: Viking Press, 1941); and Donald Fleming, *William Henry Welch and the Rise of Modern Medicine* (Boston: Little, Brown, 1954). See also Simon Flexner, "William Henry Welch," *Science* n.s. 52 (1920): 417–33; Milbank Memorial Fund, *The Eightieth Birthday of William Henry Welch: The Addresses Delivered at the Ceremonies in Memorial Continental Hall, Washington, D. C., April 8, 1930* . . . (New York: W. E. Rudge, 1930); Simon Flexner, "William Henry Welch," *Biographical Memoirs* (National Academy of Sciences) 22 (1943): 215–31.

25. See Abraham Flexner to Edwin R. Embree, 29 July 1930, JRF, folder 3.

26. Just to Welch, 23 November 1919, GEB, box 695, folder 6967; see also Just to Welch, 15 May 1921, WHW.

27. Welch to Flexner, 24 and 30 Nov. 1919, GEB, box 695, folder 6967; see also Welch to Just, 24 Nov. 1919, EEJ(H), box 125-8, folder 148.

28. Simon to Abraham Flexner, 28 Nov. 1919, GEB, box 695, folder 6967.

29. Abraham Flexner to Welch, 26 Nov. 1919, ibid.

30. McClung to Rosenwald, 7 Jan. 1920, JR, box 19, folder 3; Abraham Flexner to Graves, 23 Jan. 1920, JR, box 19, folder 3, and Graves to Rosenwald, 18 Feb. 1920, ibid.

31. See Abraham Flexner to Rosenwald, 13 Feb. and 16 March 1920, ibid.; Emmett J. Scott to Flexner, 25 August 1920, GEB, box 695, folder 6967.

32. McClung to Durkee, 9 and 25 March 1920; RF; McClung to Lillie, 26 March 1920, ibid.

33. Durkee to McClung, 17 and 30 March 1920, ibid.

34. Rosenwald to Flexner, 5 Oct. 1920, JR, box 19, folder 3.

35. Just to Heilbrunn, 8 June 1920, LVH.

36. Ibid.

37. Just to Lillie, 18 Sept. 1920, FRL(C), box 4, folder 20.

38. Just to McClung, 16 Feb. 1921, RF.

39. Just to Heilbrunn, 11 Dec. 1920, LVH.

40. Durkee to L. R. Jones, 19 Oct. 1921, RF; see also Durkee to Lillie, 21 July 1922, ibid.

41. Just to Graves, 15 Oct. 1921, JR, box 19, folder 3; L. R. Jones to Graves, 26 Oct. 1921 (with enclosures: letter from Durkee and report from Just, 19 Oct. 1921), RF; Lillie to Graves, 31 Oct. 1921, FRL(C), box 4, folder 20. The progress of Just's work in the early part of the summer is well plotted in his letters to Heilbrunn, 14 and 22 May and 4 June 1921, LVH.

42. See Just to Lillie, 18 Sept. 1920, FRL(C), box 4, folder 20.

43. Just to McClung, 18 Sept. 1920, RF.

44. Flexner to Just, 23 Oct. 1920, GEB, box 695, folder 6967.

45. Just to Flexner, 16 July 1926, RF.

46. The only adequate summary of Flexner's life and work is an auto-biography entitled *I Remember* (New York: Simon & Schuster, 1940). A good brief sketch is "Abraham Flexner," in Maxine Block, ed., *Current Biography: Who's News and Why, 1941* (New York: H. W. Wilson, 1941), pp. 289–91. See also Franklin Parker, "Abraham Flexner, 1866–1959" *History of Education Quarterly* 2 (1962): 199–209, and the obituary in the *New York Times*, 22 Sept. 1959, p. 1. A series of interviews done with Flexner as part of the New York Times Oral History Project has been published on microfiche as *Reminiscences* (Glen Rock, N.J.: Microfilming Corp., 1972).

47. For the relationship between Flexner and Quinland, see "William Quinland, 1919-22," GEB, box 702, folder 7024. See also Flexner to Graves, 21 March 1923, JR, box 19, folder 3.

48. Flexner to Rosenwald, 10 May 1930, JR, box 4, folder 2.

49. Flexner to Rosenwald, 25 July 1919, GEB, box 702, folder 7024.

50. See Just to Lillie, 15 May 1922, RF. The reference is to a letter Durkee wrote James R. Angell, National Research Council chairman (1919-20). This

letter apparently does not survive: it cannot be found in the council's files on Just, and there is no copy of it in the Angell Papers at Yale University.

51. Just to Lillie, 15 May 1922, RF; see also Just to L. R. Jones, 17 March 1922, RF.

52. Just to Lillie, 15 May 1922, RF; Just to Abraham Flexner, 21 Oct. 1920, GEB, box 695, folder 6967.

53. Details on "projects from individuals" in the Division of Biology and Agriculture can be found in the various annual reports submitted to the National Research Council executive by division chairmen. A ms. copy of the complete report for 1920–21, written by the chairman, C. E. McClung, is preserved in the Conklin papers, EGC, box 42. A summary was published in *Report of the National Academy of Sciences for the Year 1920* (Washington, D.C.: Government Printing Office, 1921), pp. 76–80.

54. Just to Lillie, 15 May 1922, RF.

55. Just to Heilbrunn, [April 1921], LVH; also Just to Rosenwald, 16 April 1922, JR, box 19, folder 3. For the reaction of another scientist, see B. H. Willier to W. J. Crozier, 15 April 1922, WJC.

56. See Lillie to Just, 20 Dec. 1920, FRL(C), box 4, folder 20.

57. Just to Heilbrunn, 4 Jan. 1922, LVH; also interview with Maribel Just Butler, 29 July 1979.

58. Just to Heilbrunn, 4 Jan. 1922, LVH.

59. Lillie to W. C. Graves, 31 Oct. 1921, FRL(C), box 4, folder 20.

60. Just to Heilbrunn, 4 Jan. 1922, LVH.

61. Just to J. R. Schramm, 16 [June] 1922, RF.

62. Lillie to L. R. Jones, 22 May 1922, RF; see also Just to Jones, 17 March 1922, ibid.

63. Jones made this comment in the margin at the bottom of the letter from Lillie cited in note 62.

64. Just to J. R. Schramm, 16 June 1922; Just to Lillie, 19 Aug. 1922, RF.

65. W. C. Graves to J. R. Schramm, 25 March 1922, RF; also Lillie to Just, 12 May 1922, FRL(C), box 4, folder 20.

66. See Lillie to Rosenwald, 14 July 1924, JR, box 19, folder 3; Graves to J. R. Schramm, 29 July 1924, ibid; W. C. Curtis to Just, 1 Jan. 1925, EEJ(H), box 125-3, folder 54; Just to the National Research Council, 13 March 1925, RF; Just to Edwin R. Embree, 16 July 1930, JRF, folder 3.

67. See Nathan I. Huggins, *Harlem Renaissance* (New York: Oxford Univ. Press, 1971); also Arna Wendell Bontemps, ed., *The Harlem Renaissance* (New York: Dodd, Mead, 1972).

68. *Crisis* 30 (1925): 270; also Logan, *Howard University*, p. 236.

69. Just to Melville J. Herskovits, 22 June 1925, MJH.

70. *Crisis* 32 (1926): 38.

71. Just to Lillie, 14 Oct. 1921, FRL(C), box 4, folder 20.

72. See Marcus F. Wheatland to Just, 4 July 1925, EEJ(H), box 125-8, folder 150.

73. Just to George L. Streeter [September 1924], CIW.

74. Streeter to Just, 24 Oct. 1924, ibid.

75. Just to Abraham Flexner, 13 Feb. 1925, GEB, box 695, folder 6967.

76. Flexner to John C. Merriam, 14 Feb. 1925; Merriam to Flexner, 20 Feb.

1925, CIW. See also Gilbert to Streeter, 20 Feb. 1925, and Gilbert to Paul Bartsch, C. B. Davenport, and Burton I. Livingston, 21 Feb. 1925, ibid.

77. Streeter to Gilbert, 24 Feb. 1925, ibid.

78. Davenport to Gilbert, 2 March 1925, ibid.

79. For Davenport's commitment to eugenics, see his correspondence with E. G. Conklin and Ross G. Harrison, EGC, box 30; RGH, box 9, folder 659. His articles on race biology include "The Racial Element in National Vitality," *Popular Science Monthly* 86 (1915): 331–33; "The Effect of Race Intermingling," *Proceedings of the American Philosophical Society* 56 (1917): 364–68; "Control of Universal Mongrelism. How a Eugenist Looks at the Matter of Marriage," *Good Health* 10 (1928): 31; "Are There Genetically Based Mental Differences between the Races?" *Science* n.s. 68 (1928): 628. His most comprehensive study was sponsored by the Carnegie Institution and published as "Race-crossing in Jamaica," *Carnegie Institution of Washington Publications* no. 395 (1929).

80. Glaser to Streeter, 25 March 1925, CIW.

81. Just to Maynard M. Metcalf, 17 April 1925, RF; also Just to Abraham Flexner, 18 April 1925, GEB, box 695, folder 6967.

82. Just to Rosenwald, 27 Aug. 1920, JR, box 19, folder 3.

83. See Just to Graves, 14 Jan. and 15 Oct. 1921; and Graves to Just, 19 Jan. and 18 Oct. 1921, ibid.

84. The recipients of Rosenwald's generosity are too numerous to mention: almost all of the fifty-plus boxes of correspondence at the University of Chicago contain several examples of personal financial transactions. One of Rosenwald's few rejections went to Carter G. Woodson, the eminent black historian: see Graves to Woodson, 14 Feb. 1918, JR, box 2, folder 24.

85. James R. Angell to Abraham Flexner, 16 April 1920, GEB, box 695, folder 6967.

86. Just to Graves, 13 May 1922; Graves to J. R. Schramm, 13 July 1922, JR, box 19, folder 3.

87. Just to Graves, 13 May 1922 (with enclosure, R. S. Wilkinson to Just, 9 May 1922), ibid.

88. Just to Rosenwald, 27 Sept. 1924, ibid.

89. Just to Graves, 29 April 1926, ibid., folder 4.

90. See Just to Abraham Flexner, 22 Feb. 1925, GEB, box 695, folder 6967.

91. Just to Graves, 25 March 1923, JR, box 19, folder 3; Just to Graves, 24 May 1926, ibid., folder 4.

92. Just to Rosenwald, 5 Jan. 1928, JR, box 19, folder 4. See JMC, box 61, for further information on Just's election to *American Men of Science*. Only 13 of 150 ballots survive; 6 of those 13 recommend Just for a star.

93. Just to Julius Rosenwald, 5 Jan. 1928, JR, box 19, folder 4.

94. Just to Abraham Flexner, 12 Oct. 1924, GEB, box 695, folder 6967.

95. See Maynard M. Metcalf to Lillie, 17 March 1925, RF.

96. Metcalf to J. Stanley Durkee, 9 May 1925, RF; also Metcalf to Just, 17 March and 7 May 1925, ibid; "Resolution Taken by the Division of Biology and Agriculture at its Annual Meeting, April 26, 1925," ibid.; minutes of that meeting, EGC, box 42. The resolution is alluded to but not reproduced in *Report of the National Academy of Sciences: Fiscal Year 1924-25* (Washington, D.C.: Government Printing Office, 1926), p. 93.

97. Metcalf to Just, 17 March 1925, JR, box 19, folder 4. For further information on the council's approach to the question of research opportunities in black colleges, see the minutes for 22 April 1923, 26 April 1924, and 26 April 1925, EGC, box 42. The issue was not mentioned in the published *Report of the National Academy of Sciences.*

98. Just to Flexner, 23 April 1925, GEB, box 695, folder 6967.

99. Abraham Flexner to Just, 27 April 1925, GEB, box 695, folder 6967.

100. W. E. B. Du Bois, "Gifts and Education," *Crisis* 29 (1925): 151; see also "The Negro College," ibid. 12 (1916): 29; editorial, ibid. 13 (1917): 111.

101. Du Bois, "Negro Education," ibid. 15 (1918): 178.

102. *General Education Board: Review and Final Report, 1902–1964* (New York: General Education Board, 1964), p. 3. The standard history of the GEB is Raymond B. Fosdick, *Adventure in Giving: The Story of the General Education Board* (New York: Harper and Row, 1962). See also *The General Education Board: An Account of its Activities, 1902–14* (New York: General Education Board, 1915). There is some historical information in the various issues of the board's *Annual Report.*

103. A clear and comprehensive early account of Washington's life is Emmett J. Scott and Lyman Beecher Stowe, *Booker T. Washington: Builder of a Civilization* (Garden City, N.Y.: Doubleday, Page, 1916). More recent biographies include Emma Lou Thornbrough, ed., *Booker T. Washington* (Englewood Cliffs, N.J.: Prentice-Hall, 1969) and Louis R. Harlan, *Booker T. Washington: The Making of a Black Leader, 1856–1901* (New York: Oxford Univ. Press, 1972). A good discussion of the intellectual issues is August Meier, *Negro Thought in America, 1880–1915: Racial Ideologies in the Age of Booker T. Washington* (Ann Arbor: Univ. of Michigan Press, 1963). See also Louis R. Harlan, "Booker T. Washington and the White Man's Burden," *American Historical Review* 71 (1966): 441–67; Allen W. Jones, "Role of Tuskegee Institute in the Education of Black Farmers," *Journal of Negro History* 60 (1975): 252–67; Alfred Young, "Educational Philosophy of Booker T. Washington: A Perspective for Black Liberation," *Phylon* 37 (1976): 224–35.

104. *Charleston News and Courier*, 4 Jan. 1899, p. 6.

105. The most scholarly biography of Carver is Linda O. McMurry, *George Washington Carver, Scientist and Symbol* (New York: Oxford Univ. Press, 1981). Other biographies include Rackham Holt, *George Washington Carver: An American Biography* (Garden City, N.Y.: Doubleday, Doran, 1943); and Lawrence Elliott, *George Washington Carver: The Man Who Overcame* (Englewood Cliffs, N.J.: Prentice-Hall, 1967). See also "Negroes Distinguished in Science," *Negro History Bulletin* 2 (1939): 69–70; and Peggy Robbins, "Gentle Genius: George Washington Carver," *American History Illustrated* 8 (June 1973): 11–21. The Pittsburgh radio program was hosted by Floyd T. Calvin, a black journalist, in 1927. A newspaper clipping on the program, source unknown but dated 25 Oct. 1927, survives in "Laura Spelman Rockefeller Memorial, Negro Problems Conference, 1927–29" (hereafter cited as NPC), GEB.

106. W. E. B. Du Bois, "The General Education Board," *Crisis* 37 (1930): 230; see also Du Bois, "Negro Education," ibid. 15 (1918): 177.

107. See Abraham Flexner to Just, 29 Dec. 1922, GEB, box 695, folder 6967.

108. Flexner to Just, 24 Feb. 1925, ibid.

109. Lillie to Metcalf, 17 Feb. 1925, RF; also, Lillie to Graves, 16 Feb. 1925, JR, box 19, folder 4.

110. Just to Heilbrunn, 16 and 29 March 1925, LVH.

111. See Just to Division of Biology and Agriculture, National Research Council, 11 Aug. 1926, RF; also Just to Graves, 17 Jan. 1927, JR, box 19, folder 4; Graves to Just, 26 Jan. 1927, ibid.; Just to L. J. Cole, 7 May 1927, RF; L. J. Cole to Maurice Holland, 4 June 1927, ibid.

112. Just to Heilbrunn, 29 March 1925, LVH.

113. Emmett J. Scott to Abraham Flexner, 9 March 1920; Flexner to Scott, 10 March 1920, GEB, box 695, folder 6967; Just to Lillie, 15 May 1922, RF.

114. Information on Young's educational background can be found in fellowship applications she submitted to W. W. Brierley at the General Education Board in 1927 and 1928, GEB, box 29, folder 271. Pat Yanni, director of pupil services in the Burgettstown Public Schools, tried unsuccessfully to locate some of Young's friends and relatives. She did, however, find a curiously precocious little story written by Young for the school newspaper: "All Aboard," *Graph* 2 (1914): 1–2.

115. Just to Graves, 14 Jan. 1921, JR, box 19, folder 3.

116. The work with Young on ultraviolet and water is discussed by Just in a report to the National Research Council, 8 April 1928, RF. For the results of the ultraviolet work, see *Anatomical Record* 27 (1927): 130–31; also, Just to Charles Packard, 11 Nov. 1927, EEJ(H), box 125-7, folder 117; F. L. Stevens to Just, 21 April 1928, ibid., box 125-8, folder 137. The first in a series of papers on the water research was published in *Physiological Zoology* 1 (1928): 122–35; see also Just to W. J. V. Osterhout, 3 Nov. 1930, EEJ(H), box 125-7, folder 116. For Young's later collaboration with Heilbrunn and Costello, see L. V. Heilbrunn and R. A. Young, "Indirect Effects of Radiation on Sea Urchin Eggs," *Biological Bulletin* 69 (1935): 274–79; D. P. Costello and R. A. Young, "The Mechanism of Membrane Elevation in the Egg of *Nereis*," *Biological Bulletin* 77 (1939): 311. Heilbrunn and Karl M. Wilbur refer to Young's unpublished work on calcium activation of *Mactra* in "Stimulation and Nuclear Breakdown in the *Nereis* Egg," *Biological Bulletin* 73 (1937): 562–63. See also Young, "The Effects of Roentgen Irradiation on Cleavage and Early Development in the Annelid, *Chaetopterus pergamentaceus*." *Biological Bulletin* 75 (1938): 378.

117. Just to Julius Rosenwald, 4 Sept. [Oct.?] 1926, JR, box 19, folder 4.

118. Just to Abraham Flexner, 26 Sept. 1925, GEB, box 695, folder 6967.

119. Just to S. O. Mast, 26 April 1925, SOM.

120. The studies in question were R. A. Young, "On the Excretory Apparatus in *Paramecium*," *Science* n.s. 60 (1924): 224; and Dmitriy Nasonov, "Der Excretionsapparat (kontraktile Vacuole) der Protozoa als homologen des Golgischen Apparatus der Metazoazellen," *Archiv für mikroskopische Anatomie und Entwicklungsmechanik* 103 (1924): 437–82. The most detailed commentary on this work was by Francis E. Lloyd, "The Contractile Vacuole," *Biological Reviews of the Cambridge Philosophical Society* 3 (1928): 329–58.

121. See Graves to N. C. Plimpton, 11 and 19 May 1925, JR, box 19, folder 4; also, Plimpton to Graves, 15 May 1925, and Emmett J. Scott to Plimpton, 9 June 1925, ibid.

122. Graves to Lillie, 2 May 1925, ibid.

123. Lillie to Graves, 5 May 1925, ibid.

124. Du Bois, "The General Education Board," *Crisis* 37 (1930): 230.

125. "Negro Problems Conference Report," p. 129, NPC, GEB. See also Just to Austin Clark, 8 Dec. 1924, AHC; Just to Abraham Flexner, 13 Feb. 1925, GEB, box 695, folder 6967; Just to Rosenwald, 13 May 1926 and 10 March 1927, JR, box 19, folder 4.

126. Just mentioned Herskovits's appointment in a letter to Maynard M. Metcalf, 13 March 1925, RF. See also correspondence between Herskovits and Locke, and Just and Herskovits, MJH. Herskovits thanked Just and Locke for their help in *The Anthropometry of the American Negro* (New York: Columbia Univ. Press, 1930), p. xiv.

127. Just to Edwin R. Embree, 27 Oct. 1931, JRF, folder 5.

128. Alain Locke, ed., *The New Negro: An Interpretation* (New York: A. and C. Boni, 1925), p. 4.

129. See Just to Herskovits, 22 June 1925, Herskovits to Just, 24 June 1925, MJH.

130. Charles H. Houston to Julian W. Mack, 9 June 1926, JR, box 19, folder 2.

131. Just to H. J. Thorkelson, 20 April 1927, GEB, box 27, folder 249; Just to Rosenwald, 10 March 1927, JR, box 19, folder 4. See also Just to Rosenwald, 17 Sept. 1926 and 21 April 1927, ibid.; Just to Abraham Flexner, 20 April 1927, GEB, box 695, folder 6968.

132. Just to Edwin R. Embree, 4 Oct. and 11 Nov. 1928, JRF, folder 1.

133. Memorandum by H. J. Thorkelson, 6 Feb. 1928, GEB, box 27, folder 248.

134. The two grants are recorded in the Minutes of the General Education Board, 18 Feb. 1927 and 24 Feb. 1928, GEB. See also H. J. Thorkelson to Just, 14 March and 6 Sept. 1927, EEJ(H), box 125-8, folder 140.

135. Just to Rosenwald, 10 March 1927, JR, box 19, folder 4.

136. Just to Rosenwald, 21 April 1927, ibid.; Graves to Mordecai W. Johnson, 6 July 1927, ibid.

137. Just to Division of Biology and Agriculture, National Research Council, 13 March 1925, RF.

138. See Graves to Just, 13 Nov. 1926; Graves to Abraham Flexner, 13 Nov. 1926, GEB, box 695, folder 6968.

139. Just to Lillie, 16 Feb. and 25 March 1923, RF; Just to Graves, 25 March 1923, JR, box 19, folder 3; see also F. A. Potts to Just, 23 Feb. 1923, EEJ(H), box 125-7, folder 119; Austin H. Clark to Just, 7 Oct. 1924, AHC; Just to Edwin R. Embree, 14 Jan. 1932, JRF, folder 6.

140. Leonard Outhwaite to Just, 12 Dec. 1927, NPC, GEB.

141. See Just to Leonard Outhwaite, 30 Dec. 1927, NPC, GEB.

142. See "Negro Problems Conference Report," NPC, GEB.

143. Stanford president to Embree, 3 Jan. 1928, ERE, box 3, folder 34.

144. Embree to Raymond Pearl, 14 March 1928, ibid., folder 30; E. G. Conklin to Embree, 10 March 1928, ibid., folder 34.

145. Just to Embree, 20 Jan. 1928, JRF, folder 1.

146. Just to Rosenwald, 5 Jan. 1928, JR, box 19, folder 4.

147. Embree, "Citizens of a Country" (ms. of address delivered at NAACP Annual Convention, Cleveland, Ohio, 27 June 1929), p. 8, ERE, box 7, folder 19.

148. Just to Embree, 26 July 1930; Embree to Just, 23 July 1930; JRF, folder 1.

149. Interview with S. M. Nabrit, 5 July 1977; interview with Roscoe L. McKinney, 5 Dec. 1976.

150. Just to Embree, 17 April 1928, JRF, folder 1.

151. Johnson to Embree, (with enclosure from Just), 11 Aug. 1928, JRF, folder 1.

152. Embree to Rosenwald, 14 Aug. 1928, ibid.

153. Embree to Johnson, 17 Sept. 1928, ibid.

154. Embree to Rosenwald, 8 May 1928, ibid.

155. George E. Vincent to Embree, 24 Sept. 1928, GEB, box 695, folder 6968.

156. Minutes, 4 Nov. 1928, JR [Addenda], box 2, folder 3; see also Embree to George E. Vincent, 18 Sept. 1928, GEB, box 695, folder 6968.

157. Embree to R. S. Lillie, 28 April 1928, JRF, folder 1.

158. R. S. Lillie to Embree, 5 May 1928, ibid. For Ralph Lillie's interest in the water research, see Just to Abraham Flexner, 27 Sept. 1926, GEB, box 695, folder 6968.

159. Minutes, 4 Nov. 1928, JR [Addenda], box 2, folder 3.

5. EUROPE: FIRST ENCOUNTERS, 1929–31

1. See Lubin to Just, 7 May 1929, EEJ(H), box 125-9, folder 159.

2. *Crisis* 23 (1922): 182; also 23 (1922): 247; 26 (1923): 272; 27 (1924): 177–78; 27 (1924): 255–58; 32 (1926): 191–92; 36 (1929): 101.

3. For Just's reluctance to attend such public functions, see Margret Boveri, *Verzweigungen: Eine Autobiographie* (Munich: R. Piper Verlag, 1977), p. 180. There is a great deal of information on Just's first four trips to Europe in these memoirs. Boveri describes in detail Just's life at the Stazione Zoologica and, later, at the Kaiser-Wilhelm-Institut in Berlin.

4. For a brief travel itinerary, see Just to Mordecai W. Johnson, 2 Oct. 1929, EEJ(H), box 125-5, folder 89; also, Just to Reinhard Dohrn, 17 Jan. 1929, SZ, A.1929.J.

5. Lillie to Dohrn, 8 Dec. 1928, FRL(W); see also Just to Dohrn, 3 Jan. 1929, SZ, A.1929.J.

6. See Heilbrunn to Margret Boveri, 12 Dec. 1928, D, B.1929-31.J.

7. This passage is from a short story, clearly autobiographical in many ways, about a foreigner's first visit to Naples. The story was written by Just on rough scraps of paper: EEJ(H), box 125-19, folder 375.

8. See Homer A. Jack, "The Biological Field Stations of Italy and Monaco," *Collecting Net* 15 (1940): 184. For further biographical information on Dohrn, see Chapter 3, note 5.

9. Otto Glaser to W. J. Crozier, 30 Sept. 1925, WJC.

10. Just to Dohrn, 17 Jan. 1929, SZ, A.1929.J.

11. Just to Embree, 27 Jan. 1929, JRF, folder 1.

12. Just to Heilbrunn, 27 May 1929, LVH; also Just to Embree, 4 April 1929, JRF, folder 2.

13. Concert program, EEJ(H), box 125-1, folder 16.

14. The intimate correspondence between Just and these scientists is preserved in the Howard collection, EEJ(H), box 125-7, folder 126, and box 125-8, folder 138.

15. Rivka Ashbel to the author, 10 June 1977.

16. Just to Aszaël, 4 Nov. 1929, EEJ(H), box 125-2, folder 27.

17. Boveri, *Verzweigungen*, p. 180.

18. Ibid., p. 182.

19. See Emmett J. Scott to Just and D. W. Woodward, 16 Dec. 1928, JRF, folder 1.

20. Fragment, EEJ(H), box 125-21, folder 427.

21. Just to Embree, 4 April 1929, JRF, folder 2.

22. Some of Just's experiences in Graz are recounted in a letter to Heilbrunn, 27 May 1929, LVH; see also Just to Friedl Weber, 15 Nov. 1929, EEJ(H), box 125-8, folder 148.

23. Fragment, EEJ.

24. Just to Dohrn, 9 May 1929, D, Be.1929–31.J.

25. Ibid.

26. The paper was published in the fall as "Breeding Habits of *Nereis dumerilii* at Naples" (1929). The earlier study was "Breeding Habits of the Heteronereis Form of *Platynereis megalops* at Woods Hole, Mass." (1914). For more on this work, including the controversy with the Woods Hole scientists, see Just's correspondence with J. Percy Moore, EEJ(H), box 125-6, folder 112; also Just to F. R. Lillie, 6 Dec. 1928, FRL(W); Just to C. M. Child, 8 Nov. 1929, EEJ(H), box 125-2, folder 49; Just to John Runnström, 9 Dec. 1929, ibid., box 125-7, folder 126.

References to the *Nereis dumerilii* article in the major domestic and foreign journals were made by (with the year of the citation): Friedrich Hempelmann (1931); Charles Pérez (1931); Silvio Ranzi (1931); Martin W. Johnson (1943); Ralph I. Smith (1950); Gunnar Thorson (1950); M. Jean Allen (1957); Ralph I. Smith (1958); R. B. Clark (1961); Yolande Boilly-Marer (1969).

27. See "The Fertilization-Reaction in *Paracentrotus* and *Echinus*" (1929). Among the scientists who cited this article in the major domestic and foreign biological journals (with year of citation) are the following: Friedrich Krüger (1931); Charles Pérez (1931); Floyd Moser (1939); Walter E. Southwick (1939); Albert Tyler (1940); Teru Hayashi (1945).

28. Dohrn to Just, 22 June 1929, EEJ(H), box 125-4, folder 65.

29. Boveri, *Verzweigungen*, p. 184.

30. EEJ.

31. Boveri, *Verzweigungen*, p. 181.

32. Just to Dohrn, 16 July 1929, D, Be.1929–31.J.

33. Affidavit by Dohrn, 25 June 1929, SZ, A.1929.J.

34. Young to Dohrn, 16 July 1929; Dohrn to Young, 17 July 1929, SZ, A.1929.J.

35. Just to Lillie, 9 Sept. 1929, FRL(W); Just to McClung, 15 Nov. 1929, D, Be.1929–31.J; Just to Conklin, 15 Nov. 1929, EGC, box 31.

36. Just to Villard, 4 Oct. 1929, D, Be.1929–31.J.

37. Lillie to Just, 17 Sept. 1929, FRL(W).

38. Conklin to Just, 22 Nov. 1929, EGC, box 31.

39. Villard to Just, 14 Oct. 1929, EEJ(H), box 125-8, folder 145; also Villard to Just, 27 Nov. 1929 (with enclosure from Stephen P. Duggan, official of the Institute of International Education), D, Be.1929–31.J.

40. Villard to Just, 2 Dec. 1929, D, Be.1929–31.J; also Just to Boveri, 13 July 1932, MB.

41. Just to Villard, 7 Dec. 1929, D, Be.1929–31.J.

42. Just to Embree, 19 Nov. and 11 Dec. 1929, JRF, folder 2.

43. Embree to Just, 13 Dec. 1929, JRF, folder 2.

44. Just to Dohrn, 7 Dec. 1929, D, Be.1929–31.J.

45. Just to Embree, 7 Oct. 1929, JRF, folder 2.

46. See Davis to Just, 25 Nov. 1929, EEJ(H), box 125-3, folder 56; A. C. Machler (Just's travel agent) to Just, 16 Dec. 1929, EEJ(H), box 125-6, folder 106.

47. EEJ(H), box 125-1, folder 16.

48. See Kuhler to Just, 14 March 1930, EEJ(H), box 125-6, folder 96.

49. For the Berlin episode, see Boveri, "Einleben in Berlin—1930 bis 1933," *Verzweigungen*, pp. 186–207.

50. See Just to J. Walter Wilson, 12 Nov. 1929, EEJ(H), box 125-9, folder 161.

51. Just to Lillie, 22 Jan. 1930, FRL(W).

52. See Just to Franz Schrader, 11 Jan. 1934, EEJ(H), box 125-7, folder 128.

53. There are few historical studies of the Kaiser-Wilhelm-Institut für Biologie. Brief descriptions may be found in "Das Kaiser-Wilhelm-Institut für Biologie," ed. Adolf von Harnack, *Handbuch der Kaiser Wilhelm-Gesellschaft zur Förderung der Wissenschaften*, 1928, pp. 103–7; and Carl Correns, "Das Kaiser-Wilhelm-Institut für Biologie in Berlin-Dahlem," in Ludolph Brauer, Albrecht Mendelssohn Bartoldy, and Adolf Meyer, eds., *Forschungsinstitute, ihre Geschichte, Organisation und Ziele*, 1930, pp. 152–53. Also useful are Richard Goldschmidt's reminiscences in his autobiography, *In and Out of the Ivory Tower* (Seattle: Washington Univ. Press, 1960). In addition, B. N. Kropp has written this author extensively (1 June 1977, 14 and 29 July 1980) about his memories of Just and the institute.

54. Holtfreter's letters to Just cover a broad range of mutual cultural interests: EEJ(H), box 125-5, folder 83.

55. Just to F. R. Lillie, 22 Jan. 1930, FRL(W); also Just to W. C. Curtis, 25 March 1931, RF.

56. In the 1930s both men were urged to consider authorizing translations of their manuals. See Camille Lhérisson to Just, 15 June 1931, EEJ(H), box 125-6, folder 100; Benjamin Kropp to Just, 18 Oct. 1930, ibid., box 125-6, folder 96.

57. Just to Embree, 28 Jan. 1930, JRF, folder 3; also Just to W. C. Curtis, 25 March 1931, RF.

58. Just to Mordecai W. Johnson, 19 Oct. 1930, EEJ(H), box 125-5, folder 89.

59. Among the scientists who cited Just's work in German journals were Wilhelm Roux (1913); W. Ruppert (1924); W. Schleip, (1925); S. Prát and K. M. Malkovský (1927); T. Péterfi (1927); John Runnström (1928, 1929); E. Bataillon (1929); E. Bataillon and Tchou-Su (1929); Leigh Hoadley (1929); A. R. Moore (1930); Josef Spek (1930); John Runnström (1930); Wilhlem Einsele (1930). Letters from German scientists, or other scientists working in Germany, include G. Ballowitz to Just, 28 May 1923, EEJ(H), box 125-2, folder 29; P. Büchner to Just, 28 July 1924 and 15 July 1925, ibid., folder 38; Josef Spek to Just, 12 April 1926 and 20 Nov. 1928, ibid., box 125-8, folder 136; see also Just to Rosenwald, 17 Sept. 1926, JR, box 19, folder 4.

60. See Pincus to W. J. Crozier, 15 March 1930, WJC.

61. Pincus to Just, 26 Oct. 1930, EEJ(H), box 125-7, folder 119.

62. Otto Glaser to W. J. Crozier, 17 March 1930; Crozier to Glaser, 19 March 1930, WJC.

63. Just to Rosenwald, 1 Aug. 1930, JRF, folder 3.

64. B. N. Kropp to the author, 29 July 1980. Lillie's correspondence alludes to Mangold's complicity with the Nazis, FRL(W).

65. Just to von Harnack, 24 Jan. 1930, KWI, Mappe 2506; also, Just to Embree, 28 Jan. 1930, JRF, folder 3.

66. Boveri, *Verzweigungen*, p. 194.

67. Just to Harnack, 3 Feb. 1930, KWI, Mappe 2506.

68. See Just's notes: EEJ(H), box 125-22, folder 426.

69. For much of the section on Just and Boveri, see Boveri, *Verzweigungen*, pp. 186–207.

70. See Johannes Holtfreter to Just, n.d., EEJ(H), box 125-5, folder 83.

71. Johannes Holtfreter to the author, 14 Nov. 1978 and 29 July 1980; also, Boveri, *Verzweigungen*, p. 202.

72. A good study of the prince and his circle is contained in Walter Henry Nelson, *The Soldier Kings: The House of Hohenzollern* (New York: Putnam, 1970).

73. See Grethe Alter to Just, 7 June 1930, EEJ(H), box 125-2. Just kept a file of notes and correspondence on Alter's choir: ibid., box 125-19, folder 380.

74. *Collecting Net* 5 (1930) no. 1: 6; also, Just to W. C. Curtis, 25 March 1931, RF.

75. *Collecting Net* 5 (1930) no. 1: 1–2; no. 2: 41.

76. Interview with S. M. Nabrit, 5 July 1977.

77. See Just to Paolo Enriques, 12 Aug. 1930, ibid., box 125-4, folder 67.

78. Just to Mordecai W. Johnson, 11 Aug. 1930, ibid., box 125-5, folder 89.

79. The congress proceedings are recounted in full in *XI° Congresso Internazionale di Zoologia. Atti*, 3 vols. (Padua, 1932).

80. See Just to Oscar Riddle, 6 Feb. 1931, EEJ(H), box 125-7, folder 122.

81. *XI° Congresso Internazionale* 1: 73.

82. See Just to Gregory Pincus, 28 Oct. 1930, EEJ(H), box 125-7, folder 119.

83. *Gazzetino di Padova*, 10 Sept. 1930, p. 4; also Just to Rosenwald, 20 Sept. 1930, JRF, folder 4.

84. Just to Bradley M. Patten, 10 March 1937, EEJ(H), box 125-7, folder 117.

85. *XI° Congresso Internazionale* 1: 68.

86. Just to Dohrn, 23 Sept. 1930, D, Be.1929–31.J.

87. *XI° Congresso Internazionale* 1: 69.

88. Boveri, *Verzweigungen*, pp. 199–200.

89. Ibid., p. 200. Boveri was mistaken about the date of their trip to the Heider chateau in Austria, placing it in 1929 (p. 183) rather than 1930. That the trip actually took place after the Padua conference is evidenced by Just's incoming correspondence, which was addressed to him at the Heider chateau at that time. For instance, Just's shipping agent, Norddeutscher Lloyd Co., wrote him there on 4 Sept. 1930 regarding his return reservations: EEJ(H), box 125-9, folder 159.

90. Just to Rosenwald, 20 Sept. 1930, JRF, folder 4.

91. Holtfreter to Just, 26 Nov. 1930, EEJ(H) box 125-5, folder 83.

92. On the Radiation Research Committee post: Curtis to Just, 1 Oct. 1930, ibid., box 125-3, folder 54. On the Oberlin lecture and consultantship: Budington to Just, 17 Oct. 1930, ibid., box 125-2, folder 41; C. G. Rogers to Just, 9 Jan. 1931, box 125-7, folder 124; C. N. Cole to Just, Feb. 1931, box 125-2, folder 51;

Just to Cole, 11 Feb. 1931, ibid. On the Society of Zoologists, see H. V. Neal to Just, 9 and 14 Jan. 1931; Just to Neal, 12 Jan. 1931, ibid., box 125-7, folder 114; M. F. Guyer to Just, 13 Jan. 1931, box 125-4, folder 75; Just to L. V. Heilbrunn, 13 Jan. 1931, LVH. For Heilbrunn's efforts to renew their intimacy, see Heilbrunn to Just, 14 Jan. 1931, EEJ(H), box 125-5, folder 80. For Lillie's, see Lillie to Just, 11 Nov. 1930, FRL(W).

93. "Annual Report from the Department of Zoology, Howard University, to Dr. E. P. Davis, Dean of the College of Arts and Sciences, submitted by E. E. Just, Professor of Zoology, March 1931." A copy was sent to the Rosenwald Fund administrators on 26 March 1931, JRF, folder 5.

94. See Just to Mordecai W. Johnson, 25 March 1931; Johnson to Just, 28 March 1931, EEJ(H), box 125-5, folder 90.

95. Interview with Lester G. Barth, 25 Jan. 1977.

96. Boveri, *Verzweigungen*, pp. 194–95.

97. See Just to Grethe Alter, 12 May 1931; Alter to Just, 19 July 1931, EEJ(H), box 125-2, folder 25.

98. Nelson, *Soldier Kings*, p. 444; see also the photograph of the prince and his sons in Nazi stormtrooper uniforms (reproduced from *Illustrated London News*) between pp. 256 and 257. See also Curt Stern to Just, 30 June 1932, EEJ(H), box 125-8, folder 138.

99. Just to Edwin R. Embree, 27 Oct. 1931, JRF, folder 5.

100. Just later wrote on Alter's behalf to Edwin R. Embree, 27 Oct. 1931, JRF, folder 5; and to Elsie W. C. Parsons (Southwest Society), 13 March 1931, and L. A. Roy (Phelps-Stokes Fund), 2 April 1931, EEJ(H), box 125-7, folder 125.

101. B. N. Kropp to Just, 18 Oct. 1930, EEJ(H), box 125-6, folder 96.

102. Boveri, *Verzweigungen*, pp. 200–201.

103. Just to Hedwig Schnetzler, 23 Feb. 1933, EEJ(M).

104. Just to Schnetzler, 1 Jan. 1934, ibid.

105. Boveri, *Verzweigungen*, pp. 200–201, details the beginning of Just and Hedwig's relationship.

6. HOWARD UNIVERSITY: CONTINUING STRUGGLE, 1929–31

1. Rayford W. Logan, *Howard University: The First Hundred Years, 1867–1967* (New York: New York Univ. Press, 1969), p. 255; see also Chapter 4, note 2.

2. For further biographical information on Johnson, see *Crisis* 32 (1926): 240; Mary White Ovington, *Portraits in Color* (New York: Viking Press, 1927), pp. 43–52; Edwin R. Embree, "Mordecai W. Johnson," in *Thirteen Against the Odds* (New York: Viking Press, 1944), pp. 175–95; "Mordecai Johnson: Twenty-five Years at Howard University," *Our World* 7 (1952): 22–27; Walter Dyson, *Howard University: The Capstone of Negro Education* (Washington, D.C.: Howard Univ., 1941), pp. 398–401.

3. Johnson to Rosenwald, 8 Oct. 1926, JR, box 19, folder 2.

4. Johnson to L. J. Cole, 28 Feb. 1927, RF.

5. For Just's positive response to Johnson in the early years, see Just to Rosenwald, 10 March 1927, JR, box 19, folder 4; Just to H. J. Thorkelson, 20 April 1927, GEB, box 27, folder 249; Just to Abraham Flexner, 20 April 1927,

ibid., 249; Just to Rosenwald, 21 April 1927, JR, box 19, folder 4. For the later negative reaction, see Just to W. C. Curtis, 24 April 1931, EEJ(H), box 125-3, folder 54; Just to Curt Stern, 20 Oct. 1931, ibid., box 125-8, folder 138.

6. Dyson, *Howard University*, p. 178. Dyson's chapter "Graduate School," pp. 178–92, provides a good overview of the early development of graduate studies at Howard. The *Howard Catalog* (published annually) is also useful for information on this aspect of the university, with many details on the graduate program, including committees, courses, students, and regulations.

7. See Melville J. Herskovits to Just, 18 Sept. 1925; Just to Herskovits, 22 Sept. 1925; Herskovits to Just, 25 Sept. 1925; MJH; Just to Abraham Flexner, 26 Sept. 1925, GEB, box 695, folder 6967; Herskovits to Just, 6 Feb. 1926, MJH.

8. Just to Rosenwald, 5 Jan. 1928, JR, box 19, folder 4.

9. See Just to H. J. Thorkelson, 17 Jan. 1927, GEB, box 695, folder 6968; Just to Rosenwald, 5 Jan. 1928, JR, box 19, folder 4. Just and four of his students were listed on the program for the annual conference of the American Society of Zoologists: see C. C. Andrews, H. Y. Chase, and T. L. Dulaney (introduced by Just), "Cytological Study of Fertilization and Mitosis in *Arbacia* Eggs Inseminated in KCN-sea-water," *Anatomical Record* 37 (1927): 161; Just and F. V. McNorton, "Mitochondria and Golgi Bodies in Mayonnaise" (1927).

10. Just to Edwin R. Embree, 17 April 1928, JRF, folder 1.

11. "The Rosenwald Institute of Zoology of Howard University," (Proposal submitted to the Julius Rosenwald Fund by Just, with covering letter to Embree from Johnson), 11 Aug. 1928, JRF, folder 1.

12. Embree to Rosenwald, 14 Aug. 1928, JRF, folder 1.

13. See Embree to Johnson, 5 Nov. 1928; Embree to Just, 5 Nov. 1928; Johnson to Embree, 9 Nov. 1928; Just to Embree, 11 Nov. 1928, JRF, folder 1.

14. For more on the conflict between Miller and Johnson, see Kelly Miller to the secretary of the General Education Board (with enclosures, Harold L. Ickes to Miller, 26 Feb. 1935; Miller to Ickes, 27 Feb. 1935; Miller to Dr. Bell, 2 March 1935), GEB, box 27, folder 251. Also, Logan, *Howard University*, p. 339.

15. The episode is recounted in a letter from Just to Hedwig Schnetzler, 26 Dec. 1933, EEJ(M).

16. *New York Times*, 3 Feb. 1929, sec. 9, p. 2.

17. Mary Church Terrell, "A Son of Howard Scales the Heights," *Washington Post*, 10 Feb. 1929; see also Roscoe C. Bruce to Just, 1 Dec. 1931, EEJ(H), box 125-2, folder 37.

18. Just to Embree, 27 Jan. 1929, JRF, folder 1.

19. Schrader to Just, 10 Sept. [1929], EEJ(H), box 125-7, folder 128.

20. Heilbrunn to Just, 2 Oct. 1929, ibid., box 125-5, folder 80.

21. See Just to Runnström, 9 Dec. 1929, ibid., box 125-7, folder 126.

22. See Runnström to Just, 11 Feb. 1930; Runnström to Just, 20 Sept. 1930; Runnström to Just, 12 Sept. 1932, ibid.; Just to Max Mason, 10 Oct. 1932, GEB, box 695, folder 6968.

23. E. P. Davis to Just, 30 Oct. 1929; Just to Davis, 30 Oct. 1929, EEJ(H), box 125-3, folder 56.

24. See Just to Davis, 23 Nov. 1929, ibid.

25. Just to Embree, 26 July 1930, JRF, folder 3.

26. Ibid.

27. Just to Emmett J. Scott, 6 Dec. 1929, EEJ(H), box 125-7, folder 129.

28. See Just to Davis, 23 and 25 Nov. 1929, ibid., box 125-3, folder 56; Emmett J. Scott to Just, 26 Nov. 1929, ibid., box 125-7, folder 129.

29. Just to Beardsley Ruml, 27 Dec. 1929, ibid., folder 126.

30. Just to James E. Shepard, 6 Dec. 1929, ibid., box 125-8, folder 133.

31. See Just to F. M. MacNaught, 10 Dec. 1929, ibid., box 125-6, folder 107.

32. W. C. Allee to Just, 30 Nov. 1929, ibid., box 125-2, folder 23.

33. F. R. Lillie to Young, 11 Jan. 1930, FRL(C), box 6, folder 27.

34. Lillie to Mordecai W. Johnson, 15 Jan. 1930, ibid.

35. Young to Lillie, [?] Jan. 1930, ibid., folder 27.

36. See Jackson Davis to H. J. Thorkelson, 15 Sept. 1928; Young to Thorkelson, 28 Jan. 1929, GEB, box 29, folder 271.

37. Interview with Donald P. Costello, 3 April 1977.

38. See Mordecai W. Johnson to Just, 10 Sept. 1927, EEJ(H), box 125-5, folder 89; Young to L. V. Heilbrunn, 5 May 1936, LVH.

39. See Heilbrunn to Lillie, 12 Jan. 1930; Lillie to Young (esp. the hand-written memo by Lillie on the carbon copy), 20 Aug. 1930, FRL(C), box 6, folder 27.

40. See Just to Hedwig Schnetzler, 16 April 1937, EEJ(M).

41. See Embree to Just, 23 July 1930; Just to Embree, 26 July 1930, JRF, Folder 3.

42. See E. P. Davis to Just, 25 November 1931, EEJ(H), box 125-3, folder 59.

43. Flexner to Embree, 29 July 1930, JRF, folder 3.

44. Lillie to Embree, 31 July 1930, ibid.

45. Just to Embree, 26 July 1930, ibid.

46. Memorandum by Warren Weaver, 3 May 1934, GEB, box 27, folder 251. A good deal of the information on the condition of the various science departments comes from this memorandum.

47. Just to Embree, 26 July 1930, JRF, folder 3.

48. See Just to E. P. Davis, 18 Nov. 1931; Davis to Just, 25 Nov. 1931, EEJ(H), box 125-3, folder 59.

49. Shohan to Just, 7 June 1931, ibid., box 125-8, folder 133. For another reference to the Shohan controversy, see Kelly Miller to Dr. Bell, 2 May 1935, GEB, box 27, folder 251.

50. See Just to Curt Stern, 20 Oct. 1931, EEJ(H), box 125-8, folder 138.

51. Shohan to Just, 7 June 1931, ibid., folder 133.

52. Just to Emanuel Celler, 15 March 1932, ibid., box 125-2, folder 46; Just to Annie Nathan Meyer, 15 March 1932, ibid., box 125-6, folder 110. See also Just to Margret Boveri, 17 Oct. 1931, MB.

53. For more on the Culemann controversy, see Culemann to Just, 19 March 1933, EEJ(H), box 125-3.

54. Just to Warren Weaver, 6 Oct. 1933, GEB, box 695, folder 6968.

55. See Just to George Arthur, 12 Nov. 1931, JRF, folder 6.

56. See Just to Douglas Whitaker, 22 Jan. 1932; Whitaker to Just, 17 Feb. 1932, EEJ(H), box 125-8, folder 151; Just to George Arthur, 15 March 1932, JRF, folder 6; Douglas Whitaker to Just, 24 Oct. 1932, EEJ(H), box 125-8, folder 151.

57. See Just to E. P. Davis, 9 Dec. 1931, EEJ(H), box 125-3, folder 59.

58. See Julia Bailey Edmonds to Just, [1931], ibid., box 125-4, folder 67.

59. See B. Hinton Brown to Just, 24 March 1931, ibid., box 125-2, folder 37; Raymond Osborn to Just, 20 April 1931, ibid., box 125-7, folder 116.

60. Muller to Just, 16 May 1931, ibid., box 125-6, folder 110.

61. Just to Embree, 26 July 1930, JRF, folder 3.

62. A. D. Mead to F. R. Lillie, 29 Nov. 1929, FRL(C), Box 5, folder 11; Allee to Just, 28 Nov. 1927, EEJ(H), Box 125-2, folder 23.

63. See Schrader to Just, 5 Dec. 1930, EEJ(H), Box 125-7, folder 128.

64. See W. C. Curtis to Lillie, 18 May 1931 (with enclosure, Heilbrunn to Curtis, 15 May 1931), FRL(W); Curtis to Just, 22 May 1931, EEJ(H), box 125-3, folder 54; Curtis to Lillie, 30 May 1931, FRL(W); Curtis to Just, 8 June 1931, EEJ(H), box 125-3, folder 54.

65. Just to Hedwig Schnetzler, 26 December 1933, EEJ(M).

66. See Schrader to Just, 12 Nov. 1929, 18 Dec. 1929, EEJ(H), box 125-7, folder 128.

67. See J. W. Wilson to Just, 26 Oct. [1931], ibid., box 125-9, folder 154.

68. Just to Hedwig Schnetzler, 11 Nov. 1933, EEJ(M).

69. Just to Hedwig Schnetzler, 1 Jan. 1934, ibid.

70. Just to Kimball Union Academy, 27 Aug. 1930, EEJ(H), box 125-9, folder 159.

71. Just to Hedwig Schnetzler, 22 Feb. 1935, EEJ(M). Also Edmund B. Spaeth to Just, 15 Jan. 1932; Just to Spaeth, 18 Jan. 1932, EEJ(H), box 125-8, folder 136.

72. See Albert L. Barrows to W. C. Curtis, 1 June 1931, RF; F. R. Lillie to Edwin R. Embree, 2 June 1931, JRF, folder 5; Embree to Johnson, 9 June 1931, ibid.; W. W. Brierley to Hermann A. Spoehr, 17 June 1931, GEB, box 695, folder 6968; Vernon Kellogg to Curtis, 18 June 1931, RF.

73. See W. C. Curtis to Just, 15 July 1931, EEJ(H), box 125-3, folder 54; Johnson to Edwin R. Embree, 19 June 1931, JRF, folder 5.

74. Just to W. C. Curtis, 24 April 1931, EEJ(H), box 125-3, folder 54.

75. *Washington Post*, 4 April 1931; *Afro-American*, 23 Jan. 1932; ibid., 20 Feb. 1932; *Dunbar News*, 23 March 1932. A folder of these and other clippings on the Howard situation is preserved at the Rockefeller Archive Center: GEB, box 27, folder 254.

76. Embree to Rosenwald, 10 April 1931, JR, box 33, folder 9.

77. See Just to Flexner, 28 Sept. 1931; Flexner to Just, 29 Sept. 1931; Just to Flexner, 2 Oct. 1931; Flexner to Just, 3 Oct. 1931, EEJ(H), box 125-4, folder 70.

78. Just to Lillie, 10 Feb. 1932, FRL(W).

79. See Just to Schwartz, 26 Sept. 1931, EEJ(H), box 125-7, folder 127; Herbert Friedmann to Leigh Hoadley, 14 Nov. 1931, LVH; also Just to E. P. Davis, 18 Nov. 1931, EEJ(H), box 125-3, folder 59; Just to Margret Boveri, 9 Dec. 1932, MB.

80. Just to George Arthur, 12 Nov. 1931, JRF, folder 6.

81. Ibid.; Just to E. P. Davis, 18 Nov. 1931, EEJ(H), box 125-3, folder 59.

82. Just to E. P. Davis, 18 Nov. 1931, EEJ(H), box 125-3, folder 59.

83. See Just to Arthur, 30 Oct. 1931, JRF, folder 5.

84. Arthur to Just, 4 Nov. 1931, ibid., folder 6; see also Just to Arthur, 2 Nov. 1931, EEJ(H), box 125-2, folder 28.

85. For a concise statement of this principle, see Just to George W. Wade, 20 Jan. 1932, EEJ(H), box 125-8, folder 147.

86. See Just to Wilkinson, 19 Oct. 1931; Wilkinson to Just, 26 Oct. 1931, ibid., box 125-9, folder 154.

87. See Just to Reginald G. Harris, March 1932; Harris to Just, 10 May 1932, CSH; Just to B. H. Willier, 31 Oct. 1931; Willier to Just, 2 Nov. 1931, EEJ(H), box 125-9, folder 154; Just to J. W. Buchanan, 15 March 1932; Buchanan to Just, 17 March 1932, ibid., box 125-2, folder 39; Just to George Arthur, 15 March 1932, JRF, folder 6.

88. See Austin Clark to Just, 25 Jan. 1932, EEJ(H), box 125-2, folder 50.

89. Just to Lillie, 10 Feb. 1932, FRL(W).

90. See Just to Embree, 3 Nov. 1930, EEJ(H), box 125-4, folder 68; Just to Embree, 9 Nov. 1930, JRF, folder 4; Just to Embree, 19 Nov. 1931, EEJ(H), box 125-4, folder 68; Just to Embree, 4 Jan. 1932, JRF, folder 6; Embree to Just, 6 Jan. 1932, ibid.

91. Just to Embree, 8 Jan. 1932, JRF, folder 6.

92. Embree to Just, 12 Jan. 1932, ibid.

93. Just to Embree, 14 Jan. 1932, ibid.

94. Just to Embree, 20 Feb. 1932, EEJ(H), box 125-4, folder 68.

95. Just to Lillie, 10 Feb. 1932, FRL(W).

96. Just to Embree, 28 March 1932, JRF, folder 6.

97. Thirkield to Just, 28 March 1932, EEJ(H), box 125-8, folder 140.

98. Just to Lillie, 12 July 1932, FRL(W).

99. Lillie to Just, 29 July 1932, FRL(W); also Franz Schrader to Embree, 18 April 1932; Embree to Schrader, 21 April 1932, JRF, folder 6.

100. Embree to Just, 7 Sept. 1932, JRF, folder 6; also Embree to Just, 3 Oct. 1932, ibid., folder 7.

101. Runnström to Just, 12 Sept. 1932, EEJ(H), box 125-7, folder 126.

7. THE SEARCH FOR A NEW LIFE, 1931–38

1. "Report of Professor E. E. Just on His Work under the Five-year Grant of the Julius Rosenwald Fund, 1928–33," JRF, folder 7.

2. The *Naturwissenschaften* article was cited by Per Eric Lindahl (1932); [Fritz] Weyer (1932); K. Linsbauer (1933); Ernst Wertheimer (1933).

Citations of "On the Origin of Mutations" were made by L. Cuénot (1932); [W. F.] Reinig (1932); Ernst Wertheimer (1933).

For Citations of "Cortical Cytoplasm and Evolution," see Ernst Wertheimer (1933); Arnold Pictet (1934); [Bernhard] Rensch (1934).

3. "Report of Professor E. E. Just," JRF, folder 7.

4. Letters between Just and Hedwig Schnetzler, 1931 to 1941, are scattered in public and private collections in Europe and America: for example, Just to Schnetzler, 4 Dec. 1932, SZ, A.1932.J; Just to Schnetzler, 5 Jan. 1934, EEJ(M); Just to Schnetzler, 21 April 1935, EEJ; Just to Schnetzler, 8 Feb. 1937, EEJ(B); Just to Hedwig Just, 16 Feb. 1941, EEJ(B).

5. Klaus Dohrn to the author, 30 Nov. 1977, and Christiane Groeben to the author, 17 Jan. 1978, outlined some of the genealogy.

6. Klaus Dohrn to the author, 30 Nov. 1977; Walter Griesshaber to the author, n.d.

7. Embree to Just, 28 April and 26 June 1933, JRF, folder 7.

8. Embree to Johnson, 4 May 1933; Johnson to Embree, 3 June 1933, JRF, folder 7.

9. See Lessing Rosenwald to Just, 12 April 1935; Just to Lessing Rosenwald, 13 April 1935 and 26 Jan. 1939, EEJ(H), box 125-7, folder 125.

10. A good analysis of this trend, with full bibliographic references to major studies of federal and private research support, is Stanley Coben, "Foundation Officials and Fellowships: Innovation in the Patronage of Science," *Minerva* 14 (1976): 225–40. Little has been published on Weaver, but his memoir, *Scene of Change: A Lifetime in American Science* (New York: Scribner, 1970), is interesting. See also the obituaries in *Physics Today* 32 (1979): 72 and *Science* n.s. 203 (1979): 534; also Robert E. Kohler, "The Management of Science: The Experience of Warren Weaver and the Rockefeller Foundation Programme in Molecular Biology," *Minerva* 14 (1976): 279–306.

11. Just to Max Mason, 10 Oct. 1932, GEB, box 695, folder 6968.

12. Memorandum by Weaver, 28 Oct. 1932, ibid.

13. Weaver to F. R. Lillie, 1 March 1933, ibid.

14. Lillie to Weaver, 6 March 1933, ibid.

15. Just to Gross, 15 May 1933, EEJ(H), box 4, folder 74; Just to Curt Stern, 16 May 1933, ibid., box 8, folder 138; Herbert Friedmann to Leigh Hoadley, 29 May 1933, LH; and Benjamin Schwartz to Just, 1 June 1933, EEJ(H), box 7, folder 127.

16. Johnson to Weaver, 9 May 1933; Weaver to Johnson, 7 July 1933, GEB, box 695, folder 6968.

17. See Weaver to E. P. Davis, 23 Sept. 1933, ibid.

18. Just to Lillie, 12 June 1933, FRL(W).

19. Lillie to Just, 12 July 1933, ibid.

20. Weaver to E. P. Davis, 23 Sept. 1933, GEB, box 695, folder 6968.

21. The evolution of the Carnegie philanthropies is well documented in several biographies of Andrew Carnegie, including Burton J. Hendrick, *The Life of Andrew Carnegie* (Garden City, N.Y.: Doubleday, Doran, 1932) and Joseph Frazier Wall, *Andrew Carnegie* (New York: Oxford Univ. Press, 1970). A useful summary of the early years of the Carnegie philanthropies is Robert MacDonald Lester, *Forty Years of Carnegie Giving: A Summary of the Benefactions of Andrew Carnegie and of the Work of the Philanthropic Trusts which he Created* (New York: Scribner's Sons, 1941). There is no comprehensive history of the Carnegie Corporation or the Carnegie Institution, but their yearbooks and annual reports contain detailed summaries of each year's work and also, on occasion, valuable historical overviews.

22. *Carnegie Institution of Washington Yearbook* no. 33 (1933–34), p. 11

23. See Just to Keppel, 19 Jan. 1927, EEJ(H), box 125-6, folder 96.

24. See Just to Keppel, 30 Dec. 1933, ibid.

25. Just to Hedwig Schnetzler, 9 Nov. 1933, EEJ(M).

26. Just to Riddle, 10 Nov. 1933, EEJ(H), box 125-7, folder 122.

27. Riddle to W. M. Gilbert, 21 Nov. 1933, CC(W).

28. Just to Keppel, 19 Nov. 1933, ibid.

29. Streeter to Gilbert, 22 Nov. 1933, ibid.

30. See John C. Merriam to Just, 27 Dec. 1933, ibid.

31. Merriam to Gilbert, 25 Jan. 1934, ibid.

32. Just to Hedwig Schnetzler, 5 Jan. 1934, EEJ(M).

33. Streeter to Merriam, 2 Feb. 1934, CC(W).

34. See Just to Hedwig Schnetzler, 5 and 9 Jan. 1934, EEJ(M).

35. *The Biology of the Cell Surface* (1939). For critical reviews and discussions of the book, see Cedric Dover, "The Significance of the Cell Surface," *Journal of the Zoological Society of India* 6 (1954): 3–42; on its philosophical implications, see W. T. Fontaine, "Philosophical Implications of the Biology of Dr. Just," *Journal of Negro History* 24 (1939): 281–90. Among the scientists who cited Just's book (with the year of the citation) are: Raymond Pearl (1939); Johannes Pfeiffer (1939); Hans Pfeiffer (1940); Ralph S. Lillie (1941); Albert Tyler (1941); S. C. Brooks (1943); Johannes Holtfreter (1943); Charles Wilber (1945): Katsuma Dan and Jean Clark Dan (1947); Jean Clark Dan (1948); Roberta Lovelace (1949); Masao Sugiyama (1951); Walter L. Wilson (1951); John Runnström and G. Kriszat (1952); B. Hagström and Britt Hagström (1954); Y. Hiramoto (1956); Albert Tyler, Alberto Monroy, and Charles B. Metz (1956); Yukio Hiramoto (1957); R. D. Allen and J. A. Griffin (1958); Naoko Kawamura and Katsuma Dan (1958); B. E. Hagström and J. Runnström (1959); A. A. Humphries (1961); David R. Burgess and Thomas E. Schroeder (1977); Thomas E. Schroeder (1978); Yuchiro Tanaka (1979); Parris Kidd and Daniel Mazia (1980); Yuchiro Tanaka (1981).

36. L. V. Heilbrunn, *An Outline of General Physiology* (Philadelphia: W. B. Saunders, 1937).

37. *Biology of the Cell Surface*, p. 7.

38. Ibid., p. 8.

39. Ibid., p. 14.

40. Ibid., pp. 11–13.

41. Ibid., p. 16.

42. Ibid., p. 27. For a discussion of Morgan's ignorance of mathematics, see also Garland E. Allen, *Thomas Hunt Morgan: The Man and His Science* (Princeton: Princeton Univ. Press, 1978) pp. 311–12.

43. *Biology of the Cell Surface*, p. 27.

44. Ibid., p. 37.

45. Ibid., pp. 47–48.

46. Ibid., pp. 65–67.

47. Ibid., pp. 47–48.

48. Ibid., pp. 75–76.

49. Ibid., p. 83. Harrison, who was the originator of the method of growing tissues outside of the body of an animal, published his pioneering work in "Outgrowth of the Nerve Fiber as a Mode of Protoplasmic Movement," *Journal of Experimental Zoology* 9 (1910): 787–848.

50. *Biology of the Cell Surface*, p. 103.

51. Ibid., pp. 104–23.

52. Ibid., pp. 105–7. This process was first described in Just, "The Fertilization-Reaction in *Echinarachnius parma*. I." (1919).

53. *Biology of the Cell Surface*, pp. 124–46.

54. Ibid., pp. 147–205.

55. Ibid., pp. 204–5.

56. Ibid., pp. 206–46.

57. Ibid., p. 334.

58. This discussion was included in the final draft of the book; see ibid., pp. 286–353.

59. Ibid., pp. 354–61.

60. See Peter Kropotkin, *Mutual Aid, A Factor of Evolution* (New York: McClore, Phillips, 1902).

61. *Biology of the Cell Surface*, p. 367.

62. Many of Just's letters to Hedwig begin with classical salutations; for example, 1 March 1937, EEJ(M).

63. See Margret Boveri, *Verzweigungen: Eine Autobiographie* (Munich: Piper Verlag, 1977), p. 201.

64. Just to Warren Weaver, 6 Oct. 1933, GEB, box 695, folder 6968.

65. Just to Hedwig Schnetzler, 4 Dec. 1932, EEJ(M).

66. Just to Schnetzler, 20, 21, and 22 Feb. 1935, 18 and 20 March 1936, ibid.

67. The proceedings of the conference, "Atti del Primo Congresso Internazionale di Elettro-Radio-Biologia," are contained in *Archivio internazionale di radiobiologia generale*, vols. 2–3 (1935). A summary review was published in *Archivio di radiologia* 11 (1935): 60–107. See also Just's correspondence with Giocondo Protti, 18 May to 19 June 1934, EEJ(H), box 125-7, folder 122.

68. W. E. Tisdale to Warren Weaver, 18 Sept. 1934, GEB, box 695, folder 6969.

69. Just to Schnetzler, 8 Jan. 1934, EEJ(M).

70. Just to Dohrn, 4 Sept. 1934, SZ, A.1934.J.

71. Schnetzler to Nancy Astor, 17 Dec. 1934, NA.

72. Just to Schnetzler, 24 Jan. 1935, EEJ(M).

73. Just to Schnetzler, 31 Jan. 1935, ibid.

74. Just to Schnetzler, 5 Feb. 1935, ibid.

75. Just to Schnetzler, 7 and 8 Feb. 1935, ibid. For the history of the controversy, see Rayford W. Logan, *Howard University: The First Hundred Years, 1867–1967* (New York: New York Univ. Press, 1969), pp. 284–304, 312, 333–46.

76. Just to Schnetzler, 8 Feb. 1935, EEJ(M).

77. Just to Schnetzler, 13 and 15 Feb. 1935, ibid.; Just to Warren Weaver, 26 Feb. 1935, GEB, box 695, folder 6969. See also *Washington Post* and *Washington Star*, February and March 1935.

78. Just to Schnetzler, 26 Feb. 1935, EEJ(M).

79. Quoted in Logan, *Howard University*, p. 299.

80. Ibid., p. 302.

81. Ibid., p. 300.

82. Just to Schnetzler, 12 and 15 Feb., 8 March, 2 May 1935, EEJ(M).

83. Just to Schnetzler, 5 March 1935, EEJ(M).

84. Just to Schnetzler, 2 May 1935, ibid.

85. Just to Schnetzler, 30 April 1935, ibid.

86. Just to Schnetzler, 19 and 29 April 1935, ibid.

87. Young to Just, 6 May 1935, EEJ(H), box 125-9, folder 158.

88. Just to L. V. Heilbrunn, 20 May 1936, LVH.

89. Just to Schnetzler, 15 March 1935, EEJ(M).

90. E. A. Varela to Just, 5 Feb. 1935, CC(W).

91. Just to Schnetzler, 22 Feb. 1935, EEJ(M).

92. Just to Schnetzler, 19 Feb. 1935, ibid.

93. Just to Schnetzler, 23 Feb. 1935 and 19 Oct. 1936, ibid.

94. Just to John C. Merriam, 23 Oct. 1935; George L. Streeter to Merriam, 28 Oct. 1935; Merriam to Streeter, 29 Oct. 1935, CC(W).

95. Just to F. P. Keppel, 28 Oct. 1935, CC; Just to Davis Foundation, 25 Oct. 1935, EEJ(H), box 125-3, folder 55; Just to Francis P. Garvan, 28 Nov. 1935, ibid., box 125-4, folder 71; Just to H. A. Moe, 3 Dec. 1935, ibid., box 125-6, folder 112.

96. Just to Gustav Oberlaender, 12 Dec. 1935; Oberlaender to Just, 16 Dec. 1935, EEJ(H), box 125-7, folder 116; Wilbur K. Thomas to Just, 19 Dec. 1935, ibid., box 125-8, folder 141.

97. Just to Lillie, 4 Oct. 1935, FRL(W); Just to A. P. Mathews, 28 Nov. and 18 Dec. 1935, EEJ(H), box 125-6, folder 108.

98. Lillie to Just, 29 Oct. 1935, FRL(W); see also Just to Lillie, 31 Oct. 1935, ibid.; Just to A. P. Mathews, 6 Dec. 1935, EEJ(H), box 125-6, folder 108.

99. A. P. Mathews to Just, 3 Dec. 1935, EEJ.

100. "Alcoholic Solutions as Gastric Secretogogues" (1936).

101. See Just to W. Forbes Morgan, 11 March 1937, EEJ(H), box 125-6, folder 112.

102. "Nuclear Increase during Development as a Factor in Differentiation and in Heredity" (1936).

103. "A Single Theory for the Physiology of Development and Genetics" American Naturalist 70 (1936): 267–312. An editorial note in large type indicates that Just paid to have his article published. Citations of this article were made by Arnold Pictet (1938); Addison Gulick (1938); Donald Paul Costello (1939); Donald Paul Costello (1945).

104. A picture of Just with the Morgan group can be found in Allen, Morgan, p. 200.

105. Wilhelm Johannsen, "The Genotype Conception of Heredity," American Naturalist 45 (1911): 129–59.

106. Ibid., 70 (1936): 272.

107. Ibid., p. 311.

108. T. H. Morgan, Embryology and Genetics (New York: Columbia Univ. Press, 1934); for an account of Morgan's views, see Allen, Morgan, p. 300.

109. American Naturalist 70 (1936): 273.

110. Ibid., pp. 274–75.

111. Ibid., pp. 280–81.

112. Ibid., p. 300.

113. Ibid., p. 303.

114. Just, "Unsolved Problems of General Biology" (1940).

115. Ibid.

116. Ms. notes by Just, EEJ(H), box 125-9, folder 164.

117. See Conklin to Heilbrunn, 12 Dec. 1932, EGC; Heilbrunn to Just, 14 Dec. 1932; Just to Heilbrunn, 23 Feb. 1933, EEJ(H), box 125-5, folder 80; Franz Schrader to Just, 10 Jan. 1934; Just to Schrader, 11 Jan. 1934, ibid., box 125-7, folder 128.

118. Just to Schnetzler, 11 Jan. 1936, EEJ(M); this is also the source for the following retrospective.

119. Just to Schnetzler, 19 March 1936, ibid.

120. Just to Schnetzler, 12, 13, and 19 March 1936, ibid.

121. Young to Heilbrunn, 21 April 1936, LVH; Just to Heilbrunn, 20 May 1936, ibid.

122. "Certificate of Incorporation of the Carl Schurz Memorial Foundation" (Philadelphia, 1930), p. 3. The standard histories of the foundation and the trust are Eugene E. Doll, *Twenty-five Years of Service, 1930–1955* (Philadelphia: Carl Schurz Memorial Foundation, n.d.), and Hans Gramm, *The Oberlaender Trust, 1931–1953* (Philadelphia: Carl Schurz Memorial Foundation, 1953).

123. "Certificate of Incorporation," pp. 26, 29.

124. Doll, *Twenty-five Years*, p. 6; Gramm, *Oberlaender Trust*, p. 39.

125. Gramm, *Oberlaender Trust*, p. 48; see also Donald P. Kent, *The Refugee Intellectual: The Americanization of the Immigrants of 1933–1941* (New York: Columbia Univ. Press, 1953), pp. 115–16. An excellent article about physicists is Charles Weiner, "A New Site for the Seminar: The Refugees and American Physics in the Thirties," in *The Intellectual Migration: Europe and America, 1930–1960*, ed. Donald Fleming and Bernard Bailyn (Cambridge: Harvard Univ. Press, 1969), pp. 190–234.

126. *Sixth Annual Report of the Carl Schurz Memorial Foundation, Inc., May 1, 1935 to April 30, 1936* (Philadelphia: Carl Schurz Memorial Foundation, 1936), p. 8.

127. See Wilbur K. Thomas to Just, 13 Jan. 1936, EEJ(H), box 125-8, folder 141.

128. Just to Keppel, 28 Oct. 1935, CC.

129. See Keppel to Just, 31 Oct. 1935, EEJ(H), box 125-6, folder 96; Lessing Rosenwald to Just, 12 April 1935, ibid., box 125-7, folder 125.

130. See Just's correspondence with Mrs. Arnold Gottlieb, 8 Oct. 1932 to 28 Oct. 1935, ibid., box 125-4, folder 73; Gannett, [March] 1933, ibid., folder 71; Barrymore, Feb. 1925, ibid., box 125-2, folder 30; Sell, 12 Feb. to 30 April 1935, ibid., box 125-8, folder 133; Lazarus, 12 Feb. to 12 April 1935, ibid., box 125-6, folder 97; Davis, 21 to 25 Oct. 1935, ibid., box 125-3, folder 55; Garvan, 28 Nov. 1935 to 4 Jan. 1936, ibid., box 125-4, folder 71; Norris, Oct. 1935 to Feb. 1939, ibid., box 125-7, folder 115; [January] 1937 to [February] 1938, EEJ(M).

131. *Fortune* 10 (1934): 60.

132. See Just to Norris, 21 Oct. 1935, EEJ(H), box 125-7, folder 115.

133. See Just's correspondence with Wolf, 20 Jan. 1936 to 25 Jan. 1939, EEJ(H), box 125-9, folder 157.

134. Just to Patterson, 30 May 1937, ibid., box 125-7, folder 117; Max Wolf to McLean, 7 Sept. 1938, ibid., box 125-9, folder 163.

135. For Just's participation on the NAACP Harlem Hospital Investigating Committee, see his correspondence with Walter White, 19 Jan. 1933 to 27 Feb. 1934, ibid., box 125-8, folder 153.

136. "Hydration and Dehydration in the Living Cell. I-IV." (1928–30). References to these articles in the major domestic and foreign biological journals were made by the following scientists (with the year of citation): Matouschek (1929); Carl Schlieper (1930); Josef Spek (1930); Friederick Krüger (1931); Charles Pérez (1931); Albert Tyler (1931); L. Genevois (1932); T. Cunliffe Barnes and Theo. L. Jahn (1934); W. W. Lepeschkin (1936); Albert Tyler and Bradley T. Scheer (1937); Sven Hörstadius and Stina Stromberg (1940); Roberta Lovelace (1949); R. Phillips Dales (1950); Georg Kriszat (1954); Donald P. Costello (1958);

P. E. Gibbs (1971); Dierdre L. Kuhl and Larry C. Oglesby (1979); Thomas E. Schroeder (1981).

137. Interview with Donald P. Costello, 3 April 1977.

138. Just to Max Wolf, 22 Feb. 1937, EEJ(M).

139. Diana Long Hall, "Science and Social Engineering: Sex Research Policies in the 1920s" (Paper delivered at MIT, March 1978).

140. Just to Rockefeller, 22 July 1936, EEJ(H), box 125-7, folder 124.

141. Nelson Beeman to Just, 6 Aug. 1936, ibid., box 125-2, folder 31.

142. Wolf to Just, 12 Oct. 1936, ibid., box 125-9, folder 157.

143. Just to Schnetzler, 18 April 1936, EEJ(M).

144. Just to di Vecchi, 6 June 1936, EEJ(H), box 125-4, folder 64.

145. Italian Ministry of Education to Just, 4 and 30 July 1936, ibid., box 125-9, folder 159.

146. Just to Mussolini, 13 Aug. 1936, ibid., box 125-6, folder 113.

147. Just to Norris, 17 Nov. 1936, ibid., box 125-7, folder 115; see also Wolf to Just, 28 Dec. 1936, ibid., box 125-9, folder 157. Only one of Just's abstracts was published in the preconference program: "Ultra-violet Radiations as Experimental Means for the Investigation of Protoplasmic Behavior" (1937). His other talk was published later as "Phenomena of Embryogenesis and Their Significance for a Theory of Development and Heredity" (1937).

148. Just to Schnetzler, 19 Feb. 1937, EEJ(M).

149. Harrison, "Embryology and its Relations," *Science* n.s. 85 (1937): 369–74; see also RGH, box 92, folder 60; box 117, folder 123.

150. Just to Schnetzler, 19 Feb. 1937, EEJ(M).

151. Monthly receipts signed "Dr. H. Schnetzler," CC(W).

152. Just to Schnetzler, 20 April 1937, EEJ(M).

153. Among the numerous studies of Lady Astor and her circle are Maurice Collis, *Nancy Astor: An Informal Biography* (London: Farber, 1960); Christopher Sykes, *Nancy: The Life of Lady Astor* (London: Collins, 1972); Elizabeth Langhorne, *Nancy Astor and Her Friends* (New York: Praeger, 1974); Anthony Masters, *Nancy Astor: A Biography* (New York: McGraw-Hill, 1981).

154. See Otto Schnetzler to Nancy Astor, 3 Dec. 1933; Karl Schnetzler to Astor, 14 June 1934, NA.

155. See Hedwig Schnetzler to Astor, 15 Jan. 1934, NA; Elisabeth Schnetzler to Astor, March and 8 May 1937, NA.

156. Hedwig Schnetzler to Astor, 15 April 1937, NA.

157. Just to Schnetzler, 9 Jan. 1934, EEJ(M).

158. See Schnetzler to Astor, 2 Sept. 1937, NA.

159. See the correspondence between Just and Valensi, 20 July 1937 to 26 January 1938, EEJ(H), box 125-8, folder 145.

160. May to Just, 28 July 1937, ibid., box 125-6, folder 108.

161. See Just to C. C. Little, 1 and 9 May 1937, ibid., folder 100; Weaver to Little, 17 May 1937; Little to Just, 19 May 1937, ibid.

162. See Marcel Prénant to the Rockefeller Foundation, 17 Dec. 1937, GEB, box 695, folder 6969; Just to Warren Weaver, Dec. 1937, ibid.; F. B. Hanson to Just, 7 Jan. 1938, ibid.

163. Hedwig Schnetzler to Astor, 18 Aug. 1937, NA.

164. Just to Astor, 18 Aug. 1937, EEJ(M).

165. Receipts dated 29 Aug. to 4 Sept. 1937, Grand Hotel Riccione, EEJ(H), box 125-1, folder 7.

166. See Hedwig Schnetzler to Astor, 2 Sept. 1937, NA.

167. Just to Astor, 28 Sept. 1937, NA.

168. Aydelotte to Lothian, 8 Oct. 1937, EEJ(M).

169. Just to Astor, 28 Sept. 1937, NA.

170. Ibid.

171. Just to Wesley, 4 Dec. 1937, EEJ(M); Johnson to Just, 28 Dec. 1937, EEJ(H), box 125-5, folder 93.

172. Just to Wesley, 6 Jan. 1938, EEJ(M).

173. Just to Johnson, 17 Jan. 1938, ibid.

174. Just to Johnson, 2 Feb. 1938, ibid.

175. Just to Norris, [24 Feb. 1938], ibid.

8. THE EXILE, 1938–40

1. Just to Johnson, 17 Jan. 1938, EEJ(M).

2. Just to Johnson, 2 Feb. 1938, ibid.

3. Just to Murray, 29 March 1938, EEJ(H), box 125-6, folder 113.

4. See Garvin to Just, 16 May 1938, ibid., box 125-4, folder 71.

5. Garvin to Just, 14 Sept. 1938, ibid.

6. Garvin to Just, 2 Nov. 1938, ibid.

7. Just to V. D. Johnston, 26 Feb. 1938, ibid., box 125-5, folder 95; Just to Johnson, 1 March 1938, ibid., folder 93; Just to V. D. Johnston, 2 Dec. 1938, ibid., folder 95.

8. Just to Harrison, 9 Dec. 1937, RGH, box 15.

9. Just to Harrison, 17 Aug. 1938, ibid.

10. Ibid.

11. Ibid.

12. Harrison to Flexner, 10 Sept. 1938, RGH, box 15.

13. Flexner to Harrison, 12 Sept. 1938, ibid.

14. Harrison to Lillie, 9 Sept. 1938, ibid., box 15.

15. Lillie to Harrison, 12 Sept. 1938, ibid.

16. Many of the brief articles in *Biological Bulletin* 95 (1948), the Lillie memorial volume, are insightful, especially B. H. Willier, "The Work and Accomplishments of Frank R. Lillie at Chicago" (pp. 151–53); R. G. Harrison, "Dr. Lillie's Relations with the National Academy of Sciences and the National Research Council" (pp. 154–57); E. G. Conklin, "Frank R. Lillie and the Marine Biological Laboratory" (pp. 158–62). A good comprehensive article is B. H. Willier, "Frank Rattray Lillie, 1870–1947," *Biographical Memoirs. National Academy of Sciences* 30 (1957): 179–236. Also useful are two general articles, E. G. Conklin, "Science and Scientists at the MBL Fifty Years Ago," *Collecting Net* 13 (1938): 101–6; and Gerhard Fankhauser, "Memories of Great Embryologists," *American Scientist* 60 (1972): 46–55. Lillie's daughter, Mary Prentice Lillie Barrows, has put out an interesting little book of personal reminiscences entitled *Moon Out of the Well* (1970). Ray Watterson's sketch in

the *Dictionary of Scientific Biography*, 8: 354–60, contains a good summary of the major bibliographic sources relating to Lillie.

17. Just to Lillie, 12 Aug. 1936, FRL(W); Just to Schnetzler, 26 Feb. 1937, EEJ(M).

18. Interview with Alfred C. Redfield, 28 July 1977.

19. The concept was first presented in "The Fertilization-Reaction in *Echinarachnius parma*. I." (1919). For citations of this work, see Chapter 3, note 45.

20. Lillie to Just, 11 Nov. 1930, FRL(W).

21. Just to Lillie, 10 Feb. 1932, ibid.

22. See Lillie's correspondence with Hsi-Wang, ibid.

23. Ms. fragment, ibid.

24. Lillie to Just, 18 Feb. 1920, FRL(C), box 4, folder 20.

25. *Zeitschrift für Zellforschung und mikroskopische Anatomie* 17 (1933): 39.

26. Just to Lillie, 12 July 1932; Lillie to Just, 29 July 1932; Just to Lillie, 15 March 1933; Lillie to Just, 18 March 1933, FRL(W).

27. Lillie to Just, 26 Aug. 1936, ibid.

28. Just to Lillie, 12 Aug. 1936, ibid.

29. Lillie to Just, 26 Aug. 1936, ibid.

30. Just to Schnetzler, 8 Feb. 1937, EEJ(M); also Just to W. C. Allee, 21 Dec. 1939, EEJ(H), box 125-2, folder 23.

31. See Johnson to Just, 11 Aug. 1939, EEJ(H), box 125-5, folder 88.

32. Karpman to Just, 8 Oct. 1938, ibid., box 125-6, folder 96.

33. Roy [Alain L. Locke] to Just, 1 Dec. 1938, ibid., box 125-9, folder 116.

34. Just to Johnson, 1 June 1939; Johnson to Just, 30 June 1938, ibid., box 125-5, folder 94.

35. Just to Johnson, 17 Jan. 1938, quoted in Johnson to Just, 30 June 1938, ibid.

36. See Richard Hurst Hill, secy., Howard Univ. Board of Trustees, "Extracts from minutes of meeting of Board of Trustees, April 11, 1939," ibid.

37. Just to Johnson, 1 June 1939, ibid.

38. Just to Lillie, 25 Jan. 1939, FRL(W); Just to Embree, 25 Jan. 1939, JRF, folder 8; Just to Heilbrunn, 25 Jan. 1939, EEJ(H), box 125-5, folder 80; Just to Flexner 25 Jan. 1939, ibid., box 125-4, folder 70; Just to Franz and Sally Schrader, 25 Jan. 1939, ibid., box 125-7, folder 128.

39. See Harrison to Heilbrunn, 16 March 1939, RGH, box 15.

40. Just to Keppel, 23 Feb. 1939; Keppel to Just, 7 March 1939, EEJ(H), box 125-6, folder 96; Keppel to Harrison, 4 May 1939, RGH.

41. Harrison to Lillie, 16 March 1939, RGH, box 15.

42. Lillie to Harrison, 20 March 1939, ibid.

43. Just to Franz and Sally Schrader, 25 Jan. 1939, EEJ(H), box 125-7, folder 128; also Schnetzler to Astor, 25 Dec. 1938, NA; G. Bobin to the author, 25 Nov. 1977.

44. Many of Just's experiences in France in 1939–40 are recounted in jottings he put down on rough scraps of paper and entitled "3 Trips to Quimper": EEJ(H), box 125-16, folder 329.

45. The most comprehensive study of the early years of the Station Biologique de Roscoff is contained in Charles Atwood Kofoid, "The Biological Stations of Europe," *United States Bureau of Education Bulletin* no. 440 (1910): 94–109.

46. Interview with Lucy Huttrer, 26 Dec. 1977.

47. Marriage certificate, EEJ(M).

48. See Astor to Hedwig Just, 13 Oct. 1939, NA.

49. Newspaper clipping, source unknown, entitled "Dr. E. E. Just unheard from in war zone," E. E. Just file, Moorland-Spingarn Research Center, Howard University.

50. "3 Trips to Quimper," EEJ(H), box 125-16, folder 329.

51. Charles Pérez to Just, 23 Oct. 1939, ibid., box 125-7, folder 118.

52. See Just to W. C. Allee, 11 June 1939, ibid., box 125-2, folder 23.

53. See "Unsolved Problems of General Biology" (1940).

54. For some of these transactions, see ·EEJ(H), box 125-1, folder 4.

55. Ms. notes, ibid.

56. Just to Heilbrunn, 19 Dec. 1940, ibid., box 125-5, folder 80.

57. Interview with Elisabeth Just Adèr, 1 Oct. 1977.

58. Lillie to Harrison, 1 Aug. 1940; Harrison to Lillie, 5 Aug. 1940; Harrison to Ernest Swift, 5 Aug. 1940, FRL(W).

59. "Passenger List," S.S. *Excambion*, EEJ.

9. AMERICA AGAIN

1. Just to Peter M. Murray, 14 Sept. 1938, EEJ(H), box 125-6, folder 113.

2. "Dr. Ernest Just Back Home," newspaper clipping, source unknown, EEJ(H).

3. Birth certificate, Elisabeth Just, 15 Oct. 1940, EEJ(M).

4. See Astor to Otto Schnetzler, 27 Jan. 1941, NA.

5. Just to Lillie, 3 Oct. 1940, FRL(W); Just to John H. Gerould, 5 Jan. 1941, JHG; Just to W. C. Allee, 5 Oct. 1940, EEJ(H), box 125-2, folder 23.

6. Just to Lillie, 3 Oct. 1940, FRL(W).

7. Lillie to Just, 7 Oct. 1940, ibid.

8. Just to W. C. Allee, 24 April 1939, EEJ(H), box 125-2, folder 23.

9. "On Abnormal Swimming Forms Induced by Treatment of Eggs of *Strongylocentrotus* with Lithium Salts and Other Means" (1940); "Egg-laying in *Nereis diversicolor* at Roscoff" (1940); "Fertilization-Reaction in Eggs of *Asterias rubens*" (1940).

10. Just to Heilbrunn, 19 Dec. 1940, EEJ(H), box 125-5, folder 80.

11. Just to Hedwig Just, 16 Feb. 1941, EEJ(B).

12. See Just to J. W. Buchanan, 15 April 1941, EEJ(H), box 125-2, folder 39. Drafts of this ms., later titled "The Origin of Man's Ethical Behavior," are preserved in the Howard collection: EEJ(H), box 125-21, folder 396.

13. Francis J. A. Neef to Just, 17 Dec. 1940; Just to Neef, 20 Dec. 1940, EEJ(H), box 125-7, folder 114.

14. Just to John H. Gerould, 5 Jan. 1941, JHG.

15. See issues of *Dartmouth Magazine*; also Just to E. E. Day, 18 Nov. 1936; Day to Just, 8 Jan. 1937, EEJ(H), box 125-3, folder 55.

16. Just to William R. Brewster, 7 June 1941, ibid., box 125-2, folder 37.

17. Just to Lloyd C. Fogg, 29 May 1941, ibid., box 125-4, folder 69.

18. Just to Converse Chellis, 7 June 1941, ibid., box 125-2, folder 47; Just to Charles E. Cushing, 8 June 1941, ibid., box 125-3; Just to C. H. Powers, 9 June

1941, ibid., box 125-7, folder 119; Just to Shongut, 9 June 1941, ibid., box 125-8, folder 133; Just to M. E. Thomas, 9 June 1941, ibid., box 125-8, folder 140; Just to Worthington, 9 June 1941, ibid., box 125-9, folder 156.

19. Just to Hedwig Just, 11 July 1941, EEJ(M).

20. See Harris to Harrison, 19 July 1941; Heilbrunn to Harrison, 24 July 1941, Schrader to Harrison, 25 July 1941, RGH, box 15.

21. Nancy Astor to Otto Schnetzler, 21 Aug. 1941, NA.

22. See Heilbrunn to Lillie, 17 Sept. 1941, FRL(W).

23. *Science* n.s. 95 (1942): 11.

24. Selig Hecht to Lillie, 8 Jan. 1942, FRL(W).

Acknowledgments

The Alfred P. Sloan Foundation granted support for the initial research on this project. I wish to thank the foundation for giving me the means to make those first steps, what were for me the most crucial ones. Arthur Singer deserves special mention. The School of Humanities and Social Science at the Massachusetts Institute of Technology supported my work in numerous ways from beginning to end. For the last two years the Class of 1922 Career Development Chair at MIT provided funds necessary to complete the biography.

My debts to individuals are numerous, and I can hope to give only some brief indication of their extent and nature. Irving Kaplan read the entire manuscript, and gave incisive and intelligent criticism over the years. Besides being a good friend and colleague, he has helped me see, appreciate, and understand science and the world that produces it. Kenneth Keniston, Leo Marx, and Charles Weiner read the manuscript along the way and always offered helpful and sensitive criticism. I owe them much as colleagues and friends. Donald Blackmer and Carl Kaysen, through their administrative acumen and wise counsel, helped provide me with the space a scholar often needs to carry out a major project. Elzbieta Chodakowska, Maurice Fox, Kenneth Hoffman, Gerald Holton, Willard Johnson, Leon Trilling, and James Young read the manuscript in its final stage, and perhaps under pressure, but still with patience and sensitivity. Garland Allen, Suzanne Berger, Evelyn Keller, Daniel Kevles, Edward Lurie, Philip Morrison, Nell Painter, Barbara Rosenkrantz, and David Rosner also read the work at its last stage, and they too offered good suggestions. William Brouwer, the late Jonathan Grandine, Antoni Jurkiewicz, Helene Jurkiewicz, David Manning, Ruth Perry, Janet Romaine, James Shreeve, Helen Slotkin, Lora Tessman, Michèle Vergne, and David Wiley gave the biography the reading of a discerning general reader, and each offered sound and sensible advice. As friends, they had no obligation to do so.

Many scientists and friends who knew Just at Woods Hole or whose teachers or parents knew him there gave me invaluable insights, memorabilia, and general encouragement. Among them were the late Lester Barth, Lucena Barth, Donald Costello, James Ebert, Sally Hughes-Schrader, Donald Lancefield, Rebecca Lancefield, Maressa Orzack, S. Milton Nabrit, Harold Plough, Constance Tolkan, the late Alfred Redfield, Paul Reznikoff, Elsa Sichel, the late H. Burr Steinbach, Susan Steinbach, and Phoebe Sturtevant. I can never convey adequate thanks to them and the Woods Hole community.

Among Just's colleagues and students at Howard who shared their detailed recollections and offered me general encouragement and advice on the project were Lillian Burwell-Lewis, Hyman Chase, Montague Cobb, Ruth Lloyd, the late Roscoe McKinney, and M. Wharton Young. In the first edition of this book, there was a transposition of names involving one student, Wallace Wormley. Wormley did in fact receive a master's degree in June 1931. I have corrected the text and wish to thank Professor Wormley for calling my attention to this matter.

Margaret Just Butcher, Maribel Just Butler, Bernice Just, Mary Just, Elisabeth Just Adèr, and Sheryl Wormley gave their support for the work and supplied many documents and illustrations. I owe them special gratitude.

Many institutions opened their archives to me, and their staffs performed specific tasks as well as provided general help. I should be ungrateful if I did not acknowledge the following archives and archivists for their services: Jane Fessenden at the Marine Biological Laboratory; Frank Portugal at the Carnegie Institution of Washington; Paul McClure at the National Academy of Sciences; Kenneth Cramer and former Provost Leonard Rieser at Dartmouth College; Albert Tannler at the University of Chicago; Esme Bhan, Thomas Battle, and Michael Winston at Howard University; the late Warren Hovius and J. William Hess at the Rockefeller Archive Center; Alexia Helsley at the South Carolina Department of Archives and History; and Christiane Groeben at the Naples Zoological Station.

Leona Capeless and Sheldon Meyer at Oxford University Press gave invaluable advice, assistance, and encouragement. Lynn Roberson typed and retyped the manuscript, and provided efficient secretarial services for the project. I owe her many thanks. My greatest gratitude is to Philip Alexander, who helped with many aspects of the research. From beginning to end, he maintained faithful and loyal dedication to the project. Frederick Parker has constantly influenced my thinking and outlook on life. My debts to him are endless.

K. R. M.

Index